D1541623

THE ECONOMY OF NATURE
AND THE EVOLUTION OF SEX

THE ECONOMY OF NATURE AND THE EVOLUTION OF SEX

Michael T. Ghiselin

UNIVERSITY OF CALIFORNIA PRESS
BERKELEY, LOS ANGELES, LONDON

UNIVERSITY OF CALIFORNIA PRESS
BERKELEY AND LOS ANGELES, CALIFORNIA
UNIVERSITY OF CALIFORNIA PRESS, LTD.
LONDON, ENGLAND
COPYRIGHT © 1974, BY
THE REGENTS OF THE UNIVERSITY OF CALIFORNIA
ISBN: 0–520–02474–5
LIBRARY OF CONGRESS CATALOG CARD NUMBER: 73–78554
PRINTED IN THE UNITED STATES OF AMERICA

FOR OLIVE GHISELIN

CONTENTS

PREFACE:
THE DYING GASPS
OF VITALISM

Quare religio pedibus subiecta vicissim
opteritur, nos exaequat victoria caelo.
T. Lucretius Carus

All scientific research presupposes a philosophy, at least in the sense of a methodology and logic. Yet rarely is it deemed necessary to discuss that philosophy when writing up the results, for the author and reader may be assumed to reason in much the same spirit. The present work constitutes an exception. The original plan was to play down the philosophical side, leaving it implicit, and perhaps developing it at some later opportunity. Yet my approach has been deliberately unconventional, to say the least. I question much that is habitually taken for granted.

We tend to treat science as strictly impersonal, the product of an objective methodology dealing with observed facts. Yet scientists are like everybody else and cannot utterly divorce their values from their professions. Nor should they do so, any more than lawyers should lack concern for justice. This habit of depersonalizing science leads to a false picture of how scientific research is done. It gives the impression that the creative aspect of scientific thought deserves no attention—that we need concern ourselves only with the finished arguments, and with the data. Yet progress is contingent upon innovation. Novelty, however, does not take care of itself. For this reason I have continued a familial tradition of concern for the mechanics of innovation (B. Ghiselin 1952, 1963). Like my father, I have been deeply influenced by Taoist philosophy, and also (as has N. R. Hanson 1958) by *Gestalt* psychology (Wertheimer 1945). I also owe a great intellectual debt to the historian Thomas Kuhn. My ideas on these topics were presented in provisional form in a book on Darwin (M. T. Ghiselin 1969a) and in greater detail later (M. T. Ghiselin 1971).

An obvious implication of such work would be that science might be rendered more effectual in its innovative processes. Could one not learn something from history, beyond the old cliché that nobody ever learns from it? I had been made aware of some conceptual difficulties in modern evolutionary theory through my efforts to see what had become of Darwin's intellectual legacy. It struck me that the study of processes going on in entities larger than organisms—species and societies, for example—might well be reconsidered. The investigation soon led to the inevitable conclusion that Western thought has reached a stage of crisis that can no longer be ignored. The traditional conception of nature was overthrown by Darwin in 1859, but the fundamental message has never really been understood. The old way of thinking and the new can never be reconciled, but efforts to do so continue to the present day and the results produce a warped and largely erroneous view of the living world. The time has come to take a strong stand on these matters, for vitalism has become almost an orthodox position for many evolutionary biologists. To be sure, the people whom I consider vitalists would spurn the label with indignation. Certainly we rarely encounter biologists willing to go along with such a vitalist as Driesch, who openly proclaimed himself as such. Yet it is one thing to say that the symptoms are not obvious, another that the malady is having no pathological effects. These days the metaphysical derangement takes many forms, but among the most pernicious is the view that organisms differ fundamentally from nonliving constellations of matter, and that the philosophy of science must be subdivided into separate compartments, one for chemistry, one for biology, perhaps a third for man or Man. (Consider Mayr 1961; Watson's 1966 critique of Simpson 1963; Dobzhansky 1968.) As Aristotle said, of course, a house is more than a pile of bricks; but it does not derive its membership in the class of houses from something other than the relations between material things. This qualification being admitted, Lucretius was right: there is nothing in the universe but atoms and the void. And to deny the philosophical ideal, however unattainable it may be, of predictive, deterministic theory is to opt out of science altogether.

Hence I have been driven to question the very foundations upon which much of modern biology still rests, and to maintain that the superstructure is not without taint. In attempting to write what almost adds up to a cosmology, one suffers from the limitations of one's ability and points of view. Indeed, by the time one managed to cover the topics here discussed in the detail that might seem called for, the accumulation of new materials would necessitate infinite revisions. But then again, one misses something in treating such matters apart from their larger context. A quantitative, rather than a verbal, intellect would have written a very different book. So would a terrestrial, rather than marine, biologist. Under such circumstances one is always tempted to play it safe—to run from controversy and to occupy positions immune from criticism. We all want to be right, but nonetheless the real satisfaction in science is that it exalts the spirit and glorifies the intellect. What is said here may or may not be true, but at any rate it is intended to elicit discussion, not to terminate it. On the other hand, I would rather not devote the rest of my life to topics upon which I have had my say. I have no intention of writing another word on the generality of topics herein discussed.

As a last preliminary remark, it may be observed that I have gone over much of the same ground as that covered by Darwin's *Descent of Man*. The full title of that work, indeed, is *The Descent of Man, and Selection in Relation to Sex*. The resemblance stands, in spite of my having reversed the order of the two main topics, and concerned myself with some additional matters. The parallel was not deliberate, in the sense of things having been planned that way. Of course, one can hardly escape being influenced by the standard of excellence in whatever field one writes about. But originally I set out with nothing more in mind than applying sexual selection to "dwarf males." The project grew into an examination of sexual selection in general. Then I saw that I could solve a lot of ecological problems that had begun to occupy my thoughts some time before. Only when I had delved into the historical and philosophical background, and had accumulated much preliminary manuscript, did the larger connections become clear to me. Traditionally *The Descent of Man* has been looked upon as two separate works, with little connection between them. I now realize that such is by no means the case. Indeed, after spending some time in the archives at Cambridge University in 1971, I am convinced that Darwin had grasped the crucial point while he was still a young man—before 1840. But that is another story.

The views expressed in this work are my sole responsibility, but I am pleased to thank those who have responded in one way or another to portions or wholes of drafts of the manuscript. I thank Herbert G. Baker, Stephen Jay Gould, William Hamner, Cadet Hand, David L. Hull, William Z. Lidicker, Ernst Mayr, Thomas Ragland, Francesco M.

Scudo, Alastair M. Stuart, Robert L. Trivers, and Perry E. Turner, Jr. T. J. M. Schopf deserves special thanks for his aid with stylistic and organizational matters, and Grant Barnes of the University of California Press went out of his way to be helpful. Jere Lipps kindly allowed me to incorporate some of our joint speculations on the evolution of bivalves. I received some help from the Committee on Research of the University of California. During the final stages of preparing the manuscript my research was supported by GS 33491 from the National Science Foundation. Much of the work was done at the Bodega Marine Laboratory.

1

INTRODUCTION:
A DIATRIBE AGAINST
THE METAPHYSICS OF
LORD VERULAM

*He writes philosophy like
a Lord Chancellor.*
William Harvey on Francis Bacon

This book is a cross between the *Kama Sutra* and the *Wealth of Nations*: it seeks to explain the diversity of reproductive phenomena in the light of some ideas derived from classical economics. The "natural economy," as it has sometimes been called, would seem to have been looked upon too much as a cooperative, rather than a competitive, association. A number of theoretical topics are reconsidered from this point of view. What is sex "for"? How can we explain the various forms of hermaphroditism? Why do animals fight, especially over mates? For what reasons are there differences in size and coloration between males and females? In whose interests do societies exist, and what has brought them into being? These questions, and many others, are examined in some detail. A large body of biological facts are presented, culled from the literature on all the major groups of organisms; these facts have been selected because they illustrate the ideas about the natural economy, and all form threads in a common fabric—one with a rather curious weave. Yet the primary concern is not with the scientific issues as such, but with our manner of dealing with them. Even if some or many of the details should prove wrong, this may not affect the more basic, philosophical, thesis.

INDIVIDUALISM

If the philosophical position here maintained admits of a label, then perhaps it should be "radical individualism." Simply stated, this means attributing great importance to individuals and relegating groups to positions of inferior status. "Radical" here means "polar"—admitting of no intermediate position. This sort of individualism is not very heterodox from a philosophical point of view, something resembling this having been advocated by philosophers such as B. Russell (1960) and Popper (1964). Likewise it is quite compatible with a broad spectrum of evolutionary thought, from that of Darwin himself, down to that of many contemporary biologists. Hence, from the point of view of biology, this position is anything but idiosyncratic. If there is anything unusual about the adoption of it here, it is the attempt to be consistent and comprehensive. But this is just the point: not only shall we consider a thesis, but we shall also follow out its implications—to the bitter end. The individualistic outlook affects the structure of theory, the canons of evidence, and the manner of presentation.

Such radical individualism profoundly affects our attitude toward the nature of the scientific endeavor. To see what a difference it can make, let us consider a common-sense treatment of a deceptively simple problem in epistemology. A widely accepted idea among philosophers of science is that the pursuit of scientific truth is a public matter—a social affair, so to speak. Only a group of scholars can engage in it, for a man working by himself lacks the criticisms of his peers and this, supposedly, is what makes him objective. Yet is this really so? Consider the example of a man alone on a desert island. He needs to know if the following hypothesis is true: "This mushroom is poisonous." Could he design and execute an experiment that would give him the answer? Indeed he could, and his desire to preserve his life would give him every reason to do the job well. Now suppose that the same man were a graduate student, whose major professor told him to demonstrate that the mushroom was not poisonous. Might not the desire to obtain a degree influence the outcome? Provine (1971) has related the case of the geneticist Weldon having his student Darbishire refute Mendel's laws, with painful consequences. Indeed, the problem of bias gets so bad in some branches of knowledge that special precautions, such as "blinds" in psychology, have become a matter of routine. To be sure, strong forces also work toward objectivity, but we cannot expect that they will inevitably prove effective, especially when we are unaware of our own ulterior motives. We all know, but rarely admit to ourselves, how strong are the pressures for conformity among scientists. Few would deny that everybody—perhaps himself excepted—finds such pressures hard to resist. Wishful thinking, the desire for fame, and the lure of grant money, tenure, and public approbation are all generated by social influences, and would never be felt on a desert island. Indeed, one might suspect that this practice of shifting the criterion of truth from the results of experiments to the opinions of the scientific community is just another manifestation of this phenomenon. To state the individualistic thesis in an extreme form, we might paraphrase Rousseau, and say that man is born wise, but society makes a fool of him.

The reader will no doubt be aware that a branch of philosophy, namely metaphysics, concerns itself with this kind of problem. Metaphysics is generally divided into two sub-branches, epistemology (the theory of knowledge) and ontology (the theory of being). Both relate to the basic theme of this book. Our position on whether groups or individuals determine what is true would be a matter of epistemology. If we were then to go on and ask "What is a group?" we would be dealing with ontology. The two branches of metaphysics are frequently treated together, for how we know can hardly be treated in complete detachment from what we know. Logic, another of the major branches of philosophy, helps provide answers to many metaphysical questions. "Logic" is usually defined as the study of valid inference; it helps us to reason, for example, by revealing fallacies (by definition invalid inferences). But it cannot tell us what determines when proper and improper modes of reasoning will be employed: that is a matter for psychology, a natural science and not a branch of philosophy at all. Nor, for that matter, can it alone tell us whether factual premises are true; only the sciences can do that.

It may readily be seen that metaphysical notions are here considered from a psychological point of view. The scientific literature ought to be read from the point of view of ulterior motives, in which it abounds. There are many ways of feigning objectivity. Some of these are mainly literary: pretend that objectivity resides in tone of voice, and banish the subject from discourse by eliminating the first-person singular. Another maneuver is to advocate metaphysical ideas that could have the same effect. Such notions have other functions, to be sure, but this is one of the more important. For the epistemological issues surrounding individualism, one idea in the philosophy of science is particularly significant. This is what may be called, following Popper (1962:137) and others, the "Baconian myth." By this is meant the notion that the truth emerges automatically from the "facts." One observes without preconceived notions, then generalizes on the basis of what one has seen. The term "Baconian" is used here advisedly, granting that Lord Verulam himself was not that simplistic. Many who have called themselves Baconians do little credit to the philosopher, who was quite aware of the value of hypotheses; and if he erred in laying too much stress upon empiricism, he did so in combatting equally, if not more, pernicious speculative habits. Still, from what he says, one gets the impression that he believed hard work more important than imaginative thinking and, on that basis, the label seems fair.

Whatever term we may wish to use for it, the fact remains that the Baconian outlook plays down the strengths of the intellect, and over-emphasizes the capabilities of perception. What we do see is strongly influenced by what we expect to see. And if we can banish one prejudice from our minds, how can we be sure that another, one of which we were not aware, has not influenced us? If, on the other hand, we have an hypothesis clearly formulated, we not only know what to look for, but can solve the problems of objectivity through sound experimental design. Even in "observing facts," hypotheses are useful: an experienced naturalist in the field, thanks to his training, sees the less conspicuous organisms that the tyro overlooks. We might almost reverse a scholastic maxim, and say that there is nothing in the senses that was not first in the intellect. When someone invokes the Baconian myth, and claims that he observes without preconceptions, we should question whether he is simply attempting to make himself more credible. But such pronouncements, even when ulteriorly motivated, should not be treated as insincere: he who believes a myth is all the more apt to gain converts.

More sophisticated philosophies of science perpetuate the Baconian myth in subtler forms. Very popular in the nineteenth century and still advocated today (Platt 1964), has been the notion of "induction by complete exclusion." This rather formidable term refers to nothing more profound than a process whereby one arrives at the truth by disproving (excluding) one hypothesis after another until but one remains. The trouble with such an approach is that it doesn't work very well in the actual conduct of research, and for various reasons. It does seem to make sense in the sort of "laboratory" exercise given to undergraduates in analytical chemistry courses. They are presented with an "unknown" and provided with a series of tests that exclude one group of compounds after another. The procedure works, because it is set up that way. In genuine research situations, however, especially those in growing fields of knowledge, success has not been preordained. One doesn't know if one has thought of all the possible hypotheses. More than one answer may be valid, and the problem may not even be soluble. Furthermore, waiting until all possibilities but one have been excluded may sacrifice, on the altar of supposed rigor, the opportunity for following up promising leads. The appropriate research strategy is rather to forge ahead with provisional hypotheses, explore their implications, and leave the more rigorous confirmation for later. Storm the citadel, and the suburbs will fall in due course.

Some overemphasis upon the principle of exclusion would seem to derive not from the Baconian myth, but rather from certain excesses surrounding advocacy of the "hypothetico-deductive" method, which is the classical picture of science as hypothesis, prediction, and refutation. Hypothetico-deductive methodology has been used by such good biologists as Darwin (Ghiselin 1969a), but the "H-D model," as it is called, takes many forms (see Popper 1965). When temperately applied, it provides a reasonable description of some basic operations used in testing hypotheses. Thus, it does seem true that refutation, rather than confirmation, of hypotheses is what scientists are up to. And Popper rightly maintains that we cannot prove an hypothesis, and that certitude therefore is not possible. We may even go so far as to say that any supposed refutation of an hypothesis may turn out to demonstrate no more than that the test was poorly designed, and hence that something other than the hypothesis has been shown to be false (see Popper 1962:238–239). Nonetheless, when the experiments don't turn out as theory has led us to expect, we know that more work is in order, and that is enough. Whatever position we take on the H-D model, this can scarcely be denied, without rejecting a fundamental principle in logic. Some philosophers reject the H-D model because scientists accept hypotheses after a series of tests that don't really refute the alternative ("null"), or which seem to provide a kind of support or confirmation. Nonetheless, as Popper says, they accept the hypothesis only after serious attempts to refute it. That isn't the same as showing that the hypothesis is true, although a superficial analysis might give that impression.

The overwhelmingly significant aspect that does get left out of the H-D model is not what it fails to do, but rather what it was never intended to do in the first place. It describes the logic of verification, not that of discovery: it tells how hypotheses are tested, rather than where they come from. To be sure, a logic of discovery would seem to be a contradiction in terms, for the invention of an hypothesis is not a process of inference, and therefore by definition it is not the concern of logic proper. Any study of how hypotheses are invented would be the province of psychological research. One could perhaps counter that the topic falls under pragmatics, or heuristic, and Caws (1969) and Meltzer (1970) actually maintain that discovery is a logical process. But such discussion tends to degenerate into an argument over the definition of the word "logic." The real issue boils down to whether creativity is or is not an orderly process. Very likely Popper, whose pleadings for libertarian ideals are quite legitimate in their proper place, is really trying to salvage the notion of free will. Others who believe that invention knows no rules may advocate a kind of authoritarianism grounded in some conception of a process that transcends reason (intuitionism). He who seeks to find order here treads upon sacred ground: we invade the sanctuary of the unconscious, and desecrate the altar of intuition. The issue will be resolved only by showing where new ideas actually do come from, and perhaps finding those laws of nature that govern their genesis.

Enormous progress occurred in the study of innovational mechanics when the historian Thomas Kuhn published his *Structure of Scientific Revolutions* (1962; 2nd ed. 1970). Kuhn treated the development of science in the light of what he called "paradigms." This term was rather loosely defined—some might consider it excessively vague —but the general idea is clear enough. A paradigm is a group of accepted ideas and practices that determine how research will be done. It can include a theory or a body of theories as its core; for example, in Newton's physics and Darwin's evolutionary biology. But even this is not necessary. The uniformitarian paradigm in geology, which replaced the catastrophist one, had mainly to do with methodology. According to Kuhn's view, one competing paradigm replaces another, but only at certain times. This leads to a distinction between what he calls "normal science" on the one hand, and "revolutionary science" on the other. In revolutionary science alone does the paradigm change. Normal science Kuhn describes as "puzzle-solving." A theory is developed and its implications are worked out in research projects. The paradigm remains the same, and each study really amounts only to variations on a theme. Clearly a somewhat limited amount of creativity is involved in some of this puzzle-solving, but Kuhn does not use "normal science" or "puzzle-solving" as opprobrious epithets. Rather, he considers normal science the proper activity for normal circumstances. The theory and the methodology are exploited so long as they give fruitful results. Only when a certain stage of development has been reached—namely when the traditional approach ceases to work— does revolutionary science become appropriate. That puzzle-solving is respectable follows from Kuhn's idea that it almost mechanically generates revolutions. He stresses the point, however, that scientists are loath to abandon their paradigms. Revolutions tend to occur long after signs of weakness or inadequacy have begun to appear, and often a school of thought is replaced largely because its adherents grow old and retire. In his most recent publications, Kuhn has emphasized that the distinction between normal and revolutionary science must not be exaggerated: there are big revolutions and small ones, and they go on all the time.

The evidence shows that, for the most part, Kuhn's ideas apply very well indeed to the Darwinian revolution (M. T. Ghiselin 1969a, 1971). The failure of biologists to modify their paradigm, even when many had come to accept at least the possibility of evolution, is a case in point. The revolution did not occur until the contradictions were rammed down their throats and an alternative paradigm was presented. Failure by many eminent biologists to embrace the new hypothesis of natural selection, or even to understand it, provides another example. The parallel discovery by Wallace—impressive for the similar pathways through which the same hypothesis was generated—substantiates at least the general thesis that a special kind of reasoning is involved.

An independent assessment of how well Kuhn's ideas apply to the Darwinian revolution has lately been published by Mayr (1972a) who reaches conclusions different from the ones presented in this book and earlier (M. T. Ghiselin 1971). Mayr presents some interesting thoughts and commentary, but his premises do not support his conclusions with respect to Kuhn. Indeed, the negative character of his evidence implies that Mayr simply has not seen the connections. He treats the Darwinian revolution as if the fact of evolution were the fundamental issue, and as if it had to do with the way species originate instead of with how one species changes into another. The basic contribution of Darwin and Wallace was not evolution, nor a mechanism of speciation, but the theory of natural selection. Grasp the mechanism that produces change and all else follows. Hence we have reasonable grounds to question Mayr's view that the unity implicit in Kuhn's view of scientific revolutions is not there. Once one understands Mayr's contentions, much of what he says can be written off as an effort to keep physics out of biology. He has repeatedly advocated notions that help widen the gulf between the sciences of life and those of inanimate matter. His opposition to "reductionism," his advocacy of "teleonomy," and his denial of the predictive power and deterministic nature of modern evolutionary theory represent over-reactions to the excesses of a philosophy of "science" that ignored biology but did contribute much of value. By a curious irony, the men who, in the past forty years, have done most to turn evolutionary biology into a "hard" science have been reluctant to take credit for that accomplishment.

On the other hand, it does seem that Kuhn's model needs to be modified and expanded in certain important respects (see M. T. Ghiselin 1971). The crucial point here relates closely to the basic theme of the present work. It turns out that Kuhn has treated innovation largely from a social, rather than an individual, point of view. It is as if science, rather than scientists, generated novelty. We will see in later chapters that Kuhn's position accords with much of modern social science. His conception of things may prove valid for normal science, and for conventional scientists, but whether it applies to revolutionary science may well be another matter. Why, for example, did Darwin invent the hypothesis of natural selection twenty years before Wallace did? If innovation were a social affair, one would expect parallel discoveries always to occur simultaneously. We might answer that Darwin's intellect transcended the limitations that society had imposed upon less unusual men. A similar case may be made for the discovery of Mendel's laws. The first discovery was a work of such originality that the scientific community could not understand it. The simultaneous rediscovery followed as the natural consequence of cytological puzzle-solving. Social influences do not foster originality, they inhibit it. The pressures for conformity, which, again, are stronger than many would like to admit are necessitated by the need for harmonious and cooperative teaching and research. A productive team of investigators is no more anxious to lose its paradigm than it is to lose its financial support.

Independent thinkers would seem to flourish in relative solitude: witness again Darwin and Mendel. Along the same lines we may note that genius is frequently associated with resistance to authority (see Caws 1969). But social independence does not by itself make anyone a genius: the village idiot can be intractable. On the other hand, it seems reasonable to infer that certain habits of thinking might help to overcome the pernicious effects of social pressures. A disposition toward thinking one's own thoughts may constitute a part of one's genetic endowment, or it may be cultivated, through either educational practices or individual choice.

We should consider another way in which originality might be fostered. Much of scientific progress results from the development of new apparatus. Some puzzles have been solved by acts no more novel than applying the latest gadget. Now if this is done routinely, why shouldn't one actively seek for a new paradigm, much as one might shop for a different kind of microscope? This has, in fact, been done by finding an alternative way of handling a problem when the usual approaches have failed. A good example is the discovery of the structure of DNA by Watson and Crick. Watson (1968) tells us that he solved the problem largely by finding out what accounted for the success of Linus Pauling, "the world's greatest chemist." Pauling, he says, used molecular models, which simplified his reasoning processes enough to handle some notions otherwise beyond anyone's grasp. Watson realized that the mathematical treatments that embellished Pauling's publications did not reflect his actual manner of conducting research. Watson's view of his exemplar's thought processes no doubt hardly does justice to their subtlety. Be this as it may, however, Watson grasped the new way of thinking and won the Nobel Prize. Some might begrudge him his reward because he did not do it the hard way, but they need not be taken seriously. The real lesson to be derived here is that the paradigm being used by Watson's competition prevented them from solving the problem, and their traditions militated against their embracing a new one; or else they were mere technicians, operating without a paradigm worthy of the name.

There would seem to exist no compelling reason for waiting until the old paradigm has broken down of its own accord before attempting to erect a new one. The process could be hastened by keeping an eye open for signs of weakness. One could go even further and seek opportunities actively. Thus, a major reason why Darwin succeeded was that he took ideas and ways of thinking from one field and transplanted them to another. His evolutionary biogeography had roots in his geological thought, and his invention of the hypothesis of natural selection owed much to his borrowing an idea from classical economics. To accomplish such transfer demands intellectual flexibility, broad knowledge, and a deep concern for fundamentals, but nothing miraculous.

Interdisciplinary transfer often involves a kind of analogical reasoning. Far-fetched comparisons have their drawbacks, not the least of which is the "analogical fallacy." Birds and airplanes, for example, resemble each other in many important respects and much useful instruction can be gained by comparing the two (for an instructive extension of this analogy to mollusks, see Gould 1971). It does not follow, however, from the mere fact of a feature being present in the one that it must occur in the other as well. We would be grossly deluded by treating eggs as the equivalents of bombs. On the other hand, analogy does serve many useful and legitimate functions. For one thing, it helps us to communicate, as may be seen from its literary applications. Perhaps more significantly for the sciences, it may facilitate our reasoning by allowing us to treat the unknown in terms of more familiar experience. We progress more rapidly when part of the phenomena has already been organized for us through our contact with like objects.

For purposes of invention, the value of analogical reasoning can scarcely be overestimated. Ideas deserving to be called "original" in the strictest sense probably do not exist. They evolve from sources not entirely disparate. It is as with organic evolution: new parts arise by the modification of ones already there. This comparison is more than just an analogy. Rather, it embodies a fundamental truth, namely that evolutionary principles may be generalized, allowing us to apprehend a far greater unity among natural phenomena than has hitherto been possible. Admittedly, such larger connections among things have been subject to speculative excess. But when we fail to look for them we adopt a defeatist attitude that prevents their being discovered at all. The historical record does in fact lend substance to the generalization. Paradigms, like organisms, have predecessors; they do not come into being through the intervention of unintelligible or supernatural agencies, but through mechanisms that are fully accessible to scientific investigation.

Extending the same basic ideas, we can see how analogical reasoning may help to reveal other important relationships the existence of which may otherwise escape us. Analogical fallacies are beside the point when we deal with more than just superficial resemblances. It does not matter that the various groups of flying animals are not identical in every respect; indeed, we learn a great deal about flight from both the convergent similarities and the differences. Likewise all entities composed of matter obey the laws of thermodynamics. But on the other hand "mathematical models" need only be logically consistent. Hence the use of "concrete models"—that is to say, comparison with something having actual physical existence—may provide a sounder guide to the real universe than the purely imaginative constructs. Calculation is precise, experience is accurate.

A great deal about the manner of pleading one's case follows. Above all else, there can be no recourse to authority. The data of experience alone determine what is true. It follows, too, that discussion shall be focused not upon the puzzles, but upon the paradigms. The fundamentals must be brought out into the open, and no longer tacitly presupposed. If a scientist adopts one paradigm and his colleague another, the two will not just disagree; in all likelihood they will find each other mutually unintelligible. That was part of Mendel's problem: the world was not ready for such a drastic shift in outlook. It may prove impossible to decide between competing paradigms unless we analyze them to the extent of comparing their basic assumptions. If the differences are not made explicit, the reader will very likely get the impression that he is being presented with nothing more than conventional solutions to traditional puzzles. And if the path to truth can be followed only by questioning everybody's motives—our own included —that step, too, becomes imperative.

Criticism also takes on a different significance. To a group of puzzle solvers, all applying a common paradigm, there is a correct solution to each puzzle, one that can be discovered by carrying out the proper operations. Criticisms are directed toward weeding out mistakes —lack of proper controls, impure reagents, and the like. Yet the paradigm determines both what operations are legitimate and what constitutes a mistake. In revolutionary science the very rules are contested. What happens to criticism then? The conclusion can scarcely be avoided that if the questions themselves can change, all answers are provisional. Experienced scientists know quite well that even the best work may ultimately be deposed by fresh attacks on the same problem; hence they tend to take a more tolerant position, say, than beginning graduate students who have never burned their fingers on genuine research. But criticism does have its legitimate revolutionary functions, particularly in pointing out weaknesses, limitations, and contradictions in a paradigm that needs to be supplanted. If inconsistencies can be shown, or if facts can be adduced for which the current thought seems inadequate, then a revolution is in order.

But such grounds of themselves do not provide sufficient reason for effecting that revolution. One doesn't abandon a theory simply because it may need to be shored up, or merely because perhaps, but only perhaps, it is not true. Effective criticism comes only when an alternative is presented. An inadequate tool is preferable to none at all, but the existence of a real option between two tools provides an opportunity for selecting the better instrument. A paradigm is an instrument in more than just a metaphorical sense, for it is something used in accomplishing a task. And the decision as to which shall be adopted is based, not upon past performance, but upon future prospects. Natural selection did not replace special creation because the latter was refuted or the former confirmed. Rather, a lot of scientists jumped at the opportunity to participate in a new kind of research. An argument for a paradigm shift, therefore, may be based on demonstrating just such advantages. To accomplish this end one needs to explain the new ideas, to show how they might be used, and to examine some of the possibilities for new research. The paradigm should be explored and illustrated. In *The Origin of Species* Darwin scarcely mentioned human biology. He said "Light will be thrown on the origin of man and his history" (Darwin 1859:488). In the context of a straightforward dispute over puzzles, he might have been taken to task for not having been more explicit. Yet he wasn't solving a puzzle. He was issuing a call to arms, in language well calculated to marshal a following.

THE PLAN OF THIS WORK

This book is, in fact, the result of a deliberate effort to overthrow a traditional paradigm, to provide an alternative, and to develop new tools for dealing with old subjects. It concerns the ultimate nature of competitive societies, and it attempts to show the advantages to biology of thinking like an economist. For anyone who has grappled with the relationships between Darwin and Malthus the idea is all too obvious. Ecologists are supposedly very much aware of the parallels between their science and economics. To judge from the current biological literature, there is now a small measure of truth in this proposition, and as time goes on we may expect the trend to continue (see, for example, Hamilton and Watt 1970; Rapport 1971; Trivers 1972). On the other hand, the degree of transfer thus far has been grossly exaggerated. Tullock (1971) lately drew attention to some correspondences that should have been obvious to anyone knowing the fundamentals of both fields. For the most part, only the economic jargon has been adopted by ecologists, and not the manner of thinking. Furthermore, the kind of economics that has been applied often turns out, for reasons to be explained later on, to have been inappropriate. The classic work on economics (A. Smith 1776) contains a number of well-known and quite elementary principles—such as extent of the market—which allows us to solve one puzzle after another. That the solutions have so long proved elusive, in spite of the fact that the theory has been available for the last two centuries, accords quite well with what one might expect from paradigm theory. People working in different fields can easily learn one another's vocabulary, but to master another way of thinking, as a rule, proves far more difficult (M. T. Ghiselin 1971).

Yet the psychological derivation of such ideas, through analogical transfer from one field to another, does not tell us about the actual constitution of the material universe. Classical economics may have provided an erroneous view of the political economy, but if so this does not alter our interpretation of the natural one. A body of theory could apply to one, to neither, or to both, and to tell which is so we must examine the economies themselves. Hence, in the present work the role of economic analogy has been threefold. It has been *metaphysical*, in providing an alternative possibility to the notion that the relations between organisms are the same as those within them. It has been *heuristical*, in suggesting possible solutions to many problems. And, finally, it has been *literary*, facilitating communication. An analogical fallacy would be involved had the parallels been taken for proofs. And merely adducing examples of similarities would provide no valid argument. Only the organisms can tell us whether the transfer has been legitimate. Nonetheless, one doesn't have to restrict discussion of the facts to this issue alone. In what follows, therefore, the questions left open are every bit as significant as those for which answers are suggested. Indeed, in many cases the point is simply that a problem does exist.

To make a case of this sort, it helps to look at a broad range of materials. It then becomes apparent that many and diverse lines of evidence all lead to the same conclusion. But one could easily lose sight of this point, and view the various examples as puzzles in isolation. Hence it should prove useful to conclude this chapter with a brief, analytical summary of those that follow.

Chapter 2 enunciates the basic thesis concerning the traditional paradigm. Namely, the study of biology as it concerns entities more inclusive than organisms has been adversely affected by teleological reasoning. That is to say, biologists have habitually assumed that this is the best of all possible worlds, as if it had been created by God and endowed with a certain kind of purposefulness. To establish this point, it is necessary to analyze the concept of teleology itself, for it turns out to be far more involved than one might think. But once one does understand what has gone wrong, the necessity for a different way of thinking becomes inescapable. This results in what, from one point of view, amounts to a digression, for doing justice to teleology demands an examinination of the ideas of Aristotle and many other philosophers and biologists. In theory, a reader who accepts this philosophical position could skip the entire chapter, and for someone who really operated with the same paradigm and who was concerned with only the puzzles as such, this would be quite reasonable. But it would be an instance of the basic premises being misunderstood; they should be stated explicitly, and from the outset. The need for a total rejection of teleology is the fundamental implication of this book and, apart from this topic, all else must be considered a side issue. An argument against these views which presupposed a teleological paradigm would beg the question. One that accepted the alternative paradigm but challenged the particular solutions to puzzles would risk taking them in a spirit that was not intended. It is shown, through an historical review, how teleological thinking has affected the study of evolution and ecology, even down to the present time. The examples drawn from the older literature illustrate the defects of the traditional paradigm; those from more recent works demonstrate that the subject has more than just historical interest. The reader should take care not to draw the wrong conclusions at this point. An orthodox critique would aim to show that, albeit the contemporary mess is as bad as ever, the millennium is at hand. An excuse would be provided for feeling offended or smug, depending upon one's position. Error is eternal, and wisdom consists in living with it, not letting our vanity tell us that it has been transcended. It is then suggested what kind of approach might be most effective in handling the same general range of problems: one should construct models of how individual organisms might be affected by selection under various conditions, and test the hypotheses by comparing organisms from different habitats. The free-enterprise economy serves as a concrete model for analogical comparison.

These general points having been enunciated, we are in a position to consider how biology can, in fact, be reconstructed upon uncompromisingly nonteleological foundations. This is done in the context of a series of exemplary puzzles. The puzzles are interrelated. All deal, more or less, with reproduction, especially with sexual dimorphism and its attendant phenomena. The choice of topics is not arbitrary. Reproduction has to do with both individuals and groups, and is therefore particularly well suited for conveying the broader message.

Chapter 3 deals with the question of what sex is "for," contrasting the older notion that envisaged sex as of use to the species with an alternative, that individuals are benefited. Sex releases variability, generates diversity, and puts the organisms that engage in it at a competitive advantage. A series of puzzles are then treated, which show how one approach not only gives better solutions, but also reveals relationships which, if the alternatives were embraced, would not even be looked for. Thus new hypotheses are proposed, explaining why organisms evolve diploidy, parthenogenesis, and various other features, the adaptive significance of which has been far from clear. In this chapter, as in succeeding ones, peripheral material is drawn in as necessary to provide background and, more importantly, to show how the same lines of reasoning might be extended: species diversity and geographical trends, for example.

Chapter 4 is the first to deal with the "economic" problems related to the differences between the sexes; ideas of division of labor, diminishing returns, and extent of the market all apply. The differences between eggs and sperm can be explained upon these principles and, with simple variants of the same basic models, one can cover the differences between males and females among organisms in general. Given suitable hypotheses and enough data, it is fairly easy to show why some animals and plants are hermaphrodites while others have separate sexes. Such matters lead naturally to the topic of sexual diversity as it relates to competition. Darwin's long-neglected idea of sexual selection, if expanded somewhat, helps to explain a great deal. Several conceivable forms of purely reproductive competition between members of the same sex and species are hypothesized. It has been possible to increase the known number of modes of sexual selection from Darwin's original two to four. This is partly a matter of terminology, but that so much was overlooked in the century since Darwin's classic on that subject fits in quite well with the underlying philosophical theme. The two modes were discovered as the outgrowth of efforts to develop a comprehensive body of theory that would explain hermaphroditism.

Chapters 5 to 7 treat the four modes of sexual selection, starting with the classical Darwinian combat between males. This topic is considered in relation to such broader issues as why animals fight, hold territories, and form various associations. Darwin's "female choice" hypothesis similarly relates to matters of courtship and family structure. A "sequestering" hypothesis is developed, which explains a number of problematic features. As a last mode of sexual selection, a form of competition involving early mating is invoked, one that clarifies the role of dispersal, age, and size. The global patterns of resources, feeding, and reproductive mechanisms, and population densities are merged into a common scheme.

Chapters 8 and 9 shift the focus somewhat, from sexual dimorphism to the origins of society. Actually the leap is not so great as one might think, since the competitive nature of all economies is implicit in what has gone before. How societies, whether bestial or human, come into being once again has to do with the individualistic theme, for it goes straight to the crux of an age-old question: in whose interests do societies exist? It bears upon the matter of teleology, because we naturally assume that society is "for" something. In an historical section, it is shown how teleological ecology and a great deal of modern social science rest upon the same false premises. Much of the criticism here is directed toward some current thought, on the evolution of social insects, which may contain a certain measure of truth, but which demonstrably is neither necessary nor sufficient to do what is claimed for it. Some alternatives based upon economics are considered. In the closing chapter, the same basic line of reasoning is extended to psychology. If purposefulness it to be found anywhere, one would think, the natural place to look for it is in human behavior. As one might expect, much has been seen which really is not there. The mind is the citadel of the old cosmology—the appropriate place to begin strewing the ruins with salt. It also serves as a fine vantage point from which to contemplate our prospects for a new vision of man and of his universe.

2

THE LEGACY OF
THE STAGIRITE, OR,
TELEOLOGY OLD
AND NEW

Long and pointed eggs are female;
those that are round, or more
rounded at the narrow end, are male.
Aristotle

In the previous chapter, it was said that we should go back to fundamentals. Taking this proposition seriously, let us trace some of the most basic ideas in evolutionary ecology all the way back to their historical origins in Greek philosophy. In so doing we can attain a greater degree of objectivity than would be possible if we confined our attention to the present. Aristotle, for example, was one of the towering geniuses of all time, and he deserves our deepest admiration for such accomplishments as the invention of formal logic. His scientific work, too, reveals the activity of a first-rate mind; but in the writings of this founder of biology, we find just the kind of mistakes to which beginners naturally fall prey. Given so much more experience, we can profit from the example of a first-rate mind going astray. We don't lose our respect for him, because his misconceptions resulted from circumstances beyond his control. We can enjoy a few laughs at his expense, and then sympathize with him when we turn to modern biology and find out that the joke is on us, for in spite of millennia of experience, we continue to suffer from the same difficulties.

TELEOLOGY AND GREEK PHILOSOPHY

Whatever else the Darwinian revolution may have overthrown it at least did away with some of the most influential philosophical ideas in our heritage from the Greeks. Darwin's reinterpretation of the living world did far more than just enrich a static conception with historical meaning and a mechanism for change. In a sense it turned the whole universe upside down, at the same time overturning notions that in his day were so fundamental to Western thought that it hardly seemed possible to question them. Indeed, the subsequent difficulties in assimilating his theory can largely be attributed to a revolution in metaphysics implicit in the theory of natural selection; implicit, but by no means obvious. To this day, even among evolutionary biologists, the revolution itself is recognized largely in name rather than in substance.

The basic feature of the Darwinian revolution has been a shift in our conception of the status of groups and individuals, and in the conception of the relations between them. Much as Copernicus moved the sun to the center of the solar system, Darwin placed the organism at the center of the biological universe. The metaphor, indeed, aptly expresses the historical situation. To Plato and Aristotle the entire cosmos appeared to partake of an order not unlike that which may be discovered in an organism. In his dialogue the *Timaeus*, Plato presents a creation myth in which the world is treated as a divine animal. Likewise, Aristotle would seem to have based much of his political philosophy on an analogy between the organism and the body politic, although it is hard to say whether he derived the basic view of things from biology or politics. A social order dominated by an aristocracy or by a monarch could easily be pressed into service as a model for both macrocosm and microcosm. The apparent orderliness so strikingly manifest in the development of an embryo, reinforced by a Greek cultural predisposition toward rationalism, could just as easily be extended from the biological world to both society and nonliving matter. Given such anthropomorphism, one might readily conceive of the whole universe as something ordered by Reason. And if such a position were granted, ideas and generalizations would take on an overwhelming significance, while concrete objects would occupy a subordinate position. And from this denigration of particular things we in turn derive our heritage of two particularly delusive philosophical ideas: essentialism and teleology.

Essentialism, or what biologists call "typology," is both a habit of thought and a doctrine concerning the nature of ultimate reality (see Popper 1963; Hull 1965). If we are to communicate, we need general terms—universals such as "chair" and "house." Such words are useful because they draw attention to certain properties deemed important. Whenever we use the word "chair" we may reasonably expect our audience to think about sitting—not food, sex, or redness. Although a chair can be red, we do not associate redness with our concept of a chair: red is not a defining property of "chair." On the other hand, proper names, the names of particular things, have no such general connotations. When someone uses the name Mary we do not think of the properties shared by everything named Mary. We are not interested in talking about all Marys but only one of them. We might suspect that the term refers to a woman or a girl, but it could just as well be the name of a town. The only thing they necessarily have in common is a label, and it would make no difference for communication if we called some of the things named Mary "Barbara" and others "Fido."

When we hear someone use a general term or universal, say, "chair," again, we may find it convenient to envision an idealized object stripped of "inessential attributes" or "nondefining characters"—to picture, as it were, "the chair" in our imaginations. Such an abstraction partakes of a certain utility in our thinking, but it stands removed from our immediate experience. We sit, not upon "the chair" but upon various particular chairs. Yet in spite of the fact that in our daily lives we are ever dealing with particulars, our organized knowledge seems to consider the generalities as occupying some privileged status. We seem to be more interested in furniture, less so in chairs, hardly, if at all, in the one before a fireplace. Hence we may come to divorce knowledge, at least scientific knowledge, from the particular objects that are known. Plato even went so far as to posit an ideal world, existing apart from particular things, and tenanted by "essences" such as the chair. "This chair" and "that chair" he degraded to the status of imperfect copies. Aristotle eliminated the ideal world, maintaining that the essences cannot exist apart from the particular things. More down to earth than Plato's, his philosophy incorporated what a Scholastic might have called an "immanent" rather than a "transcendent" sort of essence. Yet for our purposes it matters little, as the result was much the same. The ideal, the rational, and the general took on such overwhelming significance that all else was crowded out of view.

In the Peripatetic philosophy, great stress was laid upon distinctions of rank, or upon what was called "priority." Although the term as it once was used may now be unfamiliar, priority is important, because the older concept remains with us. Although "prior" in one sense means coming first in time, it also means taking first place in order of importance or of dignity, as when, in wartime, the production of guns would take priority over the production of consumer goods; or when a master would be "prior" to his slave (see Aristotle's *Categories*, chapter 12;

Metaphysics, book V, chapter 11). In his *Politics* Aristotle makes it abundantly clear that the state is prior to the citizen and the family (book I, chapter 2). This subordination was equally reflected in his biology. Indeed, his whole cosmology, from God at the top, doing nothing but contemplating himself, downward through the entire scale of beings (*scala naturae*) is permeated with it.

Such subordination is crucial to an analysis of the various notions that are loosely grouped together as "teleology." This word, as one might suspect, conceals a diversity of intensions that are not immediately apparent. Philosophers use it to cover anything remotely analogous to what biologists call "function," while biologists habitually use it as an opprobrious epithet without necessarily understanding what is wrong with the designation. The more superficial issues may be dismissed by pointing out that misunderstandings arise through the literal interpretation of metaphorical language. Of course; but the fundamental point eludes us when, as it usually does, the analysis stops at this level. A number of ambiguities are consequently overlooked, and we fail to consider the role of so-called teleological concepts in the process of investigation. When we are studying a composite thing, we sometimes need to consider the parts, sometimes the relations between those parts, often both. Here, considering the proper level takes on the utmost significance. To use Aristotle's metaphor, architecture involves more than knowledge of bricks and mortar. In examining teleology, we may observe an equally pernicious converse error. But to explain, in sufficient detail, what goes wrong, we need to subdivide "teleology" into more manageable subclasses. For future discussion it will no doubt help to provide names for these, so the following are provided; but the reader need not burden his memory with them: (1) *instrumentality*, (2) *conditional necessity*, (3) *terminating orientation*, and (4) *intelligible orderliness*. If perhaps they do not encompass the entire range of problems, at least they make the distinctions of interest to the present discussion.

1. *Instrumentality*. Among the most significant features of organized beings (whether animals, societies, machines, or what you will) is that the parts have different roles, or functions, in the activities of the whole. The various functions preponderate in specialized parts, resulting in a division of labor. Given such a brute fact, we may classify the components according to what they do: locomotion, chewing, etc. Such knowledge is useful. Neurophysiologists want to be directed to sense organs, cytogeneticists to gonads. There is nothing metaphysical about it either. A leg is, in point of fact, a member of the class of locomotory organs. Yet such an innocuous taxonomic operation should not be confused with what is called a "final cause," a term that Aristotle defines as " 'that for the sake of which' a thing exists" (*Generation*

of Animals, book I, chapter 1; cf. *Parts of Animals*, book I, chapter 1; *On the Soul*, book II, chapter 4). A tool exists for the sake of its owner; all well and good. But how far can we extend the analogy? Very far indeed, if we reason as Aristotle did. An animal, he tells us, exists for the sake of man (*Politics*, book I, chapter 8), and reproduces "in order that, as far as its nature allows, it may partake in the eternal and divine" (*On the Soul*, book II, chapter 4). Thus, an attribute appropriate to one kind of whole, to one kind of part, and to certain relations only is applied with disastrous results to an improper context.

2. *Conditional necessity* is closely related to instrumentality. In biology, it is the idea that if an organism is going to exist, or if it is to engage in some activity, then certain "necessary conditions" have to be met. For example, if an animal is going to chew it needs teeth or at least something like them. The presence of teeth, in technical jargon, would be a kind of "necessary condition" for chewing. In biological research one often asks what necessary condition it is that some part meets. That is to say, we ask what its function is, and we get a "functional explanation" as an answer: the function of a tooth is chewing. Many so-called philosophers of science would seem to believe that there is something "unclean" about functional explanations (Lehman 1965). The reason is that, in physics, an explanation in terms of a necessary condition is thought to be incomplete because it does not allow prediction. Presence of teeth tells us only that an animal can chew, not that it will inevitably do so. Physicists believe in a necessitous universe, whereas biologists are interested in a contingent one. A major reason why we biologists find it so hard to predict in any rigorous way is that the conditions of interest may be met by any one of a number of alternatives that together form a so-called "disjunctive set." Thus, the conditions for chewing are met by a group of teeth, *or* a gizzard, *or* perhaps something else. In dealing with functional relationships of this sort, we have to take organic diversity into account. Once this goal has been attained, the separate worlds of physics and biology may be envisaged as one universe.

Be this as it may, we have various ways of devising functional explanations. One way is to derive our explanation from our "understanding" of the nature of the whole, and give a plausible reason for what the parts do. It seems perfectly reasonable when we see what looks like a tooth in a jaw to assume that it is there to chew with. Yet we may be somewhat perturbed to see Aristotle, knowing that the universe is divine, inferring that man's function is to contemplate it. Here again we see that a property is attributed to a part solely on the basis of analogies. Experience does not justify such transfers from one context to another.

3. *Terminating orientation* may serve as a collective term for such

phenomena as the homing of a torpedo, tropisms and taxes, and the conscious striving we habitually attribute to intelligent animals. "Directed movements" obviously have something to do with final causes, but one may easily confuse the two. A final cause has no necessary connection with temporal sequences. The final cause of an automobile is the driver, not the junk-heap. The function of an automobile is transportation. Purpose and goals are something additional. For an organism to "home in" upon food or some other resource, a control mechanism is necessary, one that takes the place of the driver of an automobile. Cyberneticists have developed quite a body of speculative literature on this sort of topic, as one might expect (Rosenblueth, Wiener, and Bigelow 1943; Sommerhoff 1950; Moore and Lewis 1953). The main philosophical thrust of these discussions is to show that orienting mechanisms are possible. This hardly seem a profound ontological issue: since people have been constructing them for some time, we cannot reasonably question their existence. In fact, these authors have erected a straw man, which allows them to embrace final causes. The really viable issue lies in the epistemological question of how we know that a particular kind of control mechanism exists in a given system. We see order, but what tells us what sort of order we see? All sorts of ordering influences exist, and to invoke one rather than another may be a gratuitous assumption. Aristotle (*Generation of Animals*, book I, chapter 13) thought he saw evidence of reason in organic structure; he, at least, was wrong. The natural theologians of a much later period went astray too. Order is a relation between parts, and control an activity of wholes, and the situation is always ripe for confusion.

4. *Intelligible orderliness*, our final category of teleology, involves, not a slip from one level of organization to another, but rather a slip from a conceptual to a factual context. In studying how an organ functions, a physiologist habitually treats the object of investigation as if it had been designed by some intelligent agent. He might justify his procedure heuristically. That is to say, he gets results. The analogies between eyes and cameras are quite fruitful, and it helps to look for features found in the latter when studying the former. Yet we cannot validly reason that every eye must be built like a camera in all respects. The mere fact that a procedure has heuristical utility does not impose any particular structure on matter. This is especially important because of the tendency of physiologists to anticipate perfection in organic structures: their methodology leads them to overlook the imperfections. No enumeration of examples of apparent design in the living world will contradict the manifest truth that nature produces contraptions as well as contrivances (see Darwin 1877a). These days the teleology of the early naturalists is perpetuated by sophistry. We are

told that all organisms are perfectly adapted, for they are alive. We might answer that all necks must have the same length, for they extend all the way from the head to the body (M. T. Ghiselin 1966a). The superficiality of such effusions perhaps reflects an unwillingness to reason more than an inability to do so. Scientists, like everyone else, may really want to believe that this is the best of all possible worlds.

MODERN TELEOLOGICAL ECOLOGY

Let us now see how the teleological manner of thinking has affected the ecological thought of modern biologists. We can omit the intervening medieval period and the Renaissance, but it should be pointed out that the Aristotelian teleology was embraced by Scholastic philosophy, which in turn passed it on to biology in the form of what may be called the "traditional conception of the natural economy." Whatever its diversity in the minds of various thinkers, this view had one important common theme, the idea of an ultimate harmony among organisms. One kind of teleology or another and the "organicist" model of the world, which treats the universe as a great "super organism," occur in varying degrees of emphasis among naturalists of rather diverse outlook, but the basic metaphysical theme may be found in all. A few examples will suffice to give an impression of how they interpreted the relationships between organisms and their environments.

Teleological themes are particularly obvious in the ecological writings of Linnaeus. Consider a thesis that he wrote, first published in 1749, entitled *Oeconomia Naturae* (Stillingfleet 1791:123).

In short, when we follow the series of created things, and consider how providentially one is made for the sake of another, the matter comes to this, that all things are made for the sake of man; and for this and more especially, that by admiring the works of the Creator should extoll his glory, and at once enjoy all those things, of which he stands in need, in order to pass his life conveniently and pleasantly.

In the same work we find that children have lice "to consume the redundant humors," and that a snake's rattle is there to keep people from getting bitten.

Linnaeus was both a good Christian and a more or less orthodox Aristotelian. God's handiwork was very conspicuous in his world, and the taxonomic groups that he discovered mirrored the essences of things. Yet other philosophical notions could be reconciled with the traditional conception of the natural economy. God, for example, could occupy a less obtrusive position, and indeed His tendency to fade into the background is conspicuous in the scientific metaphysics of the late eighteenth century. Characteristically, laws of nature took His place: nonetheless, He still had to institute them. An interesting example of a teleological biologist in the Enlightenment is Buffon, who was decidedly a materialist. He also embraced a peculiar philosophical notion called "nominalism." That is to say, he thought that classes are in a sense arbitrary, the members sharing a name but nothing more. To nominalists, only "individuals" are "real." Yet some of Buffon's views are rather startling. He considered species to be real ("individuals"), whereas the organisms were the abstractions. He says, "We shall not consider species as a collection or succession of similar individuals, but as a whole, independent of number and of time, always active and always the same; a whole, which has been reckoned one in the works of creation, and, therefore, constitutes only a unit in Nature" (Buffon 1812:[4]:461). Here we find Buffon treating species as composite wholes, or physical units, rather than as classes defined by common possession of certain traits. His position is symptomatic of the breakdown of traditional logic, and as well of the tendency to confuse what we may call "different kinds of groups." We shall have to take up this sort of problem time and again, for the "reality" of ecological and social groups has confused scientists as much as the "reality" of taxonomic ones. Yet the nominalistic aspect to Buffon's philosophy, if providing evidence that the old ways of thinking were proving inadequate, did not fundamentally affect his conception of the natural economy, which remained quite traditional. The following passage is a good example (pp. 445–446):

> The earth, elevated above the level of the oceans, is defended against its irruptions. Its surface, enamelled with flowers, adorned with a verdure which is always renewing, and peopled with numberless species of animals is a place of perfect repose, a delightful habitation, where man, destined to aid the intentions of Nature, presides over every other being. He alone is capable of knowledge, and dignified with the faculty of admiration: God, therefore, has made him the spectator of the universe, and the witness of his perpetual miracles.

Consistent with his combination of materialism and teleology, Buffon tells us that man ought to improve upon nature. Obviously, if God has endowed him with the ability to transform the world into a garden, that is just what man should do.

The traditional conception of the natural economy flourished well into the nineteenth century, until it was utterly demolished by Darwin and Wallace in 1858 (see M. T. Ghiselin 1969b; Limoges 1970). Its refutation was implicit in the Darwinian revolution. Why this was so becomes evident when we see how much that revolution accomplished. In the first place it dethroned Reason, for the order of the universe could be accounted for by unadulterated mechanism. What previously had been the leading power in the universe was demoted to the status of a humble servant. It did away with Platonism, including, we should emphasize, its attendant mystical attitude toward numbers. The music of the spheres was replaced by the clatter of rolling dice. At the same time, and by implication, the fundamental sequence of priority was reversed. It is individuals that differ from one another, and individuals that struggle for life. Platonic Ideas, essences, types, forms, call them what you will, turned out to be hypostatized abstractions. Indeed, even considered as purely imaginative constructs, they left much to be desired. A generalization that deletes the most important features of the things generalized about is hardly sensible. Yet the kind of reasoning inherited from the ancients did just that. It left out the individuals' unique features. To understand evolution, we need to know what varies, not just what remains constant. And if we are to construct models they must be models of organisms, characterized by just such uniqueness and particularity. However profound our ideas, they are valuable only insofar as they help us to understand this animal and that plant. As was mentioned in the introductory chapter, this philosophical issue is fundamental to any real understanding of how evolutionary biologists do their research. Philosophy has never adequately reconsidered the traditional view that knowledge is of universals rather than of particulars. If this means no more than the mere fact that we must compare and generalize, all well and good. But if it means that we are not interested in that which we compare, and in that about which we generalize, we go too far. A Platonist would forsake our world for a transcendent realm of ideas. A Darwinian seeks to be at home in the land of his birth.

Reversing the priority of status for individuals and abstract groups could scarcely help but affect the teleology implicit in the traditional conception of the natural economy. Indeed, the Darwinian revolution rendered teleology superfluous. This implication of the theory of selection has been recognized, but never really assimilated. That adaptations are produced by a blind mechanism implies that the orderliness we apprehend in nature, however intelligible, does not result from conscious design. And the apparently purposeful aspects of an organism's behavior can result from quite simple mechanisms. Hence attention shifted toward the analysis of such mechanisms, away from their effects. Greater significance was attached to how the function is accomplished in particular instances, and, more importantly, to the historical chains of events that are responsible for their being accomplished that way. Yet a subtler change attracted far less consideration than it deserves. No longer could parts be viewed as mere instruments, existing for the sake of some larger being. It was as if a despotism had been overthrown, and citizens who once existed for the good of the state now looked upon society as a means for furthering their own personal welfare. In nature, only that which in a sense benefits an individual will be selected. It follows that an organism never does anything for the good of the species. A species is something that an organism uses. The economy of nature is altogether individualistic, and "altruism" is a metaphysical delusion.

The impact of this denial of instrumentality would have been overwhelming had Western thought been capable of dealing with it. It was obvious that something had happened, but the older way of thinking could still be rationalized. One result has been a tendency to interpret the individualistic outlook as an expression of nominalism (see Mayr 1959). This is unfortunate, because what really happened was that the nominalist-realist controversy simply became irrelevant in this context. A radical nominalist treats all sorts of groups as "mere abstractions," as "mental constructs" which "have no real existence in nature." Thus with respect to taxonomy, they may assert that a species is purely arbitrary, while an individual organism is the only sort of entity that exists outside our own imaginations. This position is untenable (M. T. Ghiselin 1966b, 1969b). It results from confusing different logical types, or what were earlier referred to as different kinds of groups. It overlooks the fundamental difference in status of classes on the one hand and composite wholes on the other. Both of these are groups, but they and their members may have quite different properties. As Carnap (1967:61–65) points out, treating classes as if they were composite wholes is a fallacy, and it results in such mistakes as thinking that a class must have members. Conversely, treating a composite whole as if it were nothing more than a class leads to equally absurd conclusions, such as treating something as a figment of the imagination merely because it has parts. For example, it is absurd to say that since John is a class of cells, he exists only in the mind. Such misguided nominalism leads biologists to deemphasize exactly those composite beings that are of greatest interest to evolutionists: species, communities, etc., or what are spoken of collectively as "populations," in one sense of that term.

It is not just a matter of our need to avoid being misled by words. The Darwinian revolution brought forth a new way of thinking, which demanded a new kind of logic. Rudiments of this sort of reasoning are often apparent in "cybernetics," "systems theory," and related intellectual movements. Yet the present stage of the art is far from equal to the tasks before us. Indeed, the whole point has been missed. Instead of a simplistic nominalism, many systems analysts embrace an equally simplistic realism. Every class, of whatever type, is treated as if it were a composite whole. This error is evident when Boulding (1956) treats levels as wholes. Although it is clear that some entities ranked at levels might have the ontological status of wholes, others definitely cannot be so interpreted. Thus, California and the other states of the Union together constitute a whole, the United States of America, and there is a whole-part relationship between the Republic and its subdivisions. But in the academic hierarchy, the levels of freshman, sophomore, junior, and senior merely rank autonomous individuals. The motive behind such blurring of distinctions will become increasingly evident as we proceed. Suffice it for the moment to say that the motive is ulterior: it helps to salvage the traditional metaphysics, especially its teleological aspect. Kalmus (1966) praises the Aristotelian approach, for example, and contrasts it with that of Neo-Darwinism, "the current form of Democritean biology." Yet whatever the contemporary difficulties, we do seem to be groping toward more fundamental reconstruction.

One very important historical source for the new way of thinking was the work of early economists. It is no mere coincidence that Darwin found a work on political economy useful in analyzing the structure of the natural one. Classical economics, unlike pre-Darwinian biology, focused upon the activities of individuals. It also conceived of these individuals as acting together in a system, that is, in an economy. And one such interaction, namely competition, turned out to be the driving force behind evolutionary change. Darwin's innovation consisted not so much in inventing an analogy, as it did in recognizing a context. He went on to extend his theory to encompass the whole economy of nature. In the third chapter of *The Origin of Species* he founded the science of ecology. And even in his organismal writings we find the same kind of populational outlook.

In what is called "population biology" the term "population" usually refers to a mere sample, or to collective entities in general. In evolutionary biology, which should not be confused with population biology, the word has an alternative connotation. A population is a composite whole. The parts interact with one another, and one part cannot be acted upon without influencing the others (Park 1942:121–122; M. T. Ghiselin 1969a:54–57). They form a particular kind of concrete system, but certain population biologists cannot believe in this, for, applying their simplistic nominalism, they treat the parts as mere classes.

The simplistic realists err in the converse direction, perceiving, among the parts, a unity and cohesion which really is not there. But the issue is not whether the groups are real but rather what kinds of groups they are. Some groups are integrated, some are not: their members do or do not form a composite whole. Through being so united, the parts can interact in an harmonious and coordinated manner; this makes the difference between a mob and a team, an aggregation of cells and an organism, and a populace and a state. The mere fact that a group can be distinguished does not tell us whether it is integrated, or if so how it is integrated. Some classes do not form aggregates or wholes at all: witness the class of "red-headed girls" or the trophic levels mentioned below. A crowd of people without common aim or leadership forms a collectivity, but hardly a whole in the sense of a cohesive unit. Even where some form of integration exists, it can assume various forms. Thus, men may form a single society because they happen to live in the same place, or because they voluntarily associate for common defense, or because some despot has subordinated them to his will. Now if all these possibilities exist—that a group may either form a whole or not, and that it may be variously integrated—it follows that we need criteria for deciding which possibility is true. That a given commonwealth is a state does not imply that it is a monarchy. But the old teleologists reasoned as if it did, literally viewing the natural world as God's realm and the laws of mathematics, physics, and ethics alike as His legislation. The teleologists of our present day embrace somewhat different mythologies, but reason in the same spirit.

What criteria do we need? How can we tell which possible mode of organization really characterizes the world that lies before us? We must ask a different sort of question. We no longer ask for whose sake, to what purpose, toward what goal, a thing is or acts. We do not presuppose that ours is the best of all possible worlds, or treat the universe as if it had been constructed by God. We opt for another approach, not because the old modes of thought were altogether useless, for they did well enough in their time, but because the new ones work so much better. A single question does the trick: "What has happened?" We construct an abstract model of particular events, derive the consequences, and look to the organisms for confirmation.

Now, to speak of an abstract model of *particular events*, is to enunciate a heresy. For in conventional philosophy of science, history is relegated to second rank. That laws of nature should be prior to particular events would seem implicit in the dogma that knowledge is of universals. Yet in reality this event and that event are all that we have to go on. Even in deriving laws, we need to grasp that which the laws connect. It is as with books. The table of contents is not more important than the text. For evolutionary biology, at least, history, not physics, provides the needed key to scientific knowledge. Our approach to evolutionary and populational problems should involve historical reconstruction as well as more traditional comparative and experimental techniques. We must account for what actually has occurred: what might take place under ideal conditions will never do. Otherwise we shall justify our conceptions of things from our ideas, rather than letting the way things are determine how we shall conceive of them.

When Darwin presented his most novel ideas to the world, many biologists experienced no substantial revolution in their thought. To be sure, an historical outlook became fashionable, natural selection was debated enthusiastically, and new research endeavors were launched. But the conduct of investigation was little altered, and the early evolutionists, in spite of a new jargon, for the most part remained as essentialist and teleological as ever. We see a good example in the following excerpt from a celebrated essay entitled *The Lake as a Microcosm* by Stephen A. Forbes (1887:550).

We have here an example of the triumphant beneficence of the laws of life applied to conditions seemingly the most unfavorable possible for any mutually helpful adjustment. In this lake where competitions are fierce, beyond any parallel in the worst periods of human history; where they take hold, not on goods of life merely, but on life itself; where mercy and charity and sympathy and magnanimity and all the virtues are utterly unknown; where robbery and murder and the deadly tyranny of strength over weakness are the unvarying rule; where what we call wrong-doing is always triumphant, and what we call goodness would be immediately fatal to its possessor—even here, out of these hard conditions, an order has been evolved which is the best conceivable without a total change in the conditions themselves; an equilibrium has been reached and is steadily maintained that actually accomplishes for all the parties involved the greatest good which the circumstances will at all permit. In a system where life is the universal good, but the destruction of life the well-nigh universal occupation, an order has spontaneously arisen which constantly tends to maintain life at the highest limit—a limit far higher, in fact, with respect to both quality and quantity, than would be possible in the absence of this destructive conflict. Is there not, in this reflection, solid ground for a belief in the final beneficence of the laws of organic nature? If the system of life is such that a harmonious balance of conflicting interests has been reached where every element is either hostile to or indifferent to every other, may we not trust much to the outcome where, as in human affairs, the spontaneous adjustments of nature are aided by intelligent effort, by sympathy, and by self-sacrifice?

As we shall see, this teleological outlook has continued to permeate the work of ecologists even down to the present day.

Among the strongest hindrances to the full acceptance of Darwinism has been the continued influence of idealistic philosophy. The typological approach to systematics and comparative anatomy has some curious ecological counterparts. In America, we tend to regard Ralph Waldo Emerson and Louis Agassiz as historical curiosities. Yet the same tradition is very much alive in northern Europe, and its influence upon ecology has been far stronger in that part of the world than is generally recognized. Idealistic plant ecology can be traced back to the work of Alexander von Humboldt (1806, 1807), although of course he had numerous predecessors. He is best remembered for invoking the influence of heat and other physical factors in the environment to explain the appearance of vegetation. We tend to forget that he approached the subject through "physiognomy." Physiognomy was originally the pseudo-science of determining character by studying the configuration of the face, and, to a lesser extent, the rest of the body. By analogy, the "aspect" of a landscape was thought to reveal a divine order lying beyond what was immediately perceived. Humboldt acknowledged his philosophical debt to Schelling's transcendentalism, and mentioned our old friend the "Weltorganismus" or world-organism (Humboldt 1807:90). His influence can readily be seen in the writings of the founders of plant ecology (Grisebach 1880; Kerner von Marilaun, 1863). Even quite recently we find Thinemann (1950) invoking the super-organism analogy and one of the traditional forms of teleology. The underlying metaphysics is more explicit in the work of Friederichs (1927, 1937). He refers (Friederichs, 1927:153, my translation) to a "purposefulnes of all natural phenomena, without which the unity of the Cosmos would not be conceivable." Although he does not cite his source, Friederichs here obviously refers to the teleology of Kant's *Kritik der Urteilskraft* (*Critique of the power of judgment*). German biological metaphysics often invokes Kantian notions which, although long ago refuted by analytical philosophy, retain a substantial following. Indeed, many eminent German biologists would be taken far less seriously were their metaphysics better understood.

Early in this century idealistic biology took the offensive against the theory of natural selection, arming itself with a metaphysical system called "Holism." On the surface no more than the idea that "a whole is more than the sum of its parts," Holism was basically an effort to save the teleological world view. The fundamental proposition from which it derived was nothing more profound than the mistaken notion that one may treat all classes as if they were composite wholes. It brought the super-organism analogy into high repute, and populated the world with a host of goal-seeking systems, the existence of which was dubious at best. The growing influence of Holism is readily apparent in the ecological literature. Rübel (1921) called plant associations "abstractions." The views of du Rietz gradually changed under the influence of Holism. In earlier writings (du Rietz 1921) he considered plant associations abstractions, but rejected the nominalist view that they are not real. Later (du Rietz 1929) he treated them as composite wholes. Like many opponents of Darwinism, his attitude toward teleology was to go overboard on the critical side, even rejecting functional explanation. He thus embraced a notion very popular in Neo-Kantianism, of considering teleology a good thing on a metaphysical level, but rejecting it as a form of scientific knowledge. Yet many Neo-Kantians have gone to the opposite extreme of seeing purpose everywhere.

Subsequently Holism has gradually merged with systems analysis. This latter movement has become somewhat of a fad, but the enthusiasts have largely ignored its metaphysical underpinnings, so that a great deal of naive teleology has crept into the thinking of ecologists. Thus we find Hutchinson (1948) delivering a paper entitled *Circular Causal Systems in Ecology* at a symposium on "Teleological Mechanisms." Further examples may be found in the works of Lindeman (1942), H. T. Odum and Pinkerton (1955), Margalef (1958, 1968), Patten (1959), Watt (1966), and in a number of other works which we will examine later on. The philosophical excesses of certain systems analysts are well known (Hull 1970).

Holism represents but one of many notions that the opponents of Darwinism managed to incorporate into the mainstream of ecological thought. Indeed, one can hardly even begin to understand modern ecology unless one realizes that this science developed much of its theoretical basis during the period from 1880 to 1940, when Darwin was out of fashion. It has never really recovered. The point can perhaps best be demonstrated by considering the influence of two celebrated groups of intellects, which shall here be referred to as the "Harvard crypto-vitalists" and the "Chicago School."

By the "Harvard crypto-vitalists" is meant a group of highly influential biologists, philosophers, and social theorists who often explicitly denied that they were vitalists, but whose positions amounted to much the same thing. At any rate, all to one degree or another rejected Darwinism, and attempted to resuscitate the traditional conception of the world, being strongly teleological in their outlook. We shall here consider only four of the leading figures: Lawrence J. Henderson (1878–1942), a biochemist; Walter B. Cannon (1871–1945), a physiologist; William Morton Wheeler (1865–1937), an entomologist; and Alfred North Whitehead (1861–1947), one of the most eminent philosophers of the time. It would require a full-scale historical study to find out exactly how these men influenced one another, but certainly they were more than simply colleagues. They often got together privately to share their philosophical interests (Evans and Evans 1970) and interacted as members and guests of the Society of Fellows (Barber 1970). A super-organism theme runs through their various writings, and we don't have to know exactly what went on to see that they exchanged ideas on an extensive scale.

Henderson is best known to ecologists for his book entitled *The Fitness of the Environment* (1913). In this work he proclaimed a new teleology, drawing for support upon what seem to him inexplicable examples of useful properties in water and other inorganic materials and in the structure of the universe in general. Yet his position remains ambiguous: he denied metaphysical teleology, but affirmed that something of unknown nature "organizes the universe in space and time." Henderson maintained that natural selection is somehow inadequate, citing with approval the anti-Darwinian works of Mivart and of Geddes and Thomson. As a physiological chemist he was strongly influenced by vitalistic notions, and his biochemistry assimilated Josiah Willard Gibbs's concept of a system (Parascandola 1971). His experience therefore predisposed him toward a super-organism approach (organicism) to sociology, a subject that greatly interested him in later life. It was in part due to the influence of Henderson that sociology, both animal and human, became teleological to a degree that will no doubt astound those readers who are disciplined biologists. We shall defer this topic, however, until chapter 8.

Cannon, too, wrote a highly influential book, *The Wisdom of the Body* (Cannon 1939; the first edition appeared in 1932). To him physiological ecology is rightly grateful for the term "homeostasis"— the maintenance of constant internal bodily conditions in the face of environmental change. Conversely, an excess of enthusiasm for his way of looking at things has led many physiologists to ignore the body's folly, to the detriment of their subject. Had Cannon restricted the notion of homeostasis to organisms little harm would have been done. But he went on to extend it to society, albeit more temperately than did some of his followers.

Of Wheeler (Evans and Evans 1970) we shall have more to say, for he links Harvard to Chicago. He worked on ants, and used the super-organism concept as one of his major analytical tools when dealing with insect societies. He, too, extended the analogy to human society (see E. O. Wilson 1968a).

Whitehead developed an elaborate (indeed, virtually unintelligible) "philosophy of organism" which provided much of the inspiration for vitalist speculation in the middle of our century (e.g. Agar 1938). It is hard to say how much Whitehead was influenced by biologists, including his colleagues. He obviously didn't understand natural selection, and he credits, of all bizarre sources, the "emergent evolution" of Conway Lloyd Morgan (Lloyd Morgan 1926) for having strongly influenced him (Whitehead 1925). He went so far as to call atoms "organisms." Whitehead did seek to find a better place for concrete entities in his metaphysical system, but he did not go far enough. His system is too Platonic, hence populated by bloodless and ethereal creatures, of the sort that might please a mathematician but which are ill-adapted for a truly biological philosophy.

For the history of the Chicago School an excellent source is a book which may be looked upon as its most developed expression, the so-called "Great AEPPS," after its acronym (Allee, Emerson, Park, Park, and Schmidt 1949). The seminal influence here was perhaps the morphologist Charles Otis Whitman (Lillie 1911). In several ways a link with Harvard, Whitman early came under the influence of both Louis and Alexander Agassiz, and later took his doctorate in Germany under Leuckart. At Clark University, Whitman's outstanding graduate student was perhaps William Morton Wheeler. They both later went to Chicago, where Whitman long occupied a position of considerable influence. Wheeler soon left, and ultimately made his way to Harvard. Whitman for some years served as the director of the Marine Biological Laboratory at Woods Hole. He devoted much effort in later life to research on pigeons intended to provide the basis for an attack on Darwinian views. Like many in his time, he advocated a form of "orthogenesis." This term we may crudely define as the idea that evolution proceeds like embryological development, along predetermined lines. It is an historical version of the super-organism notion. We can hardly dismiss it as coincidence that the Chicago School was most enthusiastic about a similar notion in ecology. Communities were treated as organisms, and ecological succession was like the development of an embryo. To be sure, the Chicago School was not all of one mind on this topic. Henry Chandler Cowles (1901, 1911) in his classic works on plant succession at the sand dunes near Chicago inclined toward physiological determinism, stressing physical and chemical aspects of the interplay between organisms and their surroundings. Frederick E. Clements (1916) took the super-organism idea to an extreme. His views were strongly supported in South Africa by John Phillips (1931, 1934, 1935a, 1935b), who invoked for support the holistic philosophy of his countryman General Smuts. Vernon E. Shelford, a student of Whitman, and a most influential animal ecologist, advocated a combination of the ideas of Clements and Cowles (Shelford 1911, 1915, 1931, 1932; Clements and Shelford 1939). The organicist approach was the fundamental organizing concept in the work of Chicago animal sociologists, especially W. C. Allee (a student of Shelford) and Alfred E. Emerson (see chapter 8).

Naturally the ideas of the Chicago School and its allies have met considerable resistance within ecology. Unfortunately the philosophical aspect of such controversy has left much to be desired. Some who have questioned the super-organism hypothesis would appear to have been critical realists (e.g., Nichols 1923, 1929). Tansley (1935:289–292; cf. Tansley 1921, 1929, 1947) in much the same spirit referred to a "quasi-organism." But others have embraced simplistic nominalism, dismissing every ecological group as a mere figment of the imagination. A good example is found in two papers by Gleason (1926, 1939) entitled *The Individualistic Concept of the Plant Association*. Here, ecological universals of various kinds are rejected for reasons that no longer appeal to modern systematists, such as indistinct boundaries and variation. In

a comment appended to one of these papers the younger Gleason remarked that *Homo sapiens* "does not exist except in the mind of man."
Similar notions were later enunciated by Mason and Langenheim
(1957, 1961). Yet more recently the same old metaphysical argument
was raised on the pages of *American Naturalist*. Hairston, Smith, and
Slobodkin (1960) came up with an argument for the "balance of nature" which presupposed a holistic view of the community. Ehrlich
and Birch (1967:104) asserted in response that "a 'trophic level' exists
only as an abstraction." This nominalistic thesis was rebutted by the
statement (Slobodkin, Smith, and Hairston 1967:109) that the reference had to do with "trophic levels as wholes"—which seems to imply
a naive realism, but maybe they had something else in mind. A trophic
level obviously isn't a whole, but it isn't just an abstraction either. Obviously neither faction realized they were debating medieval philosophical notions that were discarded literally centuries ago.

It may not be obvious why so many so-called "population biologists" embrace naive nominalism. The reason has to do with their
methodology, which involves various kinds of "population model"
(see Bodenheimer 1938, 1958; Andrewartha 1961). They use the word
"population" to mean a sample—the individual organisms in a given
area—because, for them, that is the object of investigation. Command-
ing an impressive, if not particularly elegant, array of mathematical
techniques, they can solve a wide variety of problems by counting the
number of individuals of different kinds. As their research can, for
their purposes at least, be conducted perfectly well without considering larger units, they see no reason to include these in their metaphysics.
In the final analysis they have simply hypostatized their methodology.

One might protest that this account exaggerates, and that ecology
isn't really all that bogged down in the middle ages. Perhaps, but even
so a very strong case can be made for the view that the super-organism
approach to community ecology remains a serious defect in both
theory and practice. In conservation ecology the facts are particularly
hard to explain upon any other basis. We find clear-cut resistance to
natural selection and a strong teleological bent in the early writings of
Charles Elton (1930:26, 56, 57). In his very influential book entitled,
The Ecology of Invasions by Animals and Plants (Elton 1958), he
attributes the stability of tropical ecosystems to the high diversity of
their inhabitants. H. T. Odum, Cantlon, and Kornicker (1960) go so
far as to maintain that such systems are diverse because the diversity
helps the systems to survive! The notion manifests itself as follows
in a recent textbook by Eugene Odum (1963:34). "It is now generally
assumed, but without much real scientific evidence, that the 'advantage'
of a diversity of species—that is, the survival value to the community
—lies in increased stability. The more species present, the greater the

possibilities for adaptation to changing conditions, whether these be short-term or long-term changes in climate or other factors." This same teleological approach, explicitly invoking the super-organism hypothesis, underlie Eugene Odum's (1969) proposals for "ecosystem development." Here we find "homeostasis" and "maturity" being used in far more than just a metaphorical sense. Undergraduate instruction and public policy, at least, are seriously threatened by ecological orthogenesis. It is as if we were teaching medicine out of *Science and Health*.

Indeed, medical research itself seems not in the best metaphysical health. We can easily understand why Linnaeus thought lice were good for children. And van Beneden (1876) was not far behind his times, when he played down the pathological aspects of parasitism and stressed the harmonious coexistence of commensalism. His avowed theological outlook could scarcely have let him do otherwise. And it is easy to see how natural selection might mitigate the antagonistic aspect of symbiotic relationships, for it in many cases would be disadvantageous for a parasite to kill its host. Yet at least some contemporary physiologists go too far. Lincicome (1971; cf. Cheng 1971) in an article entitled "The goodness of parasitism" showed that well-fed rats tended to gain weight. He doesn't tell us what happens when food is scarce, and neglects to mention that well-fed eunuchs got fat too. For some parasites, the obvious way of making a living is to treat the host like a firm being liquidated, and kill it—as do many insect parasitoids. Others maintain the host in good condition, but sterilize it, and divert the income, rather than the capital, to their use—*Sacculina*, a crustacean parasite of crabs is the textbook example of such parasitic castration. We can let Lincicome's (1971:214–215) philosophy speak for itself: "The concept of the goodness of parasitism provides an all-embracing basis for a rational philosophy and a more profound appreciation of the significance of this life expression. Goodness is, of course, an anthropomorphic descriptor but how else may we view the organic environment around us? It gives to us a sense of order and progression of nature."

Admitting, at least provisionally, that we need to get away from some of these simplistic philosophical notions, what should stand in their place? I answer that we should recognize the existence not only of classes and individuals, but more importantly of the relations between them. Communities are not just abstractions, they are things standing in relation to one another. They are systems in the sense that their component organisms interact with one another, and live in a condition of mutual interdependence and adaptation. But the components do not relate to each other as do organs in organisms; they differ most importantly in the manner in which they are integrated. We need a better analogy. Perhaps we should go along with Darwin and de Candolle, and view communities as groups of organisms continually at war with one another. At least then any cooperation between members of different species would be viewed as fundamentally selfish, like a military alliance. And harmony would be an armed peace at best. Yet such an analogy could prove equally delusive, suggesting as it does that all is destruction and bloodshed. Perhaps the figure of a commercial economy is more appropriate. It would have to be a *laissez faire* capitalist system, for nothing exists that would stand in the place of a government. Hence it might seem repugnant to liberal sentimentality. But once again we would risk slipping into metaphysics. We are not interested in passing judgment on the manner in which mushrooms conduct their daily lives. We shall use such analogies where they are appropriate for guiding our investigation or communicating an idea. We must attribute properties to systems only when we have confirmed their existence by examining the systems themselves.

TELEOLOGICAL ISSUES IN EVOLUTIONARY BIOLOGY

It is hard to say where ecology leaves off and evolutionary biology begins. Perhaps the best answer is in Hutchinson's (1965) metaphor of an ecological theater and an evolutionary play. Let us extend the analogy and ask what the play is all about. Sex, of course. And like many dramas concerned with that subject, it doesn't have a plot, and it doesn't need one. As sex provides one of the most important bonds uniting organisms into entities of larger extent, it stands at the interface between two fundamental levels of organization. Natural selection, after all, ultimately boils down to differential reproductive success. Yet an evolutionary treatment of reproductive phenomena has to deal with both the organismal and the populational aspects of biology. Individuals reproduce, populations evolve. And, largely because of this complexity, one may easily view the subject from the wrong angle. Of all biological phenomena perhaps none have been so grossly misinterpreted as the reproductive ones. For example, mammary glands are used by organisms other than those that bear them. We have to go beyond the mother's immediate and self-centered interests to appreciate their significance. Yet we cannot take the existence of such adaptations as evidence of some more ultimate altruism on the part of their possessors. If the theory of natural selection is true, they must have arisen because they provided some sort of competitive advantage (at least insofar as the theory by itself is sufficient). To be sure, we need to avoid treating the subject simplistically. Families reproduce and compete, and selection ought to perpetuate those adaptations that favor the success of family groups.

S. Wright (1938) has proposed a mechanism whereby small groups other than families might compete as units. Such competition is usually called "group selection." What is often looked upon as experiential evidence for this sort of evolution has been provided by Lewontin (1962): t alleles in mice. A gene appears in a local population, and, in a few generations, the population becomes extinct; it is a kind of lethal gene that affects a group and takes several generations to act. However the phenomenon differs from what most advocates of group selection have had in mind (let us say that it is not "group selection in the strict sense"). The distinction here will no doubt strike some readers as obscure; and, as we can see from the fact that nobody has noticed it before, it clearly is obscure, but it is nonetheless very significant. An economic analogy may help to clarify matters here. Consider two firms, one of which goes out of business. This could happen in one of two ways. In the first, a partner in one firm runs things in such a manner that prices are cut, forcing the second out of business. This is analogous to group selection in the strict sense. In the second case, the first firm does nothing, while a partner in the other one causes his firm to be run at a loss and it goes out of business. Here we have the analogue of Lewontin's mice: in both cases the system collapses of its own accord, and would do so irrespective of whether competitors were present or not. A second example (Lewontin 1970), closely akin to that of t allele, may be treated in much the same fashion. A virus infects its host. When the virus has multiplied beyond a certain point, the host is killed and the group of them is eliminated. Hence selection ought to favor those viruses that, acting together, do less damage to the host. Yet here the individual viruses all benefit from not destroying their habitat, so that this is not the usual sort of group selection either. It is worth noting that Lewontin rejects the generality of alleged cases of non-individual selection, saying that "there is virtually a complete absence of direct experimental or natural-historical verification of these interpretations" (Lewontin 1970:2). On a more global scale, group selection (in the strict sense) has been invoked by Wynne-Edwards (1962) to explain many instances of what seem to him altruistic behavior or disinterested self-sacrifice. Numerous other writers have applied it to individual cases.

W. D. Hamilton (1964, 1972) argues that close relatives may act to further one another's reproductive success (see also Williams and Williams 1957; Maynard Smith 1964, 1965). This "kin selection" is sometimes considered a third form of group selection. One might object that kin selection is anything but group selection in the strict sense, and this is a point not to be contested. Hamilton, and for that matter Darwin (see below, chapter 7) invoked it as a modified form of individualistic selection, aiming to show that what looked like acts done for the sake of other organisms could be viewed as a more ulti- mate form of self-interest. A parent caring for his offspring obviously is furthering his own reproductive success. And helping more remote relatives could be selected for in an analogous fashion.

Finally, P. J. Darlington (1972) invokes a form of differential extinction to account for various features that seem to lack an individual advantage. We should keep this mechanism conceptually distinct from the others too. Firms that go bankrupt, again, may do so irrespective of competition, and their lesser longevity will tend to populate the economy with those that have features preventing their going bankrupt, irrespective of any advantage to the more successful firms. Just as firms in which the partners cooperate with one another may be the ones that stay in business, extinction may tend to reduce the proportion of species with harmful competition between their members. This is quite different from saying, as would advocates of group selection in the strict sense, that a partner will take less than his due share of the profits in order to keep the firm in business.

The more extreme versions of such theories have been criticized by various evolutionary biologists, particularly G. C. Williams (1966). The hypotheses in question may indeed have some applicability, but at present it appears that the more orthodox selection of individuals and families is more important. For the moment, however, we should address ourselves not to whether such "groupish" mechanisms really operate, but rather to the epistemological point that they are invoked as mere plausible reasons. A phenomenon that might benefit a group is assumed to do so simply because of the possibility. Yet as we have said, so long as alternative explanations will account for the facts, and so long as the experiential data are inadequate to decide between them, we should not prefer one to the other.

Such plausible reasoning is particularly common with respect to reproductive biology. We are told that organisms do things "for the good of the species." Although Darwin, who invented both group selection in the strict sense and kin selection, categorically rejected this particular notion, ever since Weismann (1889, 1891) it has been fashionable even among professed Darwinians to attribute sex, death, reproductive cycles, and the like to the advantage for the species rather than for the individual. A particularly good recent example is Cole (1954) who writes on the "population consequences of life history phenomena." More outwardly teleological is Ayala (1970:483) who asserts, without bothering to justify this view, that "Biological evolution results from natural selection promoting or conserving the adaptedness of populations of organisms to their changing environments." Small wonder that in an explicit defense of finalistic metaphysics he says (Ayala 1968:218) that biology "requires teleology as a category of explanation."

Mayr (1961) attempts to solve the problem largely by adopting the word "teleonomy," a term invented by Pittendrigh (1958), which means the equivalent of teleology, but without the Aristotelian metaphysical taint. But the trouble lies not with our nomenclature, but with our thought habits. We are just as much teleologists when we believe that natural selection has made this the best of all possible worlds as we are when we credit God with the miracle. In the same vein, both Simpson (1958) and Pittendrigh (1958) have revived Samuel Butler's well-known simile, that a hen is just an egg's way of making another egg. They argue that classical Darwinism treated the world too much from the standpoint of individual survival: the hen makes another hen. Reversing matters stresses the perpetuation of genotypes. But this maneuver leaves out the subtlety of both the Darwinian hypothesis and the real world, which it was erected to deal with. Hens make a lot of eggs, and some of them develop into roosters. A more realistic view of the matter would be that an egg is a hen's reason for getting made by a cock. Treating a gene or a population as if it were an Aristotelian "that for the sake of which a thing exists" sweeps the whole problem under the rug.

The idea that a particular feature of some organism exists by virtue of its advantage to a population must not be lightly dismissed as unscientific. It is a perfectly reasonable hypothesis, one that can be tested by empirical research, and which should be accepted or rejected only on experiential grounds. However, the necessary operations are rarely executed. Wynne-Edwards (1962:199–200), for example, suggests that the lights of fireflies may be a mechanism for avoiding the consequences of overpopulation. The fireflies tell one another when they are too numerous, allowing them to cut their reproductive output and avoid catastrophe. He rightly notes that some of the light has nothing to do with mating. Indeed the eggs (Blair 1924) and larvae are luminous and the behavior and anatomy strongly suggest that they have a warning system that protects them from predators. But, contrary to what Wynne-Edwards asserts, the facts of structure and natural history do show that the lights are mating instruments. He says that the powerful light and receiver systems in the males of certain species are evidence that members of the same sex communicate. It does not follow, and his biology is as defective as his logic. In *Lampyris noctiluca* the male's light is nonfunctional, while the female does the signalling and is unable to fly (Schwalb 1961; Naisse 1966). In *Phausis splendidula* both sexes display, but the male flies in such a manner that the light is conspicuous to the females, not to other males, and his eyes are directed at angles that make the former, not the latter, most readily seen (Schwalb 1961). The male's ventral light in *Pyrophorus pellucidus* is extinguished after he lands near the female (Blair 1926). Further arguments along the same lines, and a good review of insect bioluminescence are provided by J. E. Lloyd (1971).

It may not be easy to find out the truth in such matters, however, for—at least at first sight—a variety of hypotheses may plausibly explain the same data, and one may be at a loss to decide among them. Consider, for example, the classical problem of "warning coloration" in Lepidoptera, originally posed by Darwin and partly solved by Wallace. A caterpillar is brightly colored. The conspicuous pattern functions in warning a predator that the caterpillar is distasteful and should not be eaten. But how did this adaptation evolve? One would think that the conspicuousness would be selected against, since the predator would kill the conspicuous individual. Several answers might be given. *First,* one might argue that although the caterpillar is both distasteful and conspicuous, the correlation is due to coincidence. *Second,* the distastefulness could have arisen because of its advantage to the distasteful individual. A bird might take a bite, taste the caterpillar, then reject it. *Third,* the coloration might be thought to attract predators to the more conspicuous individuals, but protect the family in one way or another (R. A. Fisher 1930:159). Some would be sacrificed, but the rest would be left alone. One might even want to consider the possibility of competition between more extended families. *Fourth,* and this is but a refinement of the last, one might invoke kin selection (W. D. Hamilton 1964). The individual killed would cause genes like its own to be preserved—giving a kind of aid to its siblings. *Fifth,* however, we have a possibility that is generally deemphasized. The more conspicuous individuals might actually be less likely to be those that are eaten. If a bird has learned to associate palatability with a drab appearance, and it comes across a diverse group of caterpillars, it may opt for that individual which, so to speak, "looks most like food."

For some of the above possibilities we do have good canons of evidence, such as those described by Süffert (1932), telling us, at least, that we have more than just an accidental correlation. One can find out if the animals evolve convergent patterns, if they occur only in the presence of predators, and if the prey behaves in a manner appropriate to realizing the hypothesized advantage. Records of damaged wings in butterflies show that escape is possible, but this may not tell us very much. Distinguishing between an individualistic, familial, or kinship advantage may be very difficult, for showing a correlation between gregariousness and warning coloration (Benson 1971) is not enough. All of these hypotheses entail an advantage to clustering, for one individual gets eaten and the rest are left alone. Among a group of nonrelatives, all would be safe once one had been killed, and the aggregation would be like any other herding or schooling phenomenon—a way of seeing to it that another individual gets eliminated from the gene pool (see W. D. Hamilton 1971). The evidence for an advantage to not looking like food is equivocal, especially since much of the evidence has mainly to do with butterflies rather than caterpillars. On the one hand, Brower, Brower, Stiles, Croze, and Hower (1964; see also Benson 1972) found that "uniquely colored" moths (ones with an artificial color pattern) tended to be killed more often than controls. On the other hand, some Lepidoptera have developed "intimidation displays" (see for example Blest 1957a, 1957b), suggesting that such mechanisms can work. The answers here are an open issue; the point being only that more questions are in order.

As a final complication of the warning coloration problem, we may note that an individualistic advantage may accrue to the parent rather than the offspring. Thus one might explain a tendency of distasteful caterpillars to aggregate as individuals either protecting themselves, or conferring the same advantage upon their siblings. We need not dispute these hypotheses, but it should be noted that some aggregations having this effect cannot have been produced by either pathway. The luminous firefly eggs mentioned above occur in clusters, and only the action of the mothers can have arranged them in this fashion. Another good example of the same phenomenon has been described (Pillsbury 1957) from a marine fish, *Scorpaenichthys marmoratus*. It lays poisonous eggs in large, conspicuous clusters. This phenomenon seems to be fairly common among newts and fishes (for reviews see Randall 1958; Murtha 1960; F. E. Russell 1965; Kao 1966). The topic of what we may call "parental exploitation" looms large in the present work. Parents "use" their offspring in a fashion determined by their own Darwinian fitness, and irrespective of pain or mortality inflicted upon their kin.

STRATEGY ANALYSIS

Considering the diversity of plausible hypotheses that might be invoked to explain the same phenomenon, and considering the ease with which we are led into error, it behoves us to take extraordinary precautions. The hypotheses must be systematically evaluated in the light of factual data, not just applied wherever they seem to fit. In the present study, we begin by deriving selection theory from its most basic elements. Then, and only then, it will be possible to see how the diverse modes of reproduction are appropriate to particular situations. The technique to be used throughout this work may be called a "strategy analysis," but we need to avoid reifying another metaphor here. A number of biologists have realized the potential of approaches such as one called the "theory of games" for dealing with problems in evolutionary theory (Lewontin 1961). Yet unless one goes back to fundamentals, misconceptions are all but inevitable. We must not take the term "strategy" too literally. The only real strategists here are evolutionary theorists. Carrots do not ponder differential reproductive success. Hence we must concern ourselves with strategical folly, and not presuppose that the organisms will always make the appropriate move. We have to reject outright what Watt (1966:4) calls a "basic tenet of systems analysis . . . that optimization of processes is the central aim of research." We delude ourselves if we subscribe to the notion of Sommerhoff (1969) that evolution is "goal-directed"—especially in the sense that feeding or embryological development might be said to be "goal-directed."

Going back to fundamentals also allows us to discover the "rules." Many kinds of games are possible, and we run grave risk of error if we mistake one for another. Actually evolution has no rules at all, in the sense of those that govern gentlemanly sport. The best we can say is that there are various ways to cheat, and that some kinds of cheating are hard to get away with. And just as there are no rules, there is no criterion of victory. Organisms play the game because, and only because, their ancestors did not lose. We can keep score, in terms of "fitness" or mortality, but we must not assume that the organisms are striving to gather what we consider points.

If we do not know the "rules" we may be unable even to identify the "players." Many who have written on "reproductive strategies" have been concerned, not with the actions of individuals, but with the good of the species, for no more basic reason than the tendency to view the struggle for existence as a contest between "teams" rather than "athletes." If there is any truth in this earlier work, as we should admit there may well be, it will not be discovered until we have some way of knowing what kind of spectacle really lies before us.

When denying that organisms are subject to rules in an anthropomorphic sense, we do not mean that they are unconstrained in their activities, either as individuals or as groups. Traditional ecological parameters, or as one might prefer to call them, dimensions, such as time, space, and energy, must still enter into our calculations. But for present purposes, we seek to relate these constraints to the properties of organisms, populations, and habitats. Our outlook resembles that of a functional anatomist, considering what happens to an animal or plant in various environments. This way of looking at things may be contrasted with that of more formalistic morphology, which treats diversity apart from its causes, perhaps relating it to purely descriptive schemes. A simple analogy will serve to explain what is meant. If someone wants to learn how to play chess, he derives no benefit from being told that as the game proceeds the number of pawns on the board tends to decrease. Rather, he needs to know what moves each kind of chessman can make, and what should be done in various situations. Much of community ecology is made up of rules like that which tells us about the number of pawns. This is the main reason why so much of the literature on species diversity and certain other topics has been futile.

Many of the published works on reproductive strategies ignore the analogues of rules for moving the pieces and pawns in a chess game, and fail to connect these with situations in the real world. A chess player whose first move in the game was to take his opponent's queen would gain a tremendous advantage. But that never happens, because the rules do not allow it. Ecology of this sort can usually be identified by the fact that it deals with computer models instead of organisms.

APPLICATION OF STRATEGY ANALYSIS TO NATURAL
AND POLITICAL ECONOMY

The game analogy is perhaps not the most appropriate one for discussing the actual life of organisms. Let us shift back to the sort of economic reasoning already suggested. It much more nearly approximates the organic world, and indeed accomplishes everything that the "theory of games" can do. In strategy analysis, we seek to understand how organisms with a given kind of structure are constrained by their intrinsic properties, given their extrinsic conditions of existence. An organism cannot be all creatures to all environments. Hence it must opt, as it were, for certain ones among a variety of what may be called "strategical alternatives." Much as it is impossible to construct an aircraft that combines all the mobility of a helicopter with the speed of a fighter, animals and plants often must trade an advantage for a concomitant disadvantage. Homoiotherms must burn food to keep warm, and they may starve to death under conditions where poikilotherms might not only survive, but flourish. Furthermore, an adaptive stratagem cannot work if it does not fit in with all other features of the organism's biology. If becoming sessile helps an animal to cling to a rock, but keeps the sexes from uniting, something must change if sessility is to evolve. Hence the logically possible constellations of adaptive strategies are far more numerous than those that exist in nature. Not all of them produce a biologically workable organism.

Such a functional approach to adaptive strategies provides the key with which we can unlock the secrets of organic diversity. We may reject the notion already mentioned that the variety of organisms exists to endow communities with "homeostasis." Diversity has functional reasons for existence, but these are not advantages for populations. We see its effects in human affairs, particularly in the economical advantages of a division of labor, a topic that we shall repeatedly have occasion to discuss. A craftsman supplies himself with a diversity of tools, not with a number of identical ones. He does so for the same general reason that mammals have evolved several kinds of teeth; and the advantage justifies our thinking that our heterodont dentition is an improvement over the undifferentiated, homodont type. An assembly line operates more effectively than do individual craftsmen, each doing the entire production job for himself, while a hive of bees, with its workers and reproductive individuals may reasonably be thought to enjoy a similar advantage. Similarly we may see in an ecosystem the same kind of advantage that accrues to complex industrial societies through the exercise of specialized professions.

Yet once again we need to be careful. We analyze an economy at the level of a tool, a craftsman, or a firm, or we analyze an ecosystem at the level of an organ, an organism, or a population. Given multiple levels, we readily fail to see what is really going on. The economy may benefit through a division of labor, but this in no way implies that people become specialists, or that industries use assembly lines in order to benefit the economy. Yet H. T. Odum, Cantlon, and Kornicker (1960) enunciate exactly this thesis about political economy in order to support their equally preposterous interpretation of diversity in the natural economy. Along the same lines Preston (1969) calls the cyclic fluctuations of lemmings and other organisms "strategies," expecting us to believe that they are a means of reducing the inroads of predators. He even develops his metaphysics to its own *reductio ad absurdum:* he denounces Keynesian economics, telling us that business cycles are good, and that they will correct themselves naturally.

In both our own and the natural economies, competition plays an important role. For example, it stimulates diversification. If we are deciding what vocation to follow, we avoid those fields that are saturated with practitioners. And species might reasonably be expected to change so that they come to exploit resources that are not already being used. Yet this does not mean that competition should be invoked as a cause wherever we encounter diversity; much of it results from cooperation and other influences. Our incisors do not differ from our molars to prevent one sort of tooth from impeding the function of the other. The mere correlation of diversity with a particular set of circumstances may tell us equally little about the competitive situation in ecology. It is notorious that tropical communities have a larger number of species than do those of higher latitudes (see the next chapter). Yet what may we infer from this fact about the competitive relationships between the inhabitants? Consider once more an economic analogy. Physicians in rural areas are largely general practitioners, while specialists are concentrated in the larger towns and are even more numerous in great cities. Does this necessarily mean that competition within the medical profession is more intense in urban areas, or that physicians become oculists or urologists in order to avoid competing with one another? No. Specialization is an adaptive strategy that increases production, but it is only possible under the proper circumstances. Physicians cannot specialize if there are not sufficient patients of a given kind to support the corresponding specialties. Thus, since virtually everyone needs the attention of dentists, they can make a living in small towns, but this is not true of physicians who specialize in rare ailments. And before there can be a variety of kinds of organisms, the environment must be such that a variety of adaptive strategies will work. In the case of physicians, the requisite conditions are a sufficiently large "extent of the market," as Adam Smith (1776) put it.

Now, when we invoke "the extent of the market" we use two of ecology's favorite dimensions. "Extent" is equivalent to *space*, "market" to *energy*, a term that, in a manner intriguingly like money, stands for resources in general. The third dimension is time. Others have used the term "dimension" in much the same way, except that often they imply no connection between one dimension and another. We use it here to designate a set of entities that are functionally interrelated, much as in Einstein's dictum "time is a dimension." Time, space, and energy are, at least in a number of significant contexts, interchangeable, even in everyday life. If someone asks how far it is between two towns, one can answer in three different ways. One can say "70 miles" (space), or "one hour and fifteen minutes" (time), or "my car uses about three gallons of gas" (energy). And all three enter into consideration when deciding how to travel; for example, to drive faster saves time but one uses more fuel.

Now consider how the three dimensions of ecology might be interrelated, and how they might affect the workability of adaptive stratagems. Certain biological mechanisms require a sort of leisure if if they are to be effective. If a rare plant is to utilize a rare insect as a means of transferring pollen from one individual to the next, it can only do so if the plants can wait long enough for the insects to find the flowers. Hence such flowers as the more symbiont-specific orchids must have a long blooming season. Yet where the season is brief, and where pollinators are available only for a short time during the year, such arrangements do not work very well. Hence flowers in more seasonal environments tend to attract a broader range of insects, or else they switch to pollination by wind, or to some other strategical alternative (see Baker and Hurd 1968; and below). Seasonality imposes the same kind of restriction upon the workability of many other adaptive strategies. There are fewer kinds of parasites in certain groups of organisms at higher latitudes, because it takes so long to find the hosts. However, this rule does not apply to all parasites, and one reason is space: the hosts may live at such low densities in the tropics that the advantage of time is not enough (Dogiel 1966). Economic analogies are here again instructive. In places where the main source of income is the summer tourist trade, enterprises that can only turn a profit by operating year-round tend to be excluded, since the seasonal ones can offer higher wages. Now consider another biological example. Certain kinds of feeding mechanisms will work only if the density of food organisms rises above a certain level or "standing crop" (Lipps and Mitchell 1969, MS). A "filter-feeder," such as a baleen whale that obtains its food by actively swimming through the water, will starve to death unless there are enough food organisms per unit volume of water filtered to compensate for the energy used in maintaining the animal and forcing water through the filter. The distribution and numbers of various groups of filter-feeders are closely related to the amount of food in different regions and at different times. Our whales might be compared to travelling salesmen, who lose money if their customers are so far apart that they expend their whole commissions on travelling expenses. They can compensate somewhat by selling something everyone wants or by peddling a variety of products, but efficiency is thereby reduced. Likewise it readily becomes apparent why salesmen with a particular kind of product to sell will congregate in regions of high customer density, such as urban areas and scientific meetings. The basic principle need only be extended somewhat to account for a remarkable scope of diversity patterns (see Valentine 1971). And the fact that ecology and economics can be treated from the same point of view is no accident. Time and space are resources for businessmen and animals alike. Money and ATP exist for much the same reason. Niches are professions in more than just a metaphorical sense.

We may distinguish quite a unmber of kinds of organic diversity. Ecologists have lately concerned themselves mostly with the diversity of species, and, to a lesser extent, with that of genera and other higher taxa. Yet we may also note a range of diversity within species. Local races and demes are an obvious example. Even within a local population not all the individuals are the same. Genetic recombination may give rise to individual variants, so that each organism is unique. We may make out different stages in the life cycle, often accompanied by radical metamorphoses. A variety of morphs may coexist within the same local area. Phenotypic variants are evoked by environmental influences. We have genetic polymorphism, such as the variety of camouflage patterns within single species of butterflies and snails. And finally we have sexual polymorphism: the different castes of social insects, mating types in dimorphic and trimorphic flowers, and so forth. Among these is sexual dimorphism, or the difference between male and female. All such features have their place in ecological and evolutionary theory, potentially no less important than the diversity of species.

The rest of this book examines the particulars of diversity as they relate to reproduction and to the larger issues already discussed. Alternative paradigms and hypotheses will be treated in the light of their relative ability to handle the data of experience. We shall attempt to apply comparative techniques, such as those that proved so effective in the hands of Darwin and Harvey. From the theoretical constructs it should be possible to predict under which conditions of existence a particular adaptive stratagem should be effective. We may then compare a wide variety of habitats, and see if the organisms that live there do have the expected sorts of adaptations. Or we may compare organisms living in different habitats but nonetheless sharing other features of their biology. Thus, when we find that as group after group of animals has adapted to the great depths of the sea the males have tended to become smaller, we can relate the evolutionary convergence to a common selection pressure. Likewise, when we find sessile organisms, from terrestrial plants to barnacles, displaying hermaphroditism and other convergences in their reproductive biology, we can single out the crucial selection pressure, which relates to their functional anatomy. We hold the environment constant, and vary the anatomy, or else we hold the anatomy constant, and vary the environment. The result is a system of habitat ecology.

The logic whereby hypotheses are tested in a comparative science takes the same form as that in an experimental one. Both the comparative and the experimental approaches have their own strengths and weaknesses; but the two have strengths and weaknesses in common too. Of fundamental importance to both is the principle that although we may refute, we can never verify. Our problems are most evident when we try to use correlations as tests. It is common knowledge that the mere existence of a correlation does not necessarily mean a cause and effect relationship. In natural history, one often argues for an hypothesis by showing that organisms are, in fact, distributed in a pattern that one would expect if that hypothesis were true. But no amount of data consistent with the hypothesis will show that the hypothesis really is not false. Hence the legitimate argument will be one that shows that some alternative hypothesis does not fit the data, while the favored one does withstand the test. One rejects the "null," as experimentalists often say.

In many cases, however, hypothetical explanations are provided without any effort at testing them. They are invoked simply because they fit the data, either to give the appearance of corroboration, or else to salvage some other hypothesis that seems to have been refuted by experiment or observation. The usual term "*ad hoc* hypothesis" will here be applied in both cases. Using such a label seems appropriate, in view of the fact that the two maneuvers have the same effect: they lead to untestability. A mode of reasoning that admits of no possible correction for errors has no place in good scientific research. The proper approach is to take the hypothesis and see how it deals with a variety of new materials. This technique works very well when one is actually doing the research; in general, however, it isn't easy to invoke such an experience as evidence. One can claim that one decided which data would refute which hypotheses before doing the experiment, but nobody has to be convinced. Hence the appeal must take the form "If I am not mistaken you will find that comparable studies give the same results," rather than "I have found that these data corroborate my views."

In the cases of *ad hoc* hypothesizing examined below, we will often find that multiple causes have been invoked to explain a phenomenon. An hypothesis is said to be true because it fits the facts in certain instances where it might be applied, in spite of the fact that the same hypothesis will not work for other instances. The conclusion is drawn, not that the hypothesis is false, but rather that the hypothesis is true when it is not refuted, and that different hypotheses, also true, must be invoked to cover the rest of the data. Now, is it not legitimate to ask, just what kind of science is that? We have no criterion whatever, when such procedures are deemed legitimate, for rejecting incorrect generalizations, but rather we must assume that hypotheses are true until they are refuted for each particular case! The laws of nature are universal: they apply to a range of phenomena and should be testable through their predictive capacity. If we cannot tell when a law will apply, what good is it? If an hypothesis invokes only contingencies— that is to say if it concerns particular events—then it is dishonest to treat that hypothesis as if it were universal in form and necessitous in function.

Given two hypotheses that could explain the same facts, we have three possibilities: (1) both are false, (2) one is true, the other false, (3) both are true, but they perhaps apply when different conditions are met. In the first case, our job is to find some other group of facts that will refute them. The second possibility is taken care of by refuting one of the pair. In the last case, that of different fields of applicability, we decide what conditions have to be met for each hypothesis to apply, and then show that, given a knowledge of these conditions, we can predict successfully using a combination of the two. Biologists do this all the time. There is no problem designing a set of experiments that show that a gland is both a gonad and an endocrine organ. On the other hand, we may find it difficult to decide whether we have two hypotheses that are true, or one that is true and one false, for there is no way to distinguish between a true hypothesis and one that has not been refuted. It may happen that two hypotheses generate much the same predictions, so that we have little or no reason for deciding between them. Under such circumstances, however, it is valid to prefer the hypothesis that is richer in its implications. If both serve equally well to explain a small range of phenomena, but one will cover those plus a lot else besides, it should be obvious which is better. An unnecessary hypothesis is said to violate "Ockham's razor" or the "principle of parsimony." Yet it is only too obvious that an argument from parsimony can never be a compelling one.

Such considerations lead naturally to a final point. Natural history ought to be pursued as a synthetic endeavor. We should aim at constructing a system in which the separate elements lend one another mutual support. Our task is not completed when we have taken a single organism and shown how it seems to work; neither are we finished when we have repeated the operation upon any number of examples. A biology of this kind of animal and that kind of plant is far more useful and convincing when fitted into the larger context of a biology that relates one organism to another and tells us the reasons for their similarities and differences.

3

THE LOVES OF THE PLANTS, OR, THE BIOLOGICAL ROLE OF SEX

Sexual reproduction is . . .
the master-piece of nature.
Erasmus Darwin

In the present chapter we attempt to derive the mechanisms of evolution from basic principles, and, in so doing, to explain why organisms indulge in sex. "Basic" means "general," for the fundamental model should apply to a wide diversity of organized systems. That this is indeed the case will be shown through repeated comparisons with the principles of economics, especially competition and diminishing returns.

EVOLUTION AS A CONSEQUENCE OF COMPETITIVE INTERACTION

Let us begin with an idealized case, and examine some possibilities, at first perhaps seeming to belabor the obvious. Consider an assemblage of animals that are all alike, able to give rise to offspring genetically identical to themselves, but for the moment supplied with just enough food to maintain their own bodies, but no more. Were food to become more abundant, they might then reproduce, and increase the size of the population; but since the original individuals were all alike, and since the offspring would be no different, the subsequent generations would not differ from the ancestral state. Only if there were at least two different kinds of organisms within the population would any change be possible. It does not follow, however, that given two kinds of organisms, a change is inevitable: several other necessary conditions have to be met, and nothing must interfere. If more food became available, *and if* one kind used the additional food more effectively, thereby producing more offspring than did the other, *then* the proportion of the two kinds would change. However, it is at least conceivable that the two kinds might not differ at all in how they were affected by the food supply. The population would grow, but the ratio between the two kinds would remain constant. In the former case, natural selection would have occurred, and the population, of course, would have evolved. The latter case would be an instance of variation without selection. Stated another way, variation and selection are two necessary conditions for a particular kind of evolution. Yet, at least in theory, variation without selection can produce an evolutionary change— if certain other conditions are met. If the population is small, it is relatively improbable that each kind will produce exactly the same number of offspring each generation. Hence, chance alone should tend to change the ratio between them. In other words, sampling error, including genetic drift and the founder effect, may occur. The smaller the population, the larger the random fluctuations in the proportions of variant individuals. In theory then, influences other than selection could have considerable effect, and the features known to exist in populations founded by a small number of colonizers show that the idea is no mere unsubstantiated hypothesis. We would say that variation and small population or "sample" size are both necessary conditions for this sort of change to occur. How often these conditions are met is an empirical matter of only peripheral interest to the present discussion.

Were differential growth in such a theoretical population to go on forever, the kind of organism that reproduced fastest would represent an increasingly larger proportion of the whole, although some of both kinds would always be present. Unlimited increase of this sort never happens in the real world, but something like it may be said to occur when food and other resources are in super-abundant supply. We might call such differential reproduction a form of "competition," yet this may seem to many readers a somewhat odd use of the term. Hence we ought to reconsider what we mean by "compete." Imagine two merchants selling identical goods. They might sell them at the same price and each might sell the same quantity. One merchant could obtain a larger profit if, say, he cut his overhead; and in a sense he would be more successful. Yet in this case we ordinarily would not consider him to be competing more effectively. If, on the other hand, he increased his profits by attracting customers away from some other merchant, thereby exploiting a larger proportion of the market, we would say that he was succeeding in competition. "Competition," in its stricter sense, therefore, implies an interaction between systems (trees, species, craftsmen, firms, etc.) in a context wherein *something* is in limited supply, and wherein the presence of one system affects the success of the other. Whence we get the etymology: the term literally means "to seek together." We may note too, that not all the reasons why an organism or a business succeeds or fails have to do with competition as such. A fly killed by lightning or a factory accidentally burned down are not eliminated by competition. Similarly, evolutionary change by natural selection may be a competitive process, but evolution resulting from the founder effect is something else again. In such cases of non-competitive change, it really does not matter whether competitors are present or not.

Where food happens to be unusually abundant, it is often said that there is little or no "competition for food." One might think, therefore, that under these circumstances food plays no role in competition between the organisms that are using it. When we say that there is "competition for food," we ordinarily mean that two organisms or species limit each other's food supply. When we say that two merchants "compete for customers," we mean that they vie with each other to get customers to buy at their stores. Yet the organisms and merchants are not competing successfully merely by virtue of getting food away from others or diverting customers from the "competition." What matters is reproductive success on the one hand and profits on the other, as they affect the whole interaction. When there is plenty of food, the organism that utilizes it better has a larger supply of energy with which to struggle for living space or deal with predators. When there are more customers than goods to supply their wants, a larger margin of profit may provide capital for times when the market actually does become saturated. A change in the supply of any resource will affect the manner of competition, but will not prevent it.

THE RELATIONSHIP BETWEEN COMPETITION AND SEX

How resources influence ways of competing. Let us now examine in more detail what happens as the availability of resources changes. Consider a small pool of water containing a dense population of a suitable food organism (such as a microscopic alga). A few protozoans introduced into the pool could begin to feed, rapidly growing and multiplying. So long as the food remained plentiful, the successful protozoans would be those who rapidly ingested the algae and turned them into more protozoans. When the food supply began to run low, however, reproduction would become increasingly difficult. At this stage, several alternatives are possible. For one thing, the protozoans could become dormant, expending minimal energy and awaiting better conditions. Or, they could perhaps emigrate to another pond. Another possibility would be to restructure themselves—to vary, perhaps becoming able to utilize a different kind of food. The variation might reasonably be expected to have a genetic basis. If so, it could be produced either by a reshuffling of the individual animal's own genetic material, or by a union with another individual. In either case, there would arise a form of "sexuality," as we shall here use that term. Where such phenomena occur, we should expect that the successful competitors will be those individuals who are most effective at reorganizing their genetic programs and at "getting into the gene pool." Competition, in other words, no longer depends on mere reproductive output, but upon obtaining what might be called "genetical resources."

We might ask why the protozoans do not engage in sex all along. The answer is that sex is wasteful. It uses energy, materials, and time. Hunting for a mate and reconstituting the organism interfere with feeding and growth. Furthermore, there is no advantage to becoming different when a sufficiently workable kind of organism is present. Even if some of the recombinants may be adaptively superior in one way or another, this advantage can be more than offset by the almost inevitable production of some defective ones.

We may compare this biological situation to what happens as economies develop. When the market is only beginning to open up, and demand still exceeds supply, it is expedient for manufacturers to keep on producing more and more of the same product. Retooling is expensive, and therefore it cuts into profits. Shifting to a different kind of product might even result in a loss of sales, since the customers tend to be rather conservative in what they buy. Even advertising is wasteful. Yet when the market becomes saturated, it becomes advantageous for entrepreneurs to seek larger profits at the expense of others in the same business. Different models are introduced, new products are developed, and advertising becomes profitable. Thus in both the natural and the political economy, the competitive situation changes. Yet we should avoid merely asserting that "the environment changes," or that "competition becomes more intense." Neither of these truisms gives a real insight into what is going on.

What is meant by the word "sex." We also need to be careful about how we use the term "sex," for it is ambiguous, particularly because sex happens to occur in association with various other phenomena. To be sure, everyone is free to use words as he sees fit, but our discourse will be more effective if we can be consistent and precise. It also seems reasonable to use the word to designate the most fundamental aspect of what interests us. Therefore, in the discussion to follow, "sex" will be synonymous with "mixis"—literally "mingling" but more precisely a reorganization of the genome. It is readily apparent why this phenomenon might be inadequately distinguished from others that happen to accompany it. Sex is often confused with reproduction. The two commonly go together, and the reason is that the occasion of producing a new organism is an appropriate time for producing one that is different. A working factory has to be shut down for retooling; but one still under construction is inoperative, and hence changes can be introduced with no inconvenience. The commonly used expressions "asexual reproduction" and "sexual reproduction" treat sex as if it were a subset of reproduction. We do have reproduction without sex: not only the usual sorts of vegetative growth, budding, and fission, but also reproduction with gametes but lacking genetic recombination (amictic parthenogenesis). On the other hand we also have sex uncon-nected with reproduction—what might be called "areproductive sex-uality." This occurs in the "automixis" of many protozoans (for review see Kofoid 1941; Sonneborn 1941; Wenrich 1954a; Dogiel 1965; Grell 1967). Of course, automictic organisms reproduce, but this is only because all organisms reproduce. There is no necessary connection be-tween the two phenomena.

Likewise, the physical union of individuals is often confused with what will here be called "sex" in the sense of recombination. Sex often means the genital intercourse of multicellular organisms, or the fusion of gametes or protozoa. The ambiguity these days is a problem mostly for laymen, but some confusion can be detected in the older scientific literature. For modern biology it is more important that we draw a distinction between sex and the more particular form of it called "amphimixis" (Weismann 1891), or the formation of new kinds of genomes from those of different individuals. Thus we have *mixis* and its two subclasses: *automixis* and *amphimixis*.

The difference between amphimixis and automixis is important because crossing with another individual has the fairly obvious advan-tage of increasing the variety of genetical resources that can be made available to the offspring of a single parent. Owing to amphimixis, the individuals who engage in it are members of reproductive communities, or biological species. Access to the variability has important conse-quences for the individuals who make up the species, and this in turn makes the species themselves effective reproductive entities. They pos-sess stores of latent genetical variability, and also have a capacity for mobilizing it that would not be possible were they mere collectivities of asexual or automictic individuals. Hence they each form economic units, which, like partnerships and trusts, have definite competitive advantages. Amphimixis also allows a far more rapid reorganization of the whole gene pool than would otherwise be possible. The flow of genes, like the movement of capital and technological expertise, provides for adaptive responses, and for exploiting new resources.

As a final terminological point, we need to distinguish "sex" from "gender." In common usage there is a clear ambiguity, deriving from the fact that we say there are two sexes, male and female. The technical literature suffers from the same kind of problem. Sometimes, for instance, "bisexual" is used to mean existing as populations with male and female individuals (gonochorists). At other times it means exactly the opposite, populations whose members all have the reproductive apparatus of both sexes (hermaphrodites). Such ambiguities once again suggest that problems have arisen because it is so hard to deal simultaneously with two levels of organization. Who competes with whom, for what, and how, never becomes clear unless we keep these levels conceptually distinct.

The semantics of competition. In the present analysis, sex is shown to play a crucial role in the economy of competitive systems. Yet the very idea of competition has created serious problems for many ecologists, especially those who have embraced simplistic nominalism, as when Ehrlich and Holm (1962:655) write "at most, only individuals can compete." Just as well might we argue that General Motors cannot compete with Ford. "Individual" must not be used equivocally. Of course, we have many examples of individualistic competition of the cruder sort in the biological world (see Knight-Jones and Moyse 1961) and in human economic life as well. A naive view of competition as a direct physical interaction, like the combat between gladiators, has been rightly criticized by Milne (1961), who notes how such confusion seems implicit in Darwin's metaphorical "struggle for existence." General Motors does not engage in military operations against Ford. On the other hand, the most destructive forms of competition in the political economy have been restrained through legislation; one would expect it to be carried on unbridled in the natural economy. The remorseless chemical warfare that goes on among plants (J. Bonner 1950; Whittaker and Feeny 1971) is just what Darwin would have expected. It would have been far more readily apprehended, had traditional ecology not embraced a teleological outlook.

Further problems arise from inadequately distinguishing competition from certain phenomena that accompany it. Exploitation and interference are, as Brian (1956:339) points out, important "ways" of competing, but this does not render them "forms" of competition any more than are salesmanship and employment. Some exploitation is of nonbiological resources, for example, so we should better refer to exploitation as an aspect, rather than a subset, of competition. We may likewise note an unfortunate overemphasis upon what resources are being competed "for." Thus Milne (1961:57, italics his) writes: *"Competition is the endeavour of two (or more) animals to gain the same particular thing, or to gain the measure each wants from the supply of a thing when that supply is not sufficient for both."* The result has been to treat competition excessively from the point of view of its possible results, especially limitation of growth and competitive exclusion. As Darwin (1859) originally maintained, and as was later elaborated upon by mathematical biologists (Gause 1934), organisms with similar needs tend to compete for whatever it may be that is needed. As a result, closely related species tend not to coexist very long, or else they evolve some kind of difference in niche. Competitive exclusion is analogous to one firm driving another out of business, and the diversity it produces in organisms has obvious economic analogues. Yet it seems evident that the competitive exclusion principle has been widely misconstrued. Ayala (1972) very effectively rebuts the commonly held notion that "two species competing for the same limiting resources cannot stably coexist in the same habitat" (Ayala 1972:348). Yet Gause (1934) himself asserted nothing of the sort. He only maintained that exclusion was inevitable if the two species occupied the same "Eltonian niche"—if they had the same "profession" that is. Two firms competing for the same customers may be unable to obtain a monopoly if their respective markets are not quite identical. Thus, one firm may undersell another, but not attract all the customers if the second manages to provide better service or greater convenience.

To understand competition we need to view it as a process that occurs within complex systems, one in which the interactions go on between the subunits of such systems. To compete is for a subsystem to act in a manner that tends to divert resources from another such subsystem to itself, to its own advantage, and to the detriment of the other. For the present we may as well leave the analysis at this level, and not elaborate upon such terms as "advantage." These are used in the conventional biological way, however. What really matters is that "perfect competition," in the sense used by economists, does not exist, any more than the "perfect gas" of physical chemists. The members of an ecosystem or other economy are in "imperfect competition." And whether or not two subsystems (organisms, species, firms) compete is a matter of degree.

Many biologists seem inclined to consider organisms with few resources in common as not existing in a state of competition. Of course, creatures of different species, living in different communities on different continents are as much out of competition as were the inhabitants of different parts of the world in times when there was little or no international trade. Yet it seems a mistake to treat competition as something going on only between members of a single trophic level, as Hairston, Smith, and Slobodkin (1960) would seem to do. A more extreme example is the inclination of Birch (1957) and Milne (1961) to reject the views of earlier authors that predators compete with their prey. They do so because they reason in terms of what resources are competed "for," rather than of "how" organisms compete. Milne (1961:50) even goes so far as to ask "How can two species compete to avoid a predator or a parasite or a disease?" One might answer "by running as fast as possible," but it seems more reasonable to fall back upon economic analogies once more. Granted, the man who owns a sawmill is in a more obvious, and perhaps more significant, state of competition with other owners of sawmills than he is with growers of trees and builders of houses. Yet whenever a buyer and a seller transact business, they vie with each other for an advantageous price. Deceptive advertising may just as well be intended to delude the customer into thinking that he is purchasing more than he really is, as it is to get him to patronize one firm rather than another. Buyer and seller compete for money, predator and prey compete for nutrients and energy. Likewise, two organisms or firms will benefit in some respects, yet suffer in others from competition; hence, we cannot define the word in terms of its overall effect on reproduction or profit.

As a final difficulty of competition theory, and the one toward which the present discussion is primarily directed, let us consider the idea that selection should not be viewed as competition. Birch (1957: 13) makes this point with respect to sexual selection, which for our purposes is doubly appropriate. He gives the example of male *Drosophila* of two different strains having differential success in mating, so that one type replaces the other. He argues that this is not an instance of competition, because "It is not dependent upon crowding or a shortage of resources. It takes place at quite low densities as well as high ones." Yet the conclusion follows only because he brings in the extraneous issue of limiting resources. Merchants seek to maximize their profits and attract customers away from other firms, irrespective of what happens to be in short or abundant supply. Likewise, organisms may reasonably be expected to do whatever most effectively gets them into the gene pool. The male fly that mates more often than other members of its species will, ordinarily, leave more offspring. Thus a *purely reproductive competition* is possible, in which one individual diverts genetical resources from another to his own advantage. This "sexual selection" (Darwin 1859; 1871) reveals the futility of thinking about competition as being "for" some limited resource—or of thinking in terms of what ecologists call "density dependence." Of course, the density of populations affects how organisms compete, and it must be taken into consideration. But the differences in numbers per unit of area do not affect the obvious truth that competition is universal among organisms. The economy of nature is competitive throughout, and selection, whether natural or sexual, is the most compelling manifestation of this truth.

The function of mixis. We are now in a position to specify the biological role of sex. In most general terms, sex is a means of *mobilizing genetic variability.* As such, it is not to be confused with the more ultimate sources of variability, in particular the various forms of mutation, which are the raw materials with which it works. Sex acts so as to obtain variability, and, most important, *to release it at the appropriate time.* We may infer, if this is so, that its activity in giving unity to the population is a side-effect of the circumstance that the component individuals of the population are mutually benefitted through bringing together and combining their separate genetical resources. In other words, they do not engage in sex "in order that" the population will be integrated. On this point we part company with some advocates of group selection. The situation is once again like that in commerce and trade. Markets are established because buyers and sellers satisfy their wants, and not because they seek to better the conditions of society through forming a financial community.

The proposition that sex furthers the mobilization of variability only tells us in a very general way what it does, not why. It is functionally unenlightening, like saying that a market helps trade without bothering to explain why commerce exists. We shall here defend and examine the following thesis: *a parent causes his offspring to vary under those environmental conditions in which the parent gains a competitive advantage through doing so over other members of his own species.* We should make sure that this statement is not misinterpreted, however. By "vary," is meant either to depart from the ancestral condition, or to differ from a sibling. And of crucial importance is the point that sex is advantageous to the parent, not necessarily to the offspring. Sex does "benefit" the sexual individual, because it causes him to leave more descendants. This is no tautology, for other alternatives are possible. The conclusion would seem to follow from the orthodox theory of natural selection advocated here, but not so from some of the "group selection" notions to be discussed later on.

The precise conditions under which organisms might be expected to vary have not been adequately explored. Indeed certain biologists have suggested that natural selection cannot account for the existence of sex because they find it hard to imagine such circumstances. Yet it is not hard to come up with some possibilities. We have already given an example that illustrates the basic conditions. Where a population is rapidly expanding, and where simply duplicating the old genome unchanged results in more rapid multiplication, then selection ought to eliminate the sexual individuals. If, on the other hand, the parental stock has used up those resources that allowed the population to multiply, then it should be advantageous for any individual member of the population to produce variant offspring, which might be able to exploit some other kind of resource. The idea that variable and unpredictable environments, in which it is uncertain which genotypes will be best adapted in the next generation, are likely to be conducive to sexuality has been suggested by various authors (see R. A. Fisher 1930; Mayr 1963; G. C. Williams 1966). Yet we are given little by way of concrete ideas as to how selection actually works in these cases. The danger is clear from the refutation by Schopf and Gooch (1972; Gooch and Schopf 1972) of the hypothesis that genetic variability correlates positively with environmental instability; this hypothesis, we should add, could have been tested on the basis of data published years ago (see below).

The problem becomes even more intricate when we try to account for the influence of family structure. This topic has scarcely been touched. One possibility (M. T. Ghiselin, 1969b:192) might be that if the offspring in a mammalian family were all alike genetically, the probability that an entire litter would die in a plague might be high, leaving the parents potentially able to use the available resources on raising young but with none to raise. Were only some of the offspring to die, more resources would be available for the care of the survivors, and a competitive advantage would result. Extending this line of thought, we may note that little consideration has been given to the possible benefits of individual differences within the family, and even this has mostly to do with differences between mates. A diversity of offspring might be advantageous either in reducing competition between siblings, or in furthering cooperation between them. This matter abounds in pitfalls, however. For example Trivers (1972:138) writes "Furthermore, to decrease competition among offspring, natural selection may favor females who prefer single paternity for each batch of eggs (see Hamilton 1964)." But having closely related offspring will not decrease the amount of competition between offspring and hence benefit the parent, but rather will merely set up a selection pressure rendering it advantageous to the related siblings not to harm each other. Therefore the mother would not realize the hypothesized advantage, since the reduced competition would not yet have evolved.

ON TESTING HYPOTHESES THAT MAY NOT BE
MUTUALLY EXCLUSIVE

Let us reiterate, and then apply, what was said at the end of chapter 2. We should not attempt to decide empirical issues on the basis of purely theoretical arguments, but rather let the answer be established by the data. If a theoretical argument implies that there is no sex, it is the theory that needs to be corrected, not the animals. Even so, given a set of competing hypotheses, we can approach their experiential evaluation in a number of ways. Some ways are better than others. We can show that an hypothesis is false by deriving predictions from it, and then showing that these predictions are not borne out by observation or experiment. Or we can apply the parsimony criterion and show that, given two hypotheses, one provides a broader range and greater detail of true predictions than the other. In the latter form of inference, we prefer the hypothesis with the wider field of applicability, even if the hypothesis that predicts less may be used to explain a great deal, and even if there are circumstances for which we cannot find a reason for preferring one or the other. If someone really wants to believe that the narrower hypothesis is true, we may find it very hard to convince him; and the reason is that he has opted for a form of reasoning destitute of scientific respectability. His hypotheses have become untestable, for he casts them in the form of a "particular affirmative" proposition of the sort that cannot be refuted. Instead of asserting "all x is y" or "10% of x is y" he says "some x is y." He invariably has a way out, for he can always affirm that his hypothesis is true for an indeterminate number of cases that are not contradicted by experience.

If we are going to test competing hypotheses, we need to reason so as to decide between them. It is possible, however, that two hypotheses, which on the face of it seem to exclude each other are really both true, but have different fields of applicability. The situation with respect to the utility of sex, in which perhaps group advantages might be true, by no means lacks precedent. Natural and group selection may be compared to a number of other pairs of hypotheses: strict genetic inheritance versus the inheritance of acquired characters is one, and the allopatric versus sympatric models of speciation another. Let us use the latter to illustrate how one should and should not reason about such matters. We may recollect that the "allopatric model" means that populations will not form new species unless they are split up into subunits by some kind of geographical barrier or other extrinsic isolating mechanism. The competing sympatric models deny that such a barrier is necessary, so that a species can split up into two reproductively isolated populations without the two ever having been spatially isolated. Now if we treat the allopatric model as a universal proposition, saying that if it is true and its alternative false, then we should expect to find certain patterns in the distribution of species. For example, there should be more species where barriers to dispersal have long existed. Such tests have been successfully applied, and the allopatric model is preferred to the sympatric model. The tests are not sufficiently discriminating to show that there is *no* sympatric speciation, but they do test both hypotheses, and they do show that one is preferable, even where either could be invoked *ad hoc*. Yet even if there are occasional cases of sympatric speciation, we should be able to design a test of an hypothesis that provides a mechanism for it. That is to say, we could look for more speciation in groups in which that mechanism is possible. Those who look for instances of sympatric speciation are perfectly respectable scientists; those who invoke it merely because it can be invoked do not deserve to be taken seriously.

We may test the hypotheses concerning the adaptive value of sex by arguments of like form. The hypotheses here invoked predict, in great detail, a wide range of phenomena that have long been unaccountable, and cover many phenomena which competing ones leave unexplained. Where group selection has been invoked, we can generally show that individualistic selection provides a better, more detailed explanation. But one could hardly be expected to demonstrate conclusively so vague and elusive a proposition as "there is no x, such that x has been favored by group selection." On the other hand, if someone developed a model in which a population advantage was hypothesized, he perhaps could test it. Yet if occasionally proposed, such models are rarely brought down to earth. On the basis of strictly *a priori* reasoning, Maynard Smith (1971) concluded that natural selection does not favor sexuality. Instead of rejecting his premises, he proposed a rather vague model in which sex was thought to be favored by the competition between semi-isolated groups within species. He left it at that. This hypothesis could easily be tested, for if it is true, then those populations in which sex is most frequent should be those that are broken up into small populations with little gene flow between them. There should not be so much as one instance of even a moderately panmictic species in which both sexual and asexual reproduction coexist. But the genetic systems of marine oragnisms, which are often highly panmictic thanks to the action of currents, and in which sexual and asexual reproduction do occur within the same species, but where sexuality is the general rule, contrast markedly with conditions in semi-isolated bodies of fresh water, where mixis is less frequent. The general picture developed in the rest of this chapter shows how sex benefits the individual in various habitats in a way predicted by theory. No alternative deserves to be taken seriously unless it is argued on a comparably empirical basis, and unless the logical form of the argument is respectable. However, some of the competing hypotheses can be shown to be wrong, or at least inadequate, and their shortcomings might just as well be noted. In addition, considering some other possibilities, even those of purely historical interest, puts the whole range of thinking on such matters into a more general frame of reference.

ALTERNATIVE HYPOTHESES CONCERNING THE
ADAPTIVE SIGNIFICANCE OF SEX

First let us consider an old idea that has roots in the confusion between sexuality and gender. The distinction between masculinity and femininity has served as a basis for metaphysical thinking at so many times and in so many cultures that it ranks with the stars as an object of superstition. A dualism involving active and passive forces, and often a harmony of opposites, is a recurrent theme in pre-Socratic and ancient Chinese philosophy. This antique heritage has influenced modern biology far more than one might think. In Figure 1 we reproduce a cut from a work by Lorenz Oken (1805), one of the most influential nature-philosophers of the last century. For comparison we show, in Figure 2, a not too dissimilar construct from a book on the evolution of sex by Geddes and Thomson (1901:295) published almost a century later. These authors, who contributed prolifically to the anti-Darwinian literature (e.g., Geddes and Thomson 1911), maintained that the female has an anabolic or constructive aspect, which balances the catabolic or egoistical nature of the male. We may view such untenable, vitalistic thinking as the historical precursor, at least, of some more recent speculative excesses. The effort of Hartmann to erect a general theory of sexuality incorporating a notion of bipolarity is perhaps not an unfair example (see Hartmann 1956a, also older scientific and metaphysical writings anthologized in Hartmann 1956b; for critique, Pringsheim 1964).

A second hypothesis about the function of sex is that it acts to rejuvenate the individual, the species, or both. It is maintained that the organism somehow wears out, and that the union of two individuals is needed for an adequate reconstruction. These hypotheses in part arise from inadequately distinguishing sex from reproduction. They have retained a certain measure of popularity largely because the two are so hard to separate. Maupas (1889) and Bütschli (1876:420) were perhaps overimpressed by the fact that after ciliates conjugate they seem to grow more rapidly. One could easily think that these organisms are undertaking a more limited form of sexuality when they dissolve their macronuclei and afterwards reconstitute their somatic DNA. Jennings (1913) showed that most of the cells that have conjugated actually grow less rapidly, not more rapidly, than those that have not. He concluded that crossing is advantageous because a certain minority of the exconjugants do grow faster. (See also Jennings 1910, 1929; Siegel 1958; also Beale 1954, for a review of genetics in *Paramecium aurelia*.) Unfortunately these experiments do not rigorously exclude the rejuvenation hypothesis. A mechanism that gets rid of semilethal mutations, even at the price of some mortality, would at least in theory have some

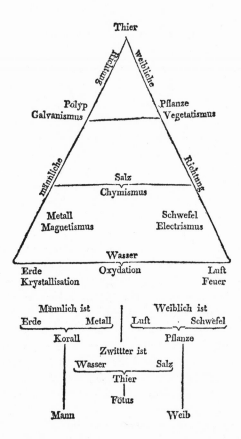

Figure 1. Diagram from a work by Oken
(1805) depicting a mystical scheme of
polarity for the sexes.

advantage too. Hence it might contain a small measure of truth, at
least, and the notion of Dougherty (1955) that sex arose as a mechanism of DNA repair at least deserves to be taken seriously. Yet mere
differential reproduction of a purely mitotic sort ought to eliminate the
deleterious mutations. A catastrophic reconstitution of the genome
would therefore appear superfluous. Indeed, one postulated advantage
of ameiotic parthenogenesis is that it allows mutations to accumulate
when they are advantageous in the heterozygous state (Suomalainen
1950). The fact that many organisms get along for generation after
generation without ever indulging in sex would seem to imply that
although it may help, sex is by no means necessary for maintaining the
quality of the nucleus. Nonetheless, the production of a whole group
of recombinants all at once does allow the parent to generate some
offspring in which lethals are not expressed and others in which they

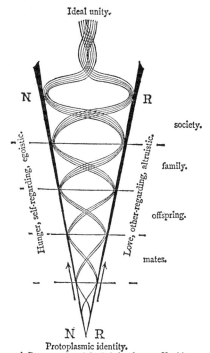

Ideal unity.

N   R

society.

Hunger, self-regarding, egoistic.

Love, other-regarding, altruistic.

family.

offspring.

mates.

N   R
Protoplasmic identity.
Diagrammatic Representation of the Relations between Nutritive,
Self-Maintaining, or Egoistic, and Reproductive, Species-
Regarding, or Altruistic Activities.

Figure 2. Diagram from a work by
Geddes and Thomson (1901) showing a
later analogue of the scheme in Figure 1.

are expressed, at a moment when new recombinants must be produced.
In this limited sense we can recognize an advantage through deleterious
genes getting culled out, but only as an accidental byproduct of advantageous genotypes being produced. The rejuvenation hypothesis,
therefore, does not necessarily contradict the one maintained here, but
only complicates the over-all picture. We shall return to this topic when
we consider the advantages of diploidy (see also Fauré-Fremiet 1953;
Corliss 1965).

For much the same reasons we may reject some older notions
(reviewed by Wenrich 1954b) that envision sex as resulting from the
physiological advantage of combining two individuals. It is true that
when genetically diverse parents are crossed the offspring benefit from
such advantages as heterozygosity. Likewise, fungi are well known to
unite and form "heterokaryons"—organisms with two kinds of nuclei,
one from each parent (see Raper 1966). It seems reasonable that, here
too, the availability of more kinds of genetic material could produce

a more physiologically vigorous individual. Such hypotheses, however, reverse the order of cause and effect. If organisms did not engage in sex, they would not develop the homozygosity that leads to inbreeding depression, and therefore they would derive no benefit from outbreeding, at least for the reasons that the hypothesis invokes. Beadle and Coonradt (1944) argued that such formation of heterokaryons may have been an intermediate step in the evolution of sexuality. Their idea does have the advantage of suggesting gradual stages whereby the recombinational apparatus could have arisen; yet so far as fungi go, the course of history would seem to have been otherwise. Heterokaryosis has been intercalated into the life cycle since meiosis originated. Hawes (1963), Mayr (1963), Stebbins (1960), and others are right in saying that sex is a very old phenomenon. We possess well-documented cases of sexuality occurring in a wide variety of organisms, including bacteria. The capacity for recombination can readily be lost, and the fact that it seems to be lacking in certain groups does not imply that it was never present in their ancestry, or that we can never find it now. Probably every creature that now exists has descended from mictic ancestors, in spite of the negative evidence.

A final alternative hypothesis is that the function of sex is to provide that variation without which natural selection could not act. It differs from the hypothesis supported here, in that sex is considered advantageous to the species, not to any individual organism. This "species-advantage" idea is known to have occurred to Darwin (see M. T. Ghiselin 1969a), but usually it is attributed to Weismann (1889, 1891). Although Weismann did not, at least in these documents, assert that sex must have originated because of this advantage, he did explain the function of various phenomena, including death as well as sex, through benefits to the species. It is difficult to reconstruct his actual views, because he does not explicitly consider the group-selection issue. At the time he was writing, it did not seem to matter (but see Lloyd Morgan 1890–1891:184–186, 193). His thesis has been reiterated by many leading evolutionary biologists (Muller 1932; Mayr 1963). The topic was taken up by C. D. Darlington (1958, first edition 1939) in a very important book entitled *The Evolution of Genetic Systems*. Darlington, too, supported a form of group selection, maintaining that not only is the individual selected, but "the community and the race" (C. D. Darlington 1958:232). This premise led him to indulge in some rather questionable criticisms of Darwin and other biologists (C. D. Darlington 1956, 1959); his attacks have been examined by G. C. Williams (1966), myself (M. T. Ghiselin 1969a), and others. His latest production (C. D. Darlington 1969) attempts to treat the human species as if it had such group adaptations. Very likely differential extinction and similar factors have had a certain influence on the evolution of genetic systems. However, when other biologists took up this topic, they were not always particularly critical when they came to the issue of selective mechanisms. They simply invoked a species benefit wherever there was a phenomenon to be explained. The result was some very significant work showing correlations between the amount of recombination and various ecological parameters (Lewis 1942; Mather 1943; Stebbins 1950, 1958; Fryxell 1957; Grant 1958; additional references below). Yet a correlation is notoriously likely to mislead, particularly in a situation such as this, where an evident benefit to a population could very likely result from individuals acting, so to speak, in their own interests. Hence, although we now have a great deal of useful material, it needs to be reinterpreted. The most recent literature is now coming to distinguish more clearly between the species and the individual advantages; yet we find a certain degree of ambiguity that suggests the issue is not taken seriously enough (see Kalmus and Smith 1960; Maynard Smith 1968; Crow and Kimura 1969). However, many of the older works need only a slight theoretical reinterpretation: they were not altogether wrong, just not quite right.

Recombination rates are, in fact, known to be subject to natural selection, and their geographical distribution patterns are what one would expect if individuals, not species or even demes, were reproducing differentially. The amount of crossing over and the degree of chromosomal polymorphism reflect the amount of genetical recombination. By comparing different species and races upon a geographical basis, it is possible to find out if features such as these are attuned to local circumstances. A great deal of pertinent work has been carried out by Dobzhansky and his collaborators (e.g., Dobzhansky, Burla, and da Cunha 1950; de Cunha and Dobzhansky 1954; Dobzhansky, Levene, Spassky, and Spassky 1959; see also Carson 1955, 1965; Bodmer and Parsons 1962; Mayr 1963) on *Drosophila*. The more widespread and ecologically diverse species recombine at a greater rate than do those with a more restricted range: hence change in the genetic system has occurred as a fairly recent adaptive response to the local circumstances and not as a result of species level extinctions. Indeed, the rate of recombination varies within populations, and species with extensive gene flow between populations do have crossing over and other manifestations of recombination. At least for some species, therefore, it seems most unlikely that sexuality is favored because of its long-term survival value for populations. Yet it would be futile to argue by enumerating single instances of individual selection producing adjustments to local circumstances, for, as we said, the advocates of group selection can always fall back upon the possibility of exceptions. Hence we might as well let this one example demonstrate that individual, rather than group, selection does control the extent of mixis in at least some

situations. To attribute the evolution of genetic systems to group selection is therefore a gratuitous assumption. To show the general adequacy of individualistic selection, we shall now proceed to demonstrate
how sex, along with other reproductive phenomena, relates to selection
pressures that would seem to affect fitness at the organismal level.

EMPIRICAL EVIDENCE: RECOMBINATION RATE AND SIZE

Among the most noteworthy features of genetic systems in unicellular organisms is that their sex appears to occur far less often than
it does in multicellular ones. It was long thought that bacteria do not
recombine at all, but we now know that the negative evidence was misleading (Lederberg 1955; Wollman, Jacob, and Hayes 1956; Jacob and
Wollman 1961). The discovery of bacterial sexuality was delayed
partly because nobody knew where to look; also the usual means for
isolating and culturing bacteria involve growing them with abundant
food, and this may have selected for asexual clones. Bacteria are peculiar in displaying what is called "meromixis"—only part of the
genome may be transferred from one individual to another. Furthermore, bacteria as a rule are haploid, each has only one chromosome,
and recombination occurs only after a period in which the cells are
physiologically incapable of engaging in recombination, the "com-
petence" being environmentally evoked (Hotchkiss 1954; Schaeffer
1956). All these features would seem to indicate that the recombination mechanism has been adapted by selection, so that the genome is
reorganized only very slowly, in the sense that it occurs over a relatively large number of generations. With haploidy there is no store of
latent variability in the form of unexpressed recessive genes. Progeny
cannot inherit different sets of chromosomes in bacteria, and actual
mixis is both partial and rare. It is difficult to be sure just what prevents
diploidy and multiple chromosomes from evolving in bacteria; but the
physiological sterility is hard to explain away as something other than
a mechanism that prevents crossing.

Various authors (see Bodmer and Parsons 1962) have noted that
as organisms become seemingly more "advanced" in evolutionary grade,
there occurs a correlated increase in the frequency of recombination,
and in the degree of elaboration of mechanisms that further it. The
"higher" protists would seem to constitute an intermediate group
between the prokaryotic bacteria with very infrequent sexuality on
the one hand, and the more "highly evolved" multicellular plants
and animals on the other. Even at this transitional level, however, Wenrich (1954b:339) points out that an "examination of the status of sexuality in the Protozoa reveals that the groups in which syngamy is
most commonly found are those which are most highly evolved." We
cannot explain the correlation away on the grounds that sexuality has

been a necessary condition of becoming highly evolved: the organisms in question have been mictic from the start, and it is only the frequency of recombination that has increased, having done so progressively, step by step along with the phylogenetic advance. We may also note that here what is called evolutionary progress has involved the origin of multicellularity, with the result that several cellular, or at least nuclear, divisions intervene before a new individual is produced. Some protozoans display an analogue of multicellularity, in that a multinucleate (polyenergid) condition is followed by multiple fission, leading to the formation of many small, uninucleate or binucleate gametes. A superficial understanding of this state of affairs led many biologists to explain the correlation between "multicellularity" and sex in terms of generation time. The idea has been that organisms that multiply very rapidly can evolve so as to keep up with environmental change by virtue of mutations alone (Stebbins 1960:208; Mayr 1963; cf. Maynard Smith 1968). (Selection for mutability is an intriguing side-issue—see, for example, Leigh 1970a.) Yet this leaves unexplained why protists don't gain an additional or an alternative evolutionary flexibility by engaging in sex. Very likely they do get such an advantage; and the reason why our predecessors have failed to see that this is so has been a failure to realize how the individuals are benefitted. Although, so far as populations are concerned mutation is an effective means of producing variation, mutation occurs more or less at random, so that the individuals are unable to control its phenotypic expression. Since, under ordinary conditions, mutations are deleterious, the appropriate adaptive strategy is to suppress them until such time as they are needed. An individual does best when it can accumulate potential variability, and can cause it to be phenotypically expressed only when the time is ripe. We may note that the advantage here derives, not from being a variant, something that the offspring cannot control, but rather from producing them: it is the parents that engage in meiosis. Such control over the release of variability can be attained with a genome in which there is a certain amount of redundancy, either through diploidy or through multiplication of genes in series. Sex can then produce diverse genotypes, and the organism need only possess some way of responding to the appropriate environmental influence to realize the full benefit of the arrangement.

Now consider an idealized case in which a unicellular organism underwent mixis at each cell division. Assuming that it would produce only two offspring, and granting that most mutations are deleterious, the probability would be low of producing a pair of which even one was more fit than itself. However, were that individual to produce a large number of progeny, the chances would be commensurately increased that at least one adaptively superior recombinant would be produced among the offspring of that individual, in spite of some defective ones being produced as well. One would expect, therefore, a genetical determinant to be selected for if it programmed the cell to divide several times, and then and only then to switch to sex. It would not matter whether these cells lived as parts of a multicellular organism or not. We may regard it as a virtual truism that there simply *must* be an upper limit to the rate of change in the structure of the genetic material. But in spite of vacuous speculation about "genetic loads" and the like, this has absolutely nothing to do with the fitness of populations. What matters is not the survival of the collective entity, but rather each individual organism's fitness, as determined by the probability that its issue will include a certain proportion of healthy and otherwise successful individuals. A very high mutation rate is a "lethal" phenotype, for it is equivalent to sterility. The upper limit of recombination will depend on the parent's ability to benefit from the advantages of recombination on the one hand, and to avoid its deleterious effects on the other. An individual that produces many offspring will be less likely to have all of them genetically defective, than will one that produces just a few. But note that we must distinguish between number of offspring produced, and number recruited into the next generation. Organisms such as man, with a low rate of populational increase, undergo considerable prenatal mortality. As shall be illustrated later on, the ability of organisms that care for their offspring to concentrate resources upon those that survive this early decimation gives the parents a considerable advantage.

We need only extend the same general line of reasoning to explain the adaptive significance of diploidy. So-called "higher" organisms, both animals and plants, are well known to display a polyphyletic evolutionary trend, in which the anagenetic level (degree of advancement) correlates with the predominance of the diploid generation. The traditional, species-advantage explanation for diploidy has been that it provides a store of variability which confers evolutionary plasticity upon the population. Genes, unexpressed in the heterozygous recessive state could be maintained as raw material for selection. Accordingly, the haploid condition of many monerans and protistans is generally explained away on the basis of their deriving flexibility through rapid multiplication. Yet it seems curious that diploidy has gradually evolved from a transient stage in haploid, unicellular organisms, to an "isomorphic alternation of generations," with both forms equally important in the life cycle, but with the variation supposedly locked up in the diploid state continually exposed to selection in the haploids.

The difficulties resolve themselves when we see that meiosis produces combinations of genes, while mitosis preserves them. When two haploid cells unite to form a (diploid) zygote, they produce a new genetic entity. If this zygote then undergoes meiosis, it will give rise to recombinant, haploid individuals. And if the reason for sex is to produce these recombinants, nothing would necessarily be gained by putting off the events responsible for it. Yet such delay might be advantageous. If conditions appropriate to variation did not eventuate until some time after fertilization had occurred, then mitotic division of the diploid cells would produce more offspring than would otherwise result. The diploid cells would themselves be recombinants, and could become evolutionarily adapted to various environmental situations, ultimately leading, perhaps, to alternating generations. The physiological advantage of having two sets of chromosomes might eventually become significant, but there is no obvious immediate benefit of this sort.

Diploidy in unicellular organisms should, then, evolve where the conditions appropriate to gametic fusion and to the production of recombinants do not coincide. Diploids can produce new combinations of genes either by union of gametes to form a zygote or by meiotic division. The advantage of the latter mechanism is that there is no need to wait for conditions to occur which are appropriate for the union of gametes. The simpler organisms are haploid, not because they are "primitive," but because they occupy "simple" environments and have "simple" ways of life. They remain genetically the same until the time comes for recombination, which they accomplish simultaneously with gametic union. But where the functional anatomy becomes more complicated, even some protistans derive advantages from a temporal division of labor.

The selective pressures favoring diploidy are accentuated in multicellular organisms. The situation is best intelligible when we consider forms that alternate between multicellular diploid and multicellular haploid generations. Two sessile, haploid individuals can cross to produce zygotes, and then reproduce mitotically to give new individuals; but until these undergo meiosis, they are all genetically alike, albeit different from haploid stock. When the diploid individuals undergo meiosis, however, they give rise to a large number and a wide diversity of offspring, and they continue to enjoy the benefits of variation when they recombine to form new individuals. Therefore, the advantage of mixis is largely lost for haploids until such time as they produce recombinants after gametic union. Diploids, on the other hand, receive the benefits of recombination in the form of varied haploid offspring. Although building up a multicellular body has some advantages for a haploid organism, the best adaptive strategy, for both them and their haploid parents, is to concentrate upon getting together for union of the sexes as soon as possible, growth notwithstanding. In consequence the haploid phase tends to be increasingly specialized for dispersal and sexual union, the diploid for the vegetative functions. (For older ideas see C. E. Allen 1937; see also Raven and Thompson 1964.)

If one wishes to gain a full appreciation of how the empirical data can be organized according to the foregoing scheme, one needs to survey life cycles throughout the plant and animal kingdoms. As this would require a staggering amount of detailed information, let us content ourselves with a brief sketch and a few illustrative examples. The main point to keep in mind is that the haploid phase is reduced where the diploid becomes larger and at the same time is able to get along without the dispersal function of a large and independent haploid. The size correlation itself is admirably illustrated by J. T. Bonner (1965), who, however, invokes the usual species advantage hypothesis to explain it.

Among unicellular organisms, haploidy throughout most of the life cycle is the general rule. The exceptions are most instructive. In diatoms of the order Pennales, two diploid individuals actually come together and unite within a mass of jelly (see Patrick 1954). Meiosis ensues, and each cell produces one or two gametes, which are exchanged to form one or more zygotes. A free-living haploid phase is not present, and clones or even individuals give rise to offspring that are not identical twins. The Foraminifera typically have an alternation of generations. *Discorbis patelliformis* (see le Calvez 1953) has a fairly large diploid phase which gives rise by meiosis to a large number of haploid individuals that do not grow to be as large as their parent. These haploid individuals crawl around and ultimately fuse together as pairs (the so-called plastogamy). Each individual then divides mitotically to form a large number of gametes, which in turn form many zygotes that once again initiate the diploid phase. In Foraminifera generally thought to be the more primitive members of the group, the haploid and diploid individuals attain the same bodily size, and plastogamy does not occur. The gametes are free-swimming, and unite in the open water. Here the chances of genetical uniformity among the offspring of a single parent are less than they are in *D. patelliformis*, and extended vegetative growth in the haploid state would not be disadvantageous. A lack of advantage to enlarging the haploid stage in *D. patelliformis* seems a better explanation than that given by le Calvez (1953:200), who suggests that the haploids are physiologically inferior to the diploids. If his hypothesis were correct, then the reduction should characterize the group as a whole; it does not. A number of protozoan taxa include groups of large, diploid organisms that split into numerous, small haploid gametes, which then unite to form new diploid individuals (Acantharia and certain sporozoans for example). This condition, which Wenrich (1954a:243) thinks was primitive among the ciliate protozoa, would produce a diversity of offspring. In some ciliates a large animal gives rise to just one or to a number of smaller gametes, which are recombinants yet are nonetheless diploid (see, for example, Finley 1943). In most ciliates, however, such small gametes are not formed. Here the advantage to diploidy may be that it allows variability to be released first at conjugation, and then later on by automixis.

The generality of multicellular algae (Chapman and Chapman 1961; Fritsch 1965) displays a clear trend in the direction of suppressing the haploid phase, although isomorphic alternation of generations is very common. A very few, especially a number of red algae, have a larger haploid generation. They differ from their isomorphic relatives, in that the diploid state is sessile on the haploid. These are not very large or very successful organisms, and the parasitic diploid stage would restore variability. Among the larger brown algae, particularly the kelps, the diploid stage is by far the larger. The haploid would appear to have lost most of its functions other than dispersal and syngamy. The diploid phase gives rise to large numbers of haploid gametes, of equal size, which settle to form microscopic but multinucleate male and female plants, which cross to form a single zygote, which in turn grows to form the large diploid organism. Here, large numbers of offspring, spread over wide areas by ocean currents, encourage a maximum amount of recombination. In some brown algae (Fucales) the haploid phase is reduced to nothing more than gametes, suggesting the usual condition of metazoans.

In the fungi too (Raper 1954, 1966), "phylogenetic advance" and morphological complexity are thought to correlate with the suppression of haploidy. Yet there occurs the very significant complication of heterokaryosis, which we have already mentioned. The haploid spores germinate to form haploid mycelia, perfectly capable of an independent existence; these may grow together and exchange nuclei. Yet the nuclei, instead of fusing remain separate for some time. The heterokaryons thus produced have two complete genomes, and therefore may be expected to enjoy some of the advantages of heterozygosity, but this does not mean that they derive no other benefit from the fusion. It is thought that some exchange of genes may take place between the separate nuclei, recalling the meromictic processes of bacteria, and perhaps explaining why certain forms (*fungi imperfecti*) seem to lack mixis. The nuclei unite and meiosis occurs, but these events take place when the haploid spores are formed, and when the cycle is about to begin anew, not during the vegetative period. An alternation of generations characterizes many parasitic fungi. The haploid stage may be free-living and saprophytic, or it may be parasitic, and, after formation of the heterokaryon, invade a new host. After some vegetative growth, sexual recombination is once again followed by dispersal. In such arrangements, the haploid phase is specialized for dispersal and union of individuals. Recombination, however, is delayed until vegetative growth has built up a large supply of haploid nuclei. Since these nuclei have diverse origins, they can release a great deal of variability in their very numerous offspring. If they had to wait until the end of the life cycle to join diverse nuclei together, they would have to expend additional resources in getting the sexual individuals together, and in nothing else. As it is, a haploid fungus grows until a suitable mate happens to be encountered, and the two remain joined together until the time appropriate for recombination. The energetical advantages here should be fairly obvious. Perhaps more significant is the fact that actual release of variability occurs with maximum body size. We might add that a minority of smaller fungi have the sort of haploid-diploid cycles known in other organisms.

In all higher plants except the Bryophyta, the haploid generation is highly reduced, and, again, has become a specialized dispersal and syngametic mechanism. The bryophytes (liverworts and mosses) are organisms of moderate size, with a small, partly parasitic diploid phase living on the larger haploid one. This arrangement should serve to ensure an adequate amount of variability, much as was suggested for the occasional red algae with which the bryophytes display a remarkable convergence. In "higher" plants the haploid dispersal stage ultimately becomes nothing more than a pollen grain. Even so, its function is limited to the transfer of genetical material from one individual to another, since transportation is now accomplished by means of seeds. The plant is able to produce a large number of highly diverse offspring.

Multicellular animals have done much the same as higher plants have, except that, ordinarily, haploidy exists only in the gametes, which are single cells, unlike pollen grains which are entire multicellular organisms. The gametes are little used in gene dispersal by animals, but those aquatic animals which do not move around very easily often retain the old mechanism. Yet even so we may note that haploidy has re-evolved in a number of metazoan lineages. It has arisen only in the males, only when these are produced through parthenogenesis, and only when the males take no part in care of the young. Here the females have become the main feeding and dispersal stages, and the males' role in the release of genetic variability gets concentrated where it is needed most.

FURTHER EMPIRICAL EVIDENCE: RECOMBINATION RATE AND FOOD

A second kind of evidence bearing upon the individualistic role of sex has to do with the environmental conditions that favor its alternatives. Sex ought to be adapted very closely to immediate environmental circumstances. If we can show that organisms switch their sexual reproductive strategies from one pattern to another as the conditions of existence change—if we can show that they do so in response to, or in correlation with, an environmental stimulus—and if we can relate what happens to individual advantage, then we have made a case for our thesis. For we can thereby demonstrate the immediate, rather than the long-term, significance of generating variability. In so doing we shall be able to get away from the somewhat groupish, and therefore misleading, interpretations of life cycles proposed by Cole (1954), Lewontin (1957), Murdock (1966), Istock (1967), Murphy (1968), and others. More individualistic models, such as those of Gadgil and Bossert (1970), are more satisfactory, and we may hope that when particular habitats are linked to their real inhabitants, a valid synthesis will emerge. Our task, however, is not easy. An organism may reproduce asexually either because sex is not advantageous, or because it is not possible. Or amictic parthenogenesis may help to prevent inbreeding. In addition, environmental challenges may be met by strategical alternatives: migration or encystment, for example, may do that job just as well as sex.

We must now return to the discussion begun earlier in the present chapter, on the difference between populations that are expanding where resources are abundant on the one hand, and those that have reached the limit that the available resources will support on the other. In the case of abundant resources there is little competition except with respect to reproductive and feeding efficacy, and therefore there is not much reason to reconstitute the genome. Where trophic resources have become limiting, we would expect competition to render variation highly advantageous. If a population is held in check by predators, yet food remains readily available, sex would still be advantageous, insofar as recombinants are able to compete better with respect to avoiding being eaten. However, rapid growth rather than crossing might still be an effective way of competing. Schmalhausen (1949) maintained that planktonic organisms low in the food chain are held in a condition of evolutionary stasis, because they supposedly are killed in a random manner by organisms that filter them unselectively out of the water. Were his hypothesis true, one might expect sexuality to be less common in such organisms, but the most we can say is that predation might put a premium on rapid growth rather than defense. From the mere fact that an organism may not be selectively killed by certain predators, we cannot infer that no selection will occur. Differential mortality and differential reproduction will still characterize whatever individuals remain after a certain percentage have been eliminated at random (Kurtén 1953:91). In fact, the oceanic plankton contains many small multicellular organisms that have become greatly altered in morphology—chaeto-gnaths and pteropods, for example. For many groups of planktonic unicellular organisms we have an abundant fossil record, showing re-peated adaptive radiations leading to great morphological complexity and considerable mass extinction. Manifestly, these organisms have been adapted by selection in a manner that suggests extensive partitioning of resources (Lipps 1970a). Asexual reproduction in the plankton is no more pronounced than it is on shore, and mainly seems to result from the difficulties of finding a mate. We need to look to areas where neither food nor predators are strongly limiting. Such conditions are realized in temporary bodies of fresh water, among parasites, and in certain other circumstances as well.

In several groups of organisms, parthenogenesis (Vandel 1931; Suomalainen 1950; Oliver 1971; Slobodchikoff and Daly 1971; S. G. Smith 1971; and other works cited below) may be shown to characterize the inhabitants of fresh water, rather than marine, environments. No instances of the reverse situation have yet been documented. Among the Gastrotricha (Remane 1935; Hyman 1951b) the largely marine Macrodasyoidea are hermaphroditic and sexual, while the predomi-nantly freshwater Chaetonotoidea are parthenogenetic and have lost the male system, except for two marine genera. In rotifers (Wesenberg-Lund 1923, 1930; Remane 1929–1933; Donner 1966; Birky and Gilbert 1971), which are mainly freshwater organisms, sexuality would seem to preponderate to a greater degree among marine members of the group. The males are distinctly suppressed in fresh water, where they become increasingly smaller than the females, and tend to appear only at the period of maximum population size. The parthenogenesis becomes most apt to involve a switch to sexuality in moderate sized bodies of water. Evidently the bloom of food resources and the subsequent shift in competitive interaction requires a growing season that is not excessively brief, while the appropriate fluctuations may not occur in the largest lakes. Certain freshwater snails that have lately moved in from the sea have evolved parthenogenesis upon invading the new habitat (Sander-son 1940; Winterbourn 1970).

Several groups of Crustacea illustrate the same principles. The Cladocera (Wesenberg-Lund 1926; Banta 1937, 1939; von Dehm 1950; Frey 1965, 1966) are well known for their cyclic parthenogenesis and occasional production of diploid males. The group is abundant in fresh water, although many live in the sea. The Ostracoda (Puri 1966; Furgeson 1967) have fewer males, hence supposedly more partheno-genesis, in freshwater populations than in marine ones. Like the rotifers and gastrotrichs, ostracods and cladocerans are small animals, and there-fore apt to flourish in temporary ponds, since their short generation time allows them to take advantage of the rich food supply.

Similarly we may note the abundance of parthenogenesis among various kinds of parasites, especially those of small body size which cannot exploit their host to the full by growing large (as some tapeworms do). Some nematodes alternate between a free-living sexual stage and a parasitic parthenogenetic one (for review see Cable 1971). The parthenogenesis of thrips (Thysanoptera), gall wasps (Doutt 1959), and aphids and other Homoptera (Kennedy and Stroyan 1959; Nur 1971) fits the same picture. It would appear that all these systems allow the animals to exploit a fairly abundant resource, yet one that is not saturated by the parasites or predators. That is to say, they adapt to fluctuating conditions, especially those of temperate climates. If the growing season is too short (as in high latitudes) or if it is too long (as in the tropics) the life cycle becomes simplified. Aphids in the tropics get along mainly on parthenogenetic reproduction (Shull 1925; Mordvilko 1928; Cognetti and Pagliai 1963). The simplification of life cycles in crustaceans and rotifers of far northern lakes has been treated extensively by S. Ekman (1905) and Olofsson (1918). Mordukhai-Boltovskoi (1967) attributes the rarity of males in cladocerans of the Caspian Sea to the more moderate conditions of this body of water. Here again, as in aphids, sexuality is adaptively appropriate to temperate conditions. Parthenogenesis among vertebrates (Beatty 1967) is exceedingly rare, and its adaptive significance remains obscure. It occurs in a few fish (Lieder 1955, 1959; Schultz 1971), amphibians (Asher and Nace 1971), and reptiles (Lantz and Cyrén 1936; Darewski and Kulikowa 1961; Maslin 1962, 1971).

As is well known, certain kinds of organisms that multiply asexually for extensive periods may be stimulated into undergoing genetical recombination by withholding food. Maupas (1889:404), who, in a classical monograph, studied this phenomenon in a wide variety of ciliates, pointed out that similar techniques had long been used by horticulturists to induce flowering. It would seem that many organisms put off sexuality until such time as their resources are exhausted to the point that a different combination of genes is likely to be advantageous. K. W. Fisher (1961) points out that bacterial recombination does not occur until late in the growth cycle of the culture. The organisms are not even "competent" (disposed to recombine) in early stages, and although it is hard on the face of it to say just why the organisms behave in this fashion, the inference is hard to avoid that sex is not advantageous, or even that it is detrimental, at this time. Among organisms with "cyclic parthenogenesis," several asexual generations are followed by a sexual one. In others, a form of self-fertilization may be followed by outbreeding. The systems vary considerably as to details, but as a general rule the sexuality tends to be concentrated at the end of the growing season, just when one might expect the environment to be closest to saturation. Only a small percentage of animals with this kind of response have been studied, but enough is known to say that "hunger" *per se* is not always responsible. Any of a variety of cues may be exploited, and if the organisms use something that allows them to prepare themselves for sexuality in advance of the opportune moment, they might gain a considerable advantage. (For response of rotifers see Gilbert 1968; Birky and Gilbert 1971.)

We should not conclude, however, that the only consideration of importance is the availability of food. Any environmental change that makes it advantageous to vary may come to evoke facultative sexuality. A most instructive example is to be found in the symbiotic zooflagellates that aid in the digestion of wood by both termites and the related wood-eating roaches (Cleveland, Hall, Sanders, and Collier 1934; Cleveland 1965; Honigberg 1970). Both the roaches and the termites lose their intestinal fauna when they moult, but it is reestablished in different ways. In the roach, the symbionts are passed out of the body via the faeces, and are taken up again in the food. The roach secretes a hormone when it moults, and this hormone causes the flagellates to indulge in sex. Such a hormone-induced sexuality is well known in various parasites, for example in *Opalina ranarum*, a protozoan that lives in the gut of amphibians (El Mofty and Smyth 1960). In termites, the flagellates are transferred when one animal feeds directly from the proctodaeum on the gut contents of another termite. The environmental change is more moderate here. Although it was long believed that sexuality is lacking in the termite flagellates, more recent work has detected it. Clearly the crucial time for recombination occurs, not with transfer from individual to individual, but with the founding of new colonies. In another group of cellulose-digesting symbionts, the rumen ciliates, epidemics of conjugation have also been observed (Dogiel 1925). Since in both the flagellates and the ciliates a large number of species conjugate simultaneously within a single host animal, it seems unlikely that we are dealing with a means of synchronizing sexual activity among members of the same species and nothing else. Many organisms have breeding seasons, but, especially among free-living tropical forms, those of different species tend not to coincide. In fact, one would only expect selection to favor separate breeding seasons as an isolating mechanism and perhaps as a means of reducing competition.

We have noted how parthenogenesis seems to preponderate in freshwater ponds and other temporary bodies of water. Parthenogenetic organisms also display intriguing patterns of geographical distribution which have been studied in great detail by Vandel (1927, 1928, 1931, 1937, 1940). Sometimes the parthenogenetic lineages can be traced to a mictic ancestral population occupying a relictual habitat (see, for instance, Mockford 1971). This suggests that the derivative forms moved from an area saturated with the ancestral stock into regions where resources are easier to come by. But other interpretations are possible. In general, the parthenogenetic forms live in higher latitudes. To some degree, but by no means always, the parthenogenesis is accompanied by polyploidy. A trend toward polyploidy in higher latitudes and in other extreme habitats has been well documented; it is usually explained in terms of the polyploids being more vigorous (Löve and Löve 1943). In some cases, the same basic trend in both parthenogenesis and ploidy is also apparent toward lower latitudes, especially where the habitat becomes dryer, suggesting that it may go along with increasingly harsher or more variable conditions. In the Notostraca for example, Longhurst (1955a, 1955b) found that the southern forms dwelling in rice fields are mainly sexual, while more northerly ones, evidently occupying transitory habitats, are increasingly parthenogenetic. Most of the foregoing work refers to terrestrial and freshwater organisms; little information is as yet available for marine ones. However, Feldmann (1957) did find that a marine alga of the genus *Cutleria* (which has an alternation of haploid with diploid generations) showed a high frequency of (haploid) males occurring in equal numbers with (haploid) females in the Mediterranean. In the English Channel the (haploid) males are less common and the (haploid) females reproduce parthenogenetically. This species, like a number of other algae (see Lewin 1960), displays a greater frequency of asexual reproduction at higher latitudes by the diploids. We may consider this phenomenon an historical accident, resulting from the concentration of dispersal and gametic union in the haploids and the greater physiological vigor of the diploids, rather than as raising problems for the hypotheses here developed concerning the adaptive significance of diploidy. Vascular plants on mountains and in high latitudes display a comparable reliance upon asexual reproduction (Bliss 1971), but in view of the complexity of sexual arrangements in this group it seems prudent not to assign causes here without further study. Similar trends have been noted in asexual reproduction of freshwater planarians through fission of the body (see Kenk 1937). The evidence seems to show that the gradients somehow reflect individual adaptations to local circumstances, and that they are not manifestations of competition between populations, as might appear at first glance. The gradients take two forms. One of these is what Vandel calls *geographical parthenogenesis:* the local populations near the extremes of the range are more often parthenogenetic than they are near the center or optimum. One could perhaps fit a group selectionist interpretation to a pattern of this sort, although a survival advantage for individuals should work just as well. On the other hand, much of the geographical difference involves what Vandel calls *spanandry:* the proportion of males within the local populations decreases toward the edge of the range. Here the individualistic advantage is manifestly reflected in differential responses to a gradient in the intensity of selection pressures.

Various hypotheses, often invoking an ability of parthenogenetic forms to colonize marginal habitats, or their supposedly greater physiological vigor, have been put forth to explain the foregoing gradients, but none seem adequate. In view of what has already been said in the present work, the basic explanation should be more or less evident. However, rather than merely state what it is, it will be more enlightening to put the whole subject of geographical gradients in broader theoretical perspective. A host of biological phenomena display latitudinal gradients. Some of these, such as Allen's rule, which relates the smaller extremities of mammals toward the poles to conservation of heat through a reduction in surface area, are obvious consequences of fairly simple environmental influences. Yet those that relate to the structure of communities and to reproductive strategies are far less easy to handle. However, many of the gradients in question have a single basic mechanism in common, for a single type of model will explain them. Let us therefore expand the present discussion, allowing us to show how the distributional patterns of sexual phenomena may be placed within the context of a more comprehensive system. As the general theory predicts far more than just the sexual phenomena, any successful test of its validity is reason for thinking that the more narrow versions are true as well.

THE GLOBAL PATTERN OF ORGANIC DIVERSITY

Species diversity gradients. The problem of latitudinal gradients in species diversity has lately exercised the pens of numerous ecologists (a useful summary is Pianka 1966; major works include: Dobzhansky 1950; Kusnezov 1957; P. J. Darlington 1959; Hairston 1959; Hutchinson 1959; Kohn 1959; Fischer 1960; Southward 1961; Paine 1962; Day 1963; Connell and Orias 1964; Simpson 1964; MacArthur 1965; Valentine 1967, 1969, 1971; Whittaker 1969; H. G. Baker 1970; R. G. Johnson 1970, 1971, 1972; Hedgpeth 1971b). However, up until very recently, little has come of this work in the way of really finding out what is going on in nature. The fact that species are more numerous in the tropics than in higher latitudes did not escape the attention of earlier naturalists. Darwin (1854), for example, provides a quantitative treatment for barnacles. We now know that there are many kinds of species diversity, but we remain pretty much at a loss as to how to deal with it in any meaningful way. As Simpson (1964) points out, a minority of taxa display reversed gradients; and some are bimodal, with diversity highest in the temperate regions and lowest near both the equator and the poles. Of the many hypotheses, some evidently are wrong, but others could all have partial validity. Such a pluralistic world—if such it be—would tend to imply that the subject deserves to be taken out of the hands of theoretical ecologists and be made once more the province of naturalists.

Yet among these speculations we may see the basic raw materials from which a synoptic theory might be constructed. Dobzhansky (1950:215) maintains that in the tropics competition seems to be more intense, for it has given rise to a wealth of specialized organisms. He relates low diversity to the fluctuation in environmental forces in seasonal habitats, which puts a premium on versatility rather than specialization. Dobzhansky and others invoke seasonality as the controlling environmental influence. Yet a flexibility in responding to some aspects of the environment does not preclude specialization with respect to others (see also Simpson 1964). As we pointed out in the preceding chapter, however, certain types of enterprise will not work under specifiable environmental conditions. We gave the diversity control of filter-feeding whales as an example. Since these animals can only obtain enough food to survive near the poles and in upwelling areas, there are naturally more of them in such places, both more species and more individuals. The whole relationship, in other words, would appear to be for the most part energetically controlled. One should not get the impression, however, that this kind of pattern affects only filter feeders. Animals that make their living through active pursuit of their prey require a certain density of food before they can turn a profit. The concentration of auks and penguins in the higher latitudes of the northern and southern hemispheres respectively was first discussed by A. R. Wallace (1876[2]:367) and has, more recently, been treated by Stehli (1968). Sea lions and ecologically comparable forms display this bimodal pattern too (see Lipps and Mitchell 1969, and below). To see that high standing crop levels rather than some other correlate of latitude is involved here, one need but observe that the various groups in question extend toward the equator on the west sides of continents, where productivity is high, and that they do so to a greater extent than can be explained upon the basis of temperature alone. Although different taxa respond in different ways, the basic influence, for many groups at least, is seasonality. An increasingly brief growing season near the poles results in sudden influxes of food into the system. Rapid growth, tending to outrun the grazing abilities of herbivores and predators, keeps the populations rapidly expanding when not being decimated through starvation. Specialized predators are at a disadvantage in so far as they waste time hunting for a particular sort of prey organism, rather than taking advantage of whatever happens to present itself. Hence, abundant energy favors those animals that scramble most effectively for food, over those adapted to catching it or using it effectively.

Now consider what happens in the more stable environments. As time becomes increasingly available, a greater number of adaptive stratagems are workable. Hence, competition between species may result, not in competitive exclusion but in the subdivision of resources, and thus in smaller niches. That constancy in the supply of resources, and not their absolute quantity, may sometimes be responsible for a high species diversity would seem to follow from the work of Hessler and Sanders (1967) on the deep-sea benthos. Here, where at once conditions are exceedingly stable yet energy is very hard to acquire, the species are remarkably diverse. However, competition *per se* is not what determines whether a community will have few species or many. The basic cause is rather a lack of competitive exclusion. In certain instances, competition in highly stable environments will actually lead to a low level of diversity. What goes on may perhaps be best elucidated if we take a fine study by Rützler (1965) as a concrete model. He counted the number of sponges on rocks of different sizes. On the smallest rocks, which are very often depopulated through being rolled or overturned by the action of waves, species were few. On larger and larger rocks, increasingly less likely to be moved, the number of species gradually increased. But a point was finally reached at which the diversity of sponges leveled off, then even diminished. On the smaller rocks, the most influential limiting factor was the availability of time, a resource necessary for the larvae to locate the rocks and to colonize them. Above a certain population density, space became ever more limiting, and competitive exclusion set in.

We need only alter the components of the foregoing model to explain the bimodal species diversity curve of the Brachiopoda (Valentine 1969). This is another group of attached animals, and one of those that is unusual for being most diverse in the temperate areas of both hemispheres, and somewhat attenuated in both high latitudes and the tropics. Interestingly enough, brachiopods also have their metropolis in waters of moderate depth, being for the most part subtidal creatures, not very abundant at great depths. For a full understanding of the phenomenon we need to take other kinds of organisms into account, however. Brachiopods (a good introductory book on them is Rudwick 1970) are two-shelled filter-feeders, convergent in both shell and feeding with the Bivalvia. They have been replaced in many habitats, especially shallow-water ones, by molluscan ecological analogues (see Stanley 1972). A similar case is that of freshwater gastropods, which have a bimodal species-diversity curve, unlike the marine ones, which peak at the equator. Hubendick (1962) discusses this freshwater distribution pattern, pointing out that the habitat is unstable, low in species diversity generally, and heavily populated by colonizers. He compares (Hubendick 1952; see also Yonge 1938) the snails of most bodies of fresh water with those of Lake Tanganyika; this old, large body of water has developed a fauna of snails remarkably convergent with different taxa of marine ones. As is well known, this sort of adaptive radiation affects many groups of animals and has taken place in numerous lakes of the same kind (there is an extensive literature—see Brooks 1950, for review). Likewise, given only a slight extension of the same principles, we can readily apprehend why aphids are less diverse in the tropics, and why sex and dispersal are diminished as well.

Indeed, we might as well go all the way, and compare what happens in a developing industry. When a new product becomes available, the automobile, for example, one firm after another begins to produce it. So long as demand exceeds supply, the number of firms should be directly proportional to the time the industry has been in existence. Yet as the market becomes increasingly saturated, one would expect to find competitive exclusion (Hupmobile goes away), character displacement (Willys succeeds only with the "Jeep"), and oligopoly (General Motors and Ford).

Returning to the natural economy, we should expect any biological phenomenon to diminish toward the poles if it works better where more time is available, provided that competition leads to exclusion in high latitudes and to niche-partitioning in low ones. We have reason to suspect that these conditions are met with respect to certain symbiotic relationships. The tropics are notorious for their abundance of parasites, although some parasitic groups are more diverse in higher latitudes. One might attribute this to a physiological delicacy, but it seems more reasonable to infer that the seasonality of high latitudes restricts the ability of parasites to home in on their hosts. Mutualism too should be affected in the same way. Especially in the tropics, marine fish and crustaceans have developed a remarkable cleaning symbiosis, in which certain species specialize in removing ectoparasites from fish (see Limbaugh 1961; Feder 1966). The cleaners have evolved a set of adaptations, such as characteristic color patterns, which become attenuated, along with the cleaning habit itself, in more temperate waters. (The existence of this gradient has been questioned, but see Losey 1971.) Whether the prevalence of the symbiosis in the tropics is due to more parasites being present there, to the symbionts having more time to locate them, or to the hosts having a greater need to avoid losing energy to the parasites, is hard to say. A significant convergence, illustrative of a trend in tropical economies, is that among the cleaners and other species, there occur analogues of both Batesian and Müllerian mimicry (see Eibl-Eibesfeldt 1955a, 1959; Wickler 1960; and, for review, Osche 1962).

Whatever may be the precise reasons for such gradients, many exist. A. R. Wallace (1895) noted the elaborate development of both warning coloration and mimicry in the tropics. A latitudinal gradient in the size and complexity of social insect communities will be treated in chapter 8. It is well to remark that many gradients are altitudinal rather than longitudinal or latitudinal. We already mentioned the deep sea; later on some high-altitude trends will be mentioned. A topic that has lately received increasing attention has been the diversity gradients of marine intertidal organisms. R. G. Johnson (1970, 1971, 1972) has elaborated a model, of some interest because we can treat it as an extension of Rützler's (1965) model for sponges, already mentioned, which has been widely ignored because of the language barrier. Johnson adds some important successional notions which might be looked upon as analogous to stages in economic development. The idea is that species are gradually added through time. At higher intertidal levels (comparable to Rützler's smaller stones) unstable environmental conditions keep the economy, so to speak, in an early stage of development. Such an hypothesis very likely has some applicability, but for the rocky shore and mudflats at least a word of caution seems in order. With increasing elevation above the lowest intertidal level, marine animals and plants are indeed exposed to ever greater environmental stress, and they are more likely to be killed, say, through the drying action of the air. Yet the rise and fall of the tides has other effects, among which are exposure to various kinds of predators, differing forms of competition, and much else of which we are perhaps quite unaware (see Castenholz 1961; Connell 1961a, 1961b, 1970; Paine 1969; J. P. Sutherland 1970; Dayton 1971; Barbour, Craig, Drysdale and M. T. Ghiselin 1973). Feeding time may prove more significant than stability. On rocky surfaces filter-feeding animals are absent in the highest intertidal levels, and at intermediate ones they are not very diverse. On mudflats filter-feeders tend to be replaced at higher levels by animals that get their food from the surface, or from the substrate, rather than from the overlying water. (Technically, it would be more accurate to say that, at higher levels, suspension feeders are replaced by deposit feeders.) A filter-feeder can only obtain food while the tide is in. The tide submerging the animals for briefer and briefer periods at higher and higher levels, should make such an enterprise increasingly less profitable at the same time—all else being equal, and it is not always equal.

Gradients in dispersal. Having suggested the broad general implications of the basic principles, let us now take up the geographical distribution of various reproductive strategies. On the whole it may be said that organisms of higher latitudes tend to adopt alternatives to sex; but so do those of lower ones under certain conditions. The exceptions are covered by our model, but even so one needs to know a lot about the basic biology of the organisms under consideration before the model can be applied.

Dispersal (see C. B. Williams 1957; Schneider 1962; C. G. Johnson 1966; Dingle 1972) may be considered one of the most important alternatives to sex. Having many of the same effects, it may function as either a supplement or a substitute. Where close relatives live together, inbreeding is probable, and organisms that go elsewhere may well be rewarded for the effort by having more vigorous and diverse offspring. On the other hand, in a place supplied with abundant resources, little is to be gained in the way of a trophic advantage by expending time and energy seeking out new pastures or reconstituting the genetic material. Given, however, a situation in which formerly abundant resources have begun to run low, the organisms are presented with a strategical alternative: compete for what remains, or find a place where conditions of existence are more favorable (see Lidicker 1962). One might leap to the conclusion that sexuality and dispersal should both diminish toward the poles. From what we have seen about parthenogenesis this notion may, crudely speaking, be correct; however, the picture is not all that simple. The expediency of any adaptive stratagem depends upon the tactical situation and the resources of the organisms. It often happens that two closely related species will coexist in more or less the same area, and nonetheless differ in their capacity for dispersal—winged *versus* wingless insects, or marine invertebrates with *versus* without a larva. Such forms even coexist within the same species, in some cases being limited to one or the other sex.

Dispersal often requires a great deal of time and energy; but the expenditure may be well repaid, especially when it allows an organism to exploit a disjunct and ephemeral supply of food or some other resource. Since an individual organism must either disperse or not, an environment in which both of the alternatives are workable may be filled by some species that specialize in one stratagem, some that specialize in its alternative, and some that are polymorphic with respect to dispersal. Yet local conditions modify the effectiveness of the alternatives, and hence influence their frequencies. Certain beetles of islands and mountaintops tend to be wingless (P. J. Darlington 1943). The basic reason is not, as Darwin thought, that wingless organisms tend to be swept to their deaths by wind. Rather, a migrant individual has little chance of finding an appropriate new habitat, and selection favors using the energy that would be dissipated in flight some other way. At higher latitudes the physical environment becomes more variable, and frequent depopulation should tend to reduce competition for food and to diminish the advantage of motility. At the same time, however, the harsher climate results in a higher instance of local catastrophe, depopulating certain areas, and rendering them particularly hospitable to immigrants. Among beetles Lindroth (1946) found an unusually high proportion of wingless individuals in Fennoscandian forms, and a balanced polymorphism in the ability to fly in *Pterostichus anthracinus*. The same kind of phenomenon occurs in freshwater Hemiptera: the power of flight is more common in those from temporary ponds than it is in those from permanent rivers and lakes (E. S. Brown 1951).

The issue here, nonetheless, can still be debated. A. R. Wallace (1876[1]:211) pointed out that some insular insects have evolved larger wings, rather than smaller ones, implying that there may be some truth in Darwin's hypothesis: stronger flyers would avoid being carried off by wind. The same conclusion was drawn by L'Héritier, Neefs, and Teissier (1937) on the basis of wing reduction in coastal insects, and these authors backed up their argument with experiments on wingless *Drosophila* competing with the wild type. The close taxonomic relationship of wingless montane forms has evidently resulted from parallel evolution; natural selection, rather than orthogenesis as advocated by Jeannel (1925), being the mechanism of course. More recently, discussion has been renewed as to what kind of evolutionary mechanism has led to the reduction of an ability to fly. Van Valen (1971) treats it as a result of group selection. P. J. Darlington (1971a) does not go so far, but thinks differential extinction has been very important in maintaining widespread adaptations. Darlington (personal communication) favors the view that the wing polymorphism reflects stages in the loss of useless organs rather than an adaptive polymorphism. On the other hand, the species that can fly and those that cannot occupy different niches. For instance, Darlington (1961) noted that in Australia most hydrophilic and arboreal carabid beetles can fly; but most of those which he considered geophiles cannot. Species number decreases with altitude, and the lowland forms tend to be polymorphic (P. J. Darlington 1970). Data from New Guinea provide good reason for thinking, as Darlington (1971b) says, that the lowland forms are adapted to a disjunct habitat. The important point here is the difference between polymorphism, and lack of any individuals that can fly. In the lowlands, only 4% of the total species have reduced wings, but 59% of these are dimorphic. In the highlands, 32% have reduced wings, but only 4% are dimorphic. Were the dimorphism not adaptive, we should hardly expect such a trend.

The geographical pattern of dispersal becomes somewhat more evident when we broaden our basis of comparison and contrast the strategies adopted in the sea with those on land. Marine waters provide an easy avenue of transport for small organisms. Dense, viscous, and rich in food, seawater readily supports small animals. They sink less readily than do larger ones, and they expend little energy keeping afloat, so that they can even grow while they disperse. Hence the eggs or young stages of marine animals are commonly released into the water, there to develop as they are moved about by currents. Some larvae can even cross entire oceans (Scheltema 1966). On land, however, things are not so easy, for air is far less dense and viscous. The aerial plankton is not a good source of energy, and the dissemules are so small that they must be inactive or suffer from desiccation. Terrestrial animals as a rule travel by air only as adults, and they expend considerable muscular energy in doing so. Spiders, which can float on their long gossamer threads are an exception. And plants, which, like many adult spiders, do not move around very much, may have very small spores or seeds, or else the seeds may be provided with parachutes. Even so, terrestrial plants, in marked contrast to marine ones, tend to evolve symbiosis with animals as a mechanism of dispersal. The same phenomenon of hitching a ride on other organisms occurs in certain freshwater animals, for whom the downhill flow of water counteracts the advantage of easy transport. In all such cases, energetical relationships thus play an overwhelming role in determining the structure of the life cycle.

Gradients in parental care. The advantage to terrestrial animals of concentrating dispersal in the adult stage has led to extensive indirect effects. Population structure has been profoundly altered. Organisms grow up in close proximity to their relatives, on the one hand creating a problem of inbreeding. On the other hand it sets the stage for a whole range of important innovations, such as cooperation within the family. Such adaptations as parental care, societies, and warning coloration are most effective where the parents and the siblings stand together as cohesive units. This may be the reason why they are far better developed on land than in the sea. From our energetical considerations we might expect a poleward suppression of the adults and elaboration of the young among terrestrial animals, at least where the young are specialized for feeding and the adults concentrate upon dispersal. Yet although such a trend ought to appear in certain taxa, we must not leave out the influence of parental care. Where motility is also a means of obtaining food for the individual—as in birds and pollinating insects —the adult stage ought to preponderate, because of the advantage of its providing energy in support of the young.

The effects of parental care on the number of eggs per clutch in birds, and on the size of the litter in mammals, have been extensively studied, and we are now beginning to get a reasonably sophisticated notion of what is going on. The basic hypothesis, which we owe to Lack (1954, 1966, 1968), has been extended, reviewed, and refined by various other workers (see, for instance, Ashmole 1963; Rowan 1965; Cody 1966; Mountford 1968; Ricklefs 1968, 1969, 1970; Tyrväinen 1969; J. Ghiselin 1970; Klomp 1970). The explicitly group selectionist views of Skutch (1949, 1967) have added much of value to the discussion, but more individualistic models increasingly prevail. It is thought that the number of offspring per family (clutch size or litter size) is regulated by the availability of food, or sometimes of other resources. The parents have as many offspring as they can raise. If they lay too many eggs, they can't provide enough food for all, and juvenile mortality offsets the advantage of starting with a larger number of nestlings. The precise selection pressures are not always evident, but it is often possible to explain the clutch size in a given species of bird as a result of what food it eats and how it utilizes it. Thus "nidicolous" birds, i.e., those in which the young remain in the nest and are fed by the parents, have smaller clutches than do "nidifugous" birds, i.e., those in which the young leave the nest and feed themselves. Lack (1966) was somewhat at a loss to explain what sets the upper limit to clutch size in nidifugous birds such as pheasants and ducks. A likely possibility is that it has to do with the number of offspring that the parents are able to care for in other ways. The parent (most often only the mother, frequently both father and mother, and but rarely the father alone) accompanies the young as they forage, locating food and dealing with predators. The family lives as a group. Individuals within larger units would compete more with one another for food, while each would receive less protection as well. A few birds have no postnatal parental care, however, so here again, something else must set the limit. Lack suggests that the ability of the mother to lay eggs limits the clutch size, but it would seem inevitable that ultimately some other influences would come into play. Even time is important: a day spent producing another egg exposes the rest to another day of possible predation, and reduces the opportunities for feeding. Clearly an infinitely large clutch would take forever to lay. Furthermore, ducks that nest in holes, and that become more numerous when artificial holes are provided for them, have larger clutches than do those that nest in more open situations (Nice 1957). Skutch (1967) would seem to reverse cause and effect, when he attributes the large broods in these birds to compensation for the difficulty of finding nesting sites. Evidently the populations are held in check by paucity of nesting sites, and more food or other resources are thereby made available, while the protected situation gives a higher return per egg. Again, there seems to be a relationship betwen the kind of food and the number of offspring (extensively reviewed by Lack 1968).

Among birds, herbivores as a rule have larger clutches and faster development than do such carnivores as hawks. The reason for this has never really been made clear. We might suggest that birds of prey have to go longer distances away from the nest if they are to increase the food supply. It is energetically more expensive to increase greatly a supply of meat than one of grain to a level that will allow the support of additional offspring. To this effect we may add the unreliability of the food supply, which results in slower development (providing a reserve of energy that prevents starvation) and, when food is scarce the likelihood that one chick will die, sparing the remainder. The notion that slow growth here reflects a need for learning has been expressed by Ashmole (1963) and criticized by Skutch (1967). The trouble is that learning occurs for the most part later on in life; and growth is faster, not slower, where the young feed themselves than it is where they are fed by their parents. The same basic idea that the properties of the young are preparation for adult existence occurs time and again in the literature, but without serious attempts to test it.

What governs clutch size has been reasonably apparent to those who have studied such phenomena. What has not been noticed is the generality of the mechanisms. They all exemplify one of the most fundamental principles of classical economics: the *law of variable proportions*, or, as it has traditionally been called, the *law of diminishing returns*. Briefly stated, this is the principle that certain resources (variable factors) can be increased so as to raise output or profits, but *only up to a certain point*, beyond which even a decrease may result. If, for example, an entrepreneur would like to increase the productivity of his factory, he can hire more workers. But he cannot go on forever simply enlarging his staff: the factory would become overcrowded and the workers would get in each other's way. At this point he would have to do something like increasing the number of shifts; but again, there are only twenty-four hours in a day. Here, the analogue of an optimal clutch size would be the number of workers giving maximal profits (not output!), while adding more shifts would be like increasing the number of broods or having the parents alternate at certain tasks. Here we cannot agree with Lack (1954) that clutch size and brood number are unrelated.

Just as the entrepreneur's strategy is determined by the price of capital and labor, so the avian reproductive strategy evolves according to the availability of resources. When the labor force is fully employed, the entrepreneur can only increase the number of workers by raising wages, and the more he seeks to hire, the more he must pay for each, the less the return on his investment, and the sooner will he reach the level of diminishing returns that makes it unprofitable to hire more workers. Yet if many workers are idle, he can easily hire more at low wages, and consequently he can enlarge his staff to a greater extent before that level will be reached. In the former case the supply is said to be "less elastic" than it is in the latter. In very stable natural economies, the community becomes saturated with organisms working at peak efficiency, and there is very little surplus energy available for them to exploit. With seasonality, however, oscillations set in, and paucity of time keeps a high standing crop of energy available for reproduction—like a pool of unemployed labor (see Ricklefs 1970:600). The organisms that live in more stable habitats therefore suffer from an energetical "inelasticity of supply," and this makes it impossible for a great increase in output to be effected by feeding a larger number of offspring. It must be stressed that what counts is profit, not efficiency. In the present work we will distinguish as unequivocally as possible between the "efficacious" (that which does the job successfully) and the "efficient" (that which does the job with minimal expenditure of resources). Wherever an entrepreneur can derive a larger return on his investment by inefficiency and waste, he will opt for inefficiency and waste. So it is with organisms: the ecosystem be damned. The idea that natural economies necessarily evolve in the direction of efficiency is every bit as much a metaphysical delusion as the notion that good will inevitably result if only government lets capitalists do as they will.

Treating these matters from the point of view of diminishing returns has the advantage that, albeit a single explanation suffices to cover all clutch-size, litter-size, and egg-size phenomena, different variable factors are rightly invoked under diverse circumstances. It also helps to clear up many uncertainties. For example, Lack explained the larger clutch sizes at higher latitudes in terms of more time being available for locating food during the longer summer days. Yet as he himself observes (Lack 1968:170) clutch size varies from west to east as well. And in some species it decreases from spring to summer. Hence, even if time is one variable factor, we have to invoke others. The obvious solution, which fits Lack's general thesis quite well, is that a higher standing crop of food is available in high latitudes, in more continental areas, and earlier in the breeding season of at least some birds. Likewise Schoener (1968) could not explain why birds with larger clutches do not have larger ranges. The answer is that what makes it possible to get along with a smaller range also makes it advantageous to have a larger clutch. Fewer offspring and larger ranges are optimal where food is sparse; where food is dense, the animals need not go far, and they can produce more young. The problem here is rather like that of why antiquarian booksellers tend to sell their goods on an international scale, while publishers of newspapers exploit local markets.

Correlation of gradients in different habitats. Given the law of variable proportions, we can derive a host of biological gradients, although its detailed application depends upon the way of life of each particular organism. High-latitude organisms live in a "boom and bust" economy, preventing, as it were, a balanced and fully employed labor force. This leaves many resources highly elastic at certain times of the year. Hence species diversity is low, and clutch size is high. The same basic pattern manifests itself when we compare the more stable, maritime climates on the west sides of the continents with the more continental ones on the eastern sides (G. G. Simpson 1964). Not only that, but many high-latitude birds migrate from east to west, rather than from north to south, after breeding in the northern hemisphere. The same general rules may be applied to habitats with seasonal rainfall. The stratagem involves working where profits are high and subsisting where expenses are low. Recalling some phenomena mentioned earlier, we may see further analogies. The hole-nesting ducks are like industries operating in areas of high rent and low wages, in which crowding the employees is more expedient than renting a larger factory when the size of the work force is enlarged. We might compare our hawks to a manufacturer who ships his goods over a long distance. It may be more lucrative for him to increase the product's quality rather than simply raise output, since the cost of transporting each unit is the same. Indeed, so general is the principle of diminishing returns that many of our traditional explanations for gradient phenomena may have to be reexamined. Bergmann's rule, which states that the larger animals live in colder habitats, should be a prime target. The empirical rule should be kept conceptually distinct from an hypothesis invoking heat conservation often invoked to explain it. These will here be called, respectively, "Bergmann's rule" and the "Bergmann effect." Big animals have smaller surface areas per unit of volume than do small ones, and hence they lose a smaller proportion through radiation. In some ways the data fit the hypothesis (see F. C. James 1970); but might we not here be dealing with a misleading correlation? As is well known, size gradients occur in directions opposite to that predicted by the hypothesis, and they also affect certain groups of marine invertebrates. It seems likely that size is determined by the length of time or the amount of other resources available for growth or other activities. We shall hazard a few suggestions in later chapters as to how such phenomena might be analyzed when dealing with gradients in sexual dimorphism.

Following our plan of comparing terrestrial situations with marine ones, we find that similar geographical trends are apparent. Thorson (1946, 1950, 1951, 1955, 1956; see Mileikovsky 1971 for critique and review of subsequent research) studied the distributional aspect of marine larval ecology in great detail, carrying out monumental researches on a large number and wide variety of marine benthic invertebrates. The developmental pathways may be classified under the following heads: (1) direct development, on the bottom, with *no larva*, (2) indirect development, in the open water, with a *planktotrophic larva* that feeds on plankton, and (3) indirect development, in the open water, but with a *lecithotrophic larva* that does not feed, but subsists upon stored yolk. Larvae of either sort vary considerably in the time they actually spend in the plankton, before switching to life on the bottom as metamorphosed juveniles. During the early stages of development, and before they enter the plankton, the young may be protected through viviparity, brood pouches, egg-cases, or other means. But sometimes, and primitively, the eggs are shed into the water, and development occurs in the plankton. As a rule the length of the planktonic stage increases toward the equator, and, conversely, the proportion of brooding, viviparous, and directly developing species increases toward the poles.

Thorson interpreted these and related facts to mean that the larvae of marine invertebrates spend time in the plankton because they can thereby exploit the food in the water. It seems, however, that although feeding is important, the larvae are basically dispersal stages, as has been suggested by various workers (e.g., Hardy 1956; Mayr 1963); the fact that larvae feed means only that they take advantage of the opportunity. Thorson attributed the suppression of larvae in high latitudes to the lack of adequate time for feeding and growth in high-latitude plankton, and to a higher "wastage" of larvae because of physical destruction and predation. It is true that the larvae do feed, but this in itself is no compelling reason to infer that they are there because of a feeding advantage. After all, people attending a football game eat hot dogs and drink beer, but their reason for being there is the sport, not the food and drink. Likewise, the supposed wastage of larvae is an illusion, which results from the fact that forms that develop entirely on the bottom divide their energetical resources among fewer offspring. For dispersal it is more efficacious to provide a little energy to a large number of young, than it is to package the same amount in a few. Success in dispersal being largely determined by chance, numbers are more important than quality. And since dispersal takes time, the time might just as well be used in feeding. We would not say of an investor who put small amounts of money into a lot of speculative enterprises, in the hope that one or a few would turn out very well, that his procedure was wasteful, at least so long as he obtained a good return for his entire investment.

Many facts are clearly out of line with Thorson's hypothesis, although he managed to save the appearances fairly well. He maintained that the food supply is not very rich in high latitudes. Yet the Antarctic, where benthic development is most common, is the most productive of all ocean waters (Sverdrup 1955; Nielsen 1963; Ryther 1963; Dell 1965; Mandelli and Burkholder 1966). Furthermore, these regions support an enormous population of organisms that are specialized for eating plankton: notably euphausids and whales. Again, we should perhaps stress the point that what matters here is standing crop, not primary productivity. These groups that are specialized for life in the plankton tend to spend their whole lives there or to retain the free-swimming phase in higher latitudes rather than to lose it. A particularly good illustration may be derived by comparing the life cycles in certain Coelenterata, particularly the Hydroidea and the Scyphozoa. Both have a basic life cycle in which they alternate between a sexual, free-swimming jellyfish stage (medusa) and an asexual, sedentary one (polyp). Allowing for some complications, which need not concern us here, we may note a clear-cut trend in the Scyphozoa for the medusa to become larger, and more diverse, and the group as a whole more prosperous in higher latitudes, where the polyp is little altered. This may reasonably be attributed to the greater standing crop of food organisms in the plankton. Among the Hydroidea, however, the benthic polyp stage becomes larger in high-latitude waters. The medusa gradually degenerates, first becoming sessile on the polyp, then rudimentary, ultimately being transformed into a specialized reproductive organ. Short-range dispersal is still effected by a planula larva. In the sterile waters around coral reefs, both groups have become highly modified. Hydroids here again suppress the medusa, and there are no really successful Scyphozoa. Thus, we contemplate the spectacle of a planktonic group doing better in the plankton, and a benthic group staying out of the plankton, increasingly, as food in both habitats becomes more abundant.

In Antarctic waters, the plankton bloom is increasingly briefer toward the south (Hart 1942; Cushing 1958). Benthic development is very common in this part of the world. Many of the larvae that do swim spend their time near the bottom rather than feeding (Pearse and Giese 1966; Pearse 1969). These are specialized for habitat selection and short-range dispersal rather than for feeding. In the Arctic, as Dunbar (1968) points out, the energy level is considerably lower. And more larvae are found there. In the deep sea, where Thorson predicted that there would be few larvae, Ockelmann (1965) found a respectable percentage with lecithotrophic larvae (see also Knudsen 1970). Again, Thorson (1950) could not explain the absence of lecithotrophic larvae in the extreme north, where food is supposedly limiting in the plankton. It seems that one would not expect them where feeding is possible or where dispersal is unnecessary. Hence we can explain the presence of lecithotrophic larvae in the tropics, where they are common, and in the deep sea, as owing to the need for dispersal and the lack of food in the water. Simon (1968) draws attention to a polychaete annelid which has a larva in the winter, and benthic development in the summer. This cycle, which resembles that of aphids, is the exact opposite of what one would expect if Thorson's hypothesis were true. We now know of numerous polymorphisms, seasonal variations, and exceptions to rules, and these complicate the originally very simple picture (see Mileikovsky 1971).

A good source of evidence for the role of larvae is the "tadpole" of ascidians ("sea-squirts") (for an introductory account of these and other protochordates see Barrington 1965). These animals are sessile filter-feeders, and surprisingly do have a very limited ability to move about as adults—but only around eight centimeters a month (Carlisle 1961). The larvae never feed: their sole function is dispersal and habitat selection (Millar 1966). The tadpole as a rule can swim quite well, and possesses good sensory apparatus and means of attachment but the free-swimming period is generally less than two days (Abbott 1955:586). In certain ascidians of the genus *Mogula* the tail has been reduced and Berrill (1931) attributes the reduction to sandy habitats where habitat selection would be less crucial. Likewise he observes that the tail has been lost in certain derivatives of this group which have taken to a permanent existence in the open water. (Additional references: Goodbody 1961; Haven 1971.)

A remarkable convergence has occurred in several groups of marine invertebrates that seem to have in common a sandy bottom habitat—but this correlation may be due to a false impression. It affects only some members of each group, and appears to be largely confined to the tropics. An unusually large planktotrophic larva has evolved, one which evidently can travel long distances. In the phylum Sipunculida (for general biology see Baltzer 1934; Hyman 1959; for comparative development see Rice 1967), there is a *Pelagosphaera* larva which may reach about a centimeter in diameter, while the ordinary forms rarely exceed a millimeter (Dawydoff 1930; Åksson 1961; Jägersten 1963). Similarly, in certain Hemichordata (for general information see [van der] Horst, 1927–1939, 1932–1956; Hyman 1959) the small *Tornaria* larva has been modified into a much larger, more complicated *Planktosphaera* ([van der] Horst 1936; [van der] Horst and Helmcke 1956b). A similar elaboration in our own phylum is the *Amphioxides* larva of certain acranians (relatives of "Amphioxus") which, according to Wickstead and Bone (1959; see also Goldschmidt 1906, 1933; Bone 1961) have an extended period of feeding and development in the plankton but evidently spend a lot of time homing in on a suitable habitat. Other, less satisfactory, or less highly developed, examples could be adduced. It would seem that under tropical conditions certain benthic forms tend to go in for a combination of feeding with dispersal and habitat selection, in contrast to many Antarctic forms which stress the habitat selection alone. The length of time available for various incompatible activities evidently sets priorities.

Knudsen (1950) tried to reconcile Thorson's ideas with data on tropical West African marine mollusks. He found that the area from the Cape Verde Islands to St. Paul de Loana has an unusually high percentage of species with nonpelagic development: 67%, as compared with 10% for Bermuda and 25% for the Persian Gulf. He argues that the lack of nutrients in the Guinea current is responsible. Actually, there is an annual season of upwelling in this area, which brings nutrients up from deeper water, increasing the growth rate of plants and coinciding with a period of good fishing (Howat 1945). Areas such as the West Coast of North America also have a reduced amount of planktonic development. Thorson (1951) argued that here the larvae tend to get swept out to sea. This *ad hoc* hypothesis is not necessary, for here again we have an upwelling area, one of the most productive in the world, and remarkable for its high algal diversity and standing crop (G. M. Smith 1945). On the whole it seems far more fruitful to attribute the larval biology to the prevailing energetical climate.

Among sedentary and sessile marine invertebrates, the forms with a narrower range of food organisms or a more limited range of substrate have a longer period of larval development, less care of the young, and a larger number of eggs. They employ a "shotgun approach" in dispersal and finding food. Albeit many are sent forth, few make it, but these are highly successful. In the low energy situation of the tropics, specialists predominate. The same is true of the deep sea, where niches are also finely partitioned, but in a different manner. The energetical gradient also nicely explains why in trophic generalists, care of the young, and fewer but larger eggs, increase toward the poles. It also explains why a certain amount of specialization occurs in the seasonal habitats: those activities for which time is not limiting can be exploited by specialists. The same basic generalization applies to at least some vertebrates, notably fishes (N. B. Marshall 1953). Although planktonic fish generally produce large numbers of very small eggs, the benthic ones display a poleward trend in the direction of fewer and larger eggs, coupled with care of the young. Egg size also increases with depth, as it does in cephalopods (Robson 1932) and other invertebrates. Benthic cephalopods likewise have smaller eggs when part of development is planktonic (Mangold-Wirz and Fioroni 1970). We have insufficient knowledge of brood care in the deep sea, but it does seem that a size relationship fits into the picture. A number of deep-sea fishes have a pelagic larva, and as adults they tend to evolve capacious mouths and other mechanisms for feeding on larger food items. A comparable trend is apparent in the abyssal bivalves, which include a higher proportion of carnivores than do shallow water faunas (Knudsen 1970). In pelagic animals it seems that feeding by the young does provide a useful store of energy. But the reason for lack of a larva and for parental care in benthic forms would appear to be the joint influence of dispersal being unnecessary and the adult being in a good position to obtain energy on the bottom. For any individual case, which stratagem gets adopted will depend upon the particular kind of organism involved and where it lives. We must reject the widespread notion that climatic stability necessarily favors such adaptations as care of the young, for the patterns adduced fit only some kinds of organisms. This is particularly evident when we realize that certain of the gradients in the sea go in directions opposite to those on land. Tropical birds have fewer eggs than do their relatives from higher latitudes, tropical marine gastropods have more. We should hesitate to generalize about organisms when all we have studied is a few aberrant creatures, such as birds and *Drosophila*.

Some caution should be exercised in relating the competitive situation to the allocation of resources (see Gadgil and Solbrig 1972; Pianka 1972). Theoretically, one would expect selection, under conditions in which there is a scramble for food, to favor those individuals who devote much of their resources to reproduction; and where the environment is closer to saturation, greater returns should be derived from other activities—such as predator avoidance. One might think that many offspring and less parental care would be appropriate in the former, few offspring but more care for each in the latter, case. This generalization may be true for some organisms, but there is no necessary connection between the proportion of resources given over to reproduction, and the number of offspring, while the relationship between size and numbers holds only if all else remains constant. The relative sizes and numbers of seeds in terrestrial plants have lately been discussed by [van der] Pijl (1969), Harper, Lovell, and Moore (1970) and Stebbins (1970, 1971). An interesting pattern, displaying much that bears comparison with the situation among invertebrates, especially marine ones, seems to be emerging. Thus, seeds in many plant species are polymorphic, dispersal mechanisms are suppressed on islands and on mountaintops, and parasites tend to have many small seeds. Yet we may make out differences, too. While terrestrial plants and invertebrates often rely upon animals for transport, the marine ones get along more often by the action of currents and their own ability to swim (but consider such exceptions as spiders). Yet whatever the connections, one cannot just read off the mode of reproduction from the habitat, or the habitat from the mode of reproduction. Thus the largest and the smallest seeds alike may be found in the inhabitants of tropical rain forests. The best we now can say is that a shotgun approach to dispersal tends to go along with more but smaller offspring; but one rarely has the opportunity to formulate reliable predictive models. To make sense out of such phenomena requires that one examine the situation in great detail, for the point at which diminishing returns set in is affected by many influences. Thus small animals would seem to derive but limited advantage from a shotgun approach, for their size already makes great numbers of offspring difficult (M. T. Ghiselin

1963). Parental care may be favored where the young feed less effectively than the adults, or where avoiding predation is important, or for any of various other reasons.

SUGGESTIONS FOR NEW RESEARCH

We have considered how, with a knowledge of the autecology of a given kind of organism, an understanding of its habitat, and a grasp of the pertinent biological principles, one may with reasonable accuracy predict the reproductive strategies and diversity patterns of organisms. Our acquaintance with the details, to be sure, is rarely so great that we can expect to add nothing after the fact. Yet in a general way, we can indeed predict, and we need not hypothesize *ad hoc*. The hypotheses here proposed account for the data better than do earlier ones; thus, the reversed gradients are implicit in one system, but must be rationalized to save the others. Most important, we find a number of previously unexpected relationships coming to light. The individualistic approach not only can be made to cover the same phenomena as its alternatives, but it proves far more effective at solving biological problems. We can argue for it on just such a basis, whether or not we attempt to refute the competing hypotheses. Let us conclude the present chapter, therefore, by showing how certain important biological phenomena might be reinvestigated from a fresh point of view.

Parasites. The life cycles of parasites could perhaps provide very illuminating materials bearing upon the interrelationships between dispersal and sex. It is widely asserted that the high reproductive rate of parasites is "necessary for the survival of the species." We have already observed that this inference is based upon fallacious reasoning. Reproductive success is virtually a synonym for natural selection. The reason why tapeworms lay so many eggs is that they have sufficient resources for doing so. Their only alternative would be to lay many small eggs or fewer large ones. Since the eggs find their way to the next host largely by chance, such organisms naturally adopt a shotgun approach to dispersal. As Jarecka (1961) has shown, however, tapeworm eggs do possess some remarkable homing mechanisms, such as mimicking the food of the host; but these tend not to be energetically expensive. This shotgun approach is characteristic of endoparasites in general, for as a rule they cannot transport their offspring from one host to another, although there are exceptions.

In those ectoparasites in which the larvae are the dispersal stage (ectoparasitic gastropods, for example), the eggs are likewise small and numerous. Where the adults provide the offspring with both transportation and energy for growth, the young may be quite few, as is the case in the viviparous tsetse fly, in which the maggots develop within the body of the mother. If the adult provides dispersal, but does not feed, then either many small or few large offspring should work, depending on the circumstances. Our analysis is further complicated by the problem of where sex fits in. Parasites that regularly pass from one host to another during their life cycle often feed on all of them. In view of the rich energy supply, selection ought to favor their taking advantage of it. Yet the system also tends to be arranged in a manner that furthers at once dispersal and the closely related function of bringing the sexual individuals together. Some of the stages in a parasite's life cycle may not involve feeding. *Trichina*, for example, just encysts and waits. Very often the purely asexual stages live in hosts that are the usual food of those host species in which sex occurs. It happens that larger, hence more active and "highly evolved" organisms tend to occupy higher levels in the trophic pyramid. The resulting correlation led certain older authors to view the relationship as a phylogenetic one —of parasites climbing, as it were, the phylogenetic tree—and the notion remains with us. But the exceptions, such as the malaria parasite, which is asexual in man and sexual in mosquitoes, show that the rule is misleading. Transfer from host to host serves to collect a wide variety of sparsely distributed and numerically scarce parasites together, concentrating them such that the probability of uniting diverse genotypes is maximized in the definitive host.

Weeds. Another body of knowledge that needs to be clarified is the peculiar genetic systems of weeds (and of certain other organisms for which we might as well use the same term). An impressive amount of research has been done along these lines, much of which bears upon the adaptive significance of sex itself. Weeds display a variety of adaptations which suggest that the amount of genetical recombination has been secondarily reduced (Stebbins 1957). Various constellations of adaptive strategies have been recorded among them (Ehrendorfer 1965). If perennial, they tend to go in for asexual reproduction; if annual, they maximize seed output at the expense of recombination. Their genetical adaptations include: selfing (with inbreeding and homozygosity), apomixis, low chromosome numbers, infrequent chiasmata, polyploidy, and asymmetrical chromosomes (Grant 1958; Stebbins 1958). These are all features that should have the effect of decreasing the amount of recombination (Stebbins 1950). Such features have been taken as evidence in favor of the Mather-Darlington hypothesis that recombination is adjusted to the needs of the species. In not one single instance, however, has anyone excluded the alternative possibility that individuals are benefitted; nor does there exist valid evidence that benefits accruing to the populations are not the fortuitous result of selection for advantages to their members. Lewontin (1965:78–79) asserts that colonizing species have the greatest possibility for interdeme (i.e. group) selection. This may be true, but an actuality cannot be inferred from mere possibility. One is more likely to find trolls under bridges than elsewhere, but the existence of bridges does not lead rational men to believe in trolls. It is perhaps not too much of a quibble to add that, in view of the asexual conditions of so many weeds and colonizers, interdeme selection is not always possible, for by definition there cannot be a deme without sex. In fact, a critical look at weeds should convince anyone that group selection has been attributed to them as gratuitously as have trolls to bridges.

The idea that weeds occupy "temporary," "disturbed," and "marginal" habitats needs to be qualified. As Lewontin (1965) rightly points out, the basic consideration is that these organisms live where resources important to them are abundant, and the effective stratagem is to maximize reproductive output rather than the ability to compete with respect to space, water, or the like. Yet it is somewhat misleading to assert (Lewontin 1965:78, italics his) that "a newly formed colony is at first low in density and the individuals are in competition with individuals of *other* species, as opposed to the situation in older populations where any competition is primarily with members of the same species." The individuals of the newly founded colony compete with other individuals irrespective of species by growing rapidly and using up resources before other individuals can exploit them. As we have already mentioned, this is the effective way to compete in such an enonomy: he who moves fast corners the market and drives the competition out of business. How plants manage to succeed under such circumstances depends upon the situation. If the habitat is regularly, but not completely, depopulated, an effective strategy is a long life with a resistant stage. Thus we get annuals with perhaps limited powers of dispersal, but seeds that live a long time; or we get perennials that really flourish only after a catastrophe. The stratagem is like switching back and forth from stocks to bonds as the market fluctuates. Where depopulation occasionally occurs and leads to a spotty distribution of habitats that soon disappear, a short life is advantageous. This results in annuals that put all their available energy into seeds with good dispersal ability. Here the strategy is like investing in a diversity of speculative companies and selling out as soon as the period of rapid expansion is over. Wourms (1972:410–411) compares the variable hatching periods within clutches of annual fishes to what happens in the germination of certain plants, and analogizes it to "a roulette game in which the player bets on all numbers."

In view of the way that the natural economy would seem to operate, we cannot accept the suggestion of McIntosh (1970) that the brevity of life in weeds exists because it makes the population evolve more rapidly. Why, at the same time, is sex suppressed? Rather, weeds die because they live in habitats in which their days are numbered, and by putting all their resources into offspring, they spend on the next generation those resources that would be wasted on their own. It is as selectively disadvantageous for a doomed individual to prolong its life to the detriment of its offspring, as it is foolish for a businessman not to liquidate a firm that can no longer turn a profit.

Some geneticists have maintained that weeds evolved a reduced sexuality as a means of conserving well-adapted combinations of genes, or, in more moderate versions, they have looked upon it as a pre-adaptation to the weedy habit. Albeit a good point, this is only one aspect of the problem. Where rapid multiplication is both expedient and possible, it is clearly disadvantageous to produce variants that are adaptively inferior to those already present. Yet in addition, there is no reason even for producing recombinants which, if different, are equally fit. Sex would only waste energy, time, and the lives of individuals. It therefore would have no positive advantage, but rather a negative one. Who would liquidate a prosperous business to try his fortune at one which had never been attempted? Stebbins (1970) notes that annuals tend to self, but attributes it to better fitness for colonizing, not to retention of genetic combinations.

It has long been known (Darwin 1877b) that cleistogamous flowers, that is to say those morphologically adapted to prevent out-crossing, will occasionally open up and allow insects to fertilize them, especially when conditions are favorable and the probability of out-breeding is therefore high. This and some analogous phenomena indicate that here, as in monerans and protistans, selection has favored a low frequency of recombination, attuned to the needs of the individuals.

Weeds tend to be physiologically adaptable in a way that suggests that they have substituted phenotypic plasticity for an ability to modify the genetic material. From this phenomenon Baker (1965:165) has developed a theory of the "general purpose genotype." An ability to handle various ecological situations is purchased at the expense of a reduced capacity to adjust to particular local circumstances. Such genotypes do not evolve, however, as has been suggested, because it is advantageous to build up a large population. All organisms, without exception, multiply as much as they can (Malthus 1798). General purpose genotypes evolve because, under the prevailing conditions of existence, the stratagem works for the individuals that use it. It seems reasonable to predict that more such genotypes will be found, if only we search for them in situations where the individual advantage is likely to be realized. The obvious place to look would be in sessile marine plants and invertebrates, such as tunicates, bryozoans and sponges. Some possibilities along these lines will be mentioned later on, when we treat other aspects of reproduction in such organisms.

4

LOVE'S LABOR DIVIDED, OR, THE UNION AND SEPARATION OF THE SEXES

*All the various species of things transform
into one another by the process of variation
in form. Their beginning and ending is like
an unbroken ring, of which it is impossible
to discover the principle.*

Chuang-Tzu

From man's anthropocentric point of view, it seems only natural that organisms should be either male or female, not both. Hermaphroditism is somehow conceived of as abnormal, or as a feature of "lower" organisms, while its contrary, "gonochorism," is viewed as the norm, perhaps even as the ideal. This conception of the order of things was expounded in Aristotle's *History of Animals*, and it remains with us. Yet when we realize the diversity and success of hermaphrodites, the error is evident. The sexes are united in flourishing groups of higher plants, in the advanced subclasses of gastropods, and in the more specialized and modified crustaceans and annelid worms. Yet even these creatures do retain a fundamental differentiation of gender at the cellular level, for each individual forms spermatozoa and eggs.

Going backward in the evolutionary sequence, we find the functional equivalent of spermatozoa and eggs, and these adapted to the same role, in unicellular organisms. In many protozoans and protophytes, an evolutionary trend seems apparent, one beginning with individuals that are physiologically specialized for crossing yet are morphologically indistinguishable (isogamy), and proceeding through various degrees of differentiation with gametes of unequal size (anisogamy), ultimately leading to a state (oogamy) with many small and highly motile "male" gametes and few, large, immobile "female" ones. Here a dichotomy between active and passive individuals would thus seem to parallel the masculinity-femininity distinction among metazoans, while undifferentiated gametes, able to cross with like individuals, might be compared to hermaphrodites. Perhaps, then, we may use the evolution of anisogamy as a concrete model, providing a simplified example that could be extended to cover various other situations.

DIVISION OF LABOR BETWEEN THE SEXES

The basic reason for the origin of distinct genders would seem to be the division of labor, a principle that goes back to Aristotle, but one that did not become important in biology until the nineteenth century. It is hard to say when the division of labor was first applied to sex, for the transfer from economics to biology seems not to have been the work of a single intellect, and it occurred but gradually and imperfectly. The idea that labor is divided between the sexes, among the castes of social insects, and between the polypoid and medusoid generations of coelenterates was clearly and explicitly developed by Haeckel (1869). The principle seems to have been applied to the gametes by several authors; yet a clear-cut enunciation of it is hard to find. Rather than risk doing injustice to history, let us quote a passage from a work by Oscar Hertwig first published in 1892 (authorized translation, O. Hertwig 1909:278).

The theory built up by Weismann, Strasburger, Maupas, Richard Hertwig, and myself may be worked out more in detail in the following manner. During fertilisation two circumstances must be considered, which work together and yet are opposed to one another. In the first place, it is necessary for the nuclear substances of the two cells to become mixed; hence the cells must be able to find one another and to unite. Secondly, fertilisation affords the starting point for a new process of development and a new cycle of cell divisions; hence it is equally important that there should be present, quite from the beginning, a sufficient quantity of developmental substance, in order to avoid wasting time in procuring it by means of the ordinary processes of nutrition.

In order to satisfy the first of these conditions, the cells must be motile, and hence active; in order to satisfy the second, they must collect these substances, and hence increase in size, and this of necessity interferes with their motility. Hence one of these causes tends to render the cells motile and active, and the other to make them non-motile and passive. Nature has solved the difficulty by dividing these properties—which cannot of necessity be united in one body, since they are opposed to one another—between the two cells which are to join in the act of fertilisation, according to the principle of division of labour. She has made one cell active and fertilising, that is to say male, and the other passive and fertilisable, or female. The female cell or egg is told off to supply the substances which are necessary for the nourishment and increase of the cell protoplasm during the rapid course of the processes of development. Hence, whilst developing in the ovary, it has stored up yolk material, and in consequence has become large and non-motile. Upon the male cell, on the other hand, the second task has devolved, namely of effecting a union with the resting egg-cell. Hence it has transformed itself into a contractile spermatozoon, in order to be able to move freely, and, to as large an extent as possible, has got rid of all substances, such as yolk material or even protoplasm itself, which would tend to interfere with this main purpose. In addition it has assumed a shape which is most suitable for penetrating through the membrane which protects the egg, and for boring its way through the yolk.

From our contemporary viewpoint, Hertwig's formulation is adequate but it does not go far enough. He does not make it clear just what is advantageous in differentiating the gametes. He suggests that a sort of interference is involved, but tells us nothing about the selective mechanisms, assuming, like so many biologists of his day, that if there is a purpose to be explained there is no problem with respect to what has happened. He describes the morphological results, but omits the most important point. The basic underlying advantage is an energetical one, and one that derives from broad general principles; but at the time these had not been applied to biology, except rather vaguely by Darwin (1871, 1874; see below, chapter 8). Indeed, even recent work by Kalmus and Smith (1960) and Scudo (1967a) aiming to demonstrate a population-level advantage to anisogamy in furthering the combination of gametes, leaves out the energetical advantages to individual organisms.

Suppose that a manufacturer of heavy goods such as machine tools wanted to set up an effective program for both selling them and getting them to his customers. He might equip each salesman with a large truck piled high with machinery to be sold on the spot, and have them scout the countryside for customers. This arrangement would have certain advantages, such as immediate delivery, but at the same time a lot of energy would be wasted hauling the stock from place to place. It would be cheaper to use a heavy vehicle only for delivery, and to provide the salesmen with automobiles in which they could cover a far wider area on a given amount of fuel. Extending the analogy, we do not expect a manufacturer to mount his entire plant on wheels and produce goods on the road. Yet even this generalization must be qualified to some degree, for it all depends on the circumstances. The captain of a fishing vessel may put the crew to work at maintenance while on the way to the fishing grounds, and while returning to port may expect them to process the catch.

So it is with sexual roles. Eggs differ from sperm because the female gametes specialize in providing the zygotes with energy and other resources. The male gametes are specialized for uniting with the female ones. Any energy used by the females in moving about would necessarily be subtracted from that passed on to the zygotes. Hence the ideal female would be an absolutely passive organism. The male, on the other hand, should concentrate upon obtaining maximal dispersal from a given quantity of energy. In the simple case of protistan anisogamy, he does this by dividing himself into many small units. By virtue of greater numbers, he raises the probability of encountering a suitable mate. At the same time, the male gametes have less energy per cell. Although as individuals they can cover less distance, as a group they can cover more, for they do not have to expend so much of their resources in transporting stored energy. In addition, they can traverse a greater distance in a given period of time. Hence, when a protozoan undergoes multiple fission, giving rise to numerous microgametes, he acts in his own interest, for although his offspring each have what we might consider low fitness, collectively they derive an advantage through numbers. Our ideal male, therefore, would be an absolutely motile organism—not an absolutely active one, as might be conjectured.

The evolution of size differences in protistan gametes can be traced fairly easily, owing to the availability of much comparative material. The series derived from such comparisons, however, should be conceived of as representing stages, rather than actual phylogenetic lineages. In phytoflagellates many such series are known (Lewin 1954). These begin with strictly isogamous forms; yet even so, they may differ behaviorally, with one (the female) being less active than the other. It is not known how the differentiation began. Perhaps the mating types originally evolved as blocks to inbreeding. Nor is it easy to discern what was the original advantage to the difference in motility. Given, however, a less active (energy sparing) subpopulation, the more active one would clearly gain an advantage if it mated with them, rather than with individuals that had expended energy finding mates, and hence had less for the next generation. In this respect the phenomenon of *relative sexuality* is noteworthy. Whether an individual is a macrogamete or a microgamete may be only a matter of degree, and an intermediate form may act as a microgamete in fertilizing a larger individual, or as a macrogamete in being fertilized by a smaller one. Evidently the specificity has evolved largely as a block that deters the microgametes from uniting with the less advantageous partners.

One can demonstrate that anisogamy has energetical advantages by establishing that there is some kind of correlation between the need for motility in bringing the gametes together and the degree of difference in size between the gametes. The trouble is that many protists feed or photosynthesize as they travel, and therefore derive less of a locomotory advantage through subdividing their bodies. However, where the breeding season is very brief, time spent travelling is limiting, and the advantage of numbers is then accentuated. In ciliates, the stage at which conjugation occurs coincides with a relatively small body size, and the conjugants are known to be hyperactive. But most ciliates are not unlike hermaphroditic metazoans, in that both conjugants are fertilized and each develops into an individual that is "new" in the sense of having a different genetic makeup. Furthermore, since conjugation is not obligatory, some of the advantages to dimorphism are lost. Extensive dimorphism has, however, arisen in attached or sedentary ciliates in the orders Peritrichida and Suctorida. Certain members of both groups bud off small, highly motile microconjugants that swim about and fertilize attached macroconjugants, then degenerate (Kormos and Kormos 1957, 1958). Sexual dimorphism, albeit to a lesser degree, also results through unequal division in some rumen ciliates (Dogiel 1925), and also among the symbiotic flagellates of wood-eating cockroaches. Although the existence of this dimorphism has been questioned, there is nonetheless good reason to accept Dogiel's evidence, for he provides abundant data.

SEX-RATIO AND COMPETITION

Thus we have the best of reasons for calling the unicellular organisms that are destined to give rise to male gametes "males" and the ones that will produce eggs "females." This is, of course, a convention of terminology, but one that makes functional sense. We may then go on to ask how many individuals of each gender one should expect to find in the entire population. In other words, what ought to be the sex ratio? We should observe from the start that, in theory at any rate, one sperm suffices to fertilize one egg, and that one male can give rise to many sperm. Hence, from the point of view of maximizing the reproductive output of the population, we should have, if not an equal number of eggs and sperm, at least a numerical preponderance of egg-producing individuals. Such a conclusion might be argued on the basis of a false analogy with a chicken farm: a few cocks service many hens. Yet the reproductive output of populations does not determine the sex ratio in nature, which, if we admit certain qualifications, ordinarily is around unity (one-to-one). (See Darwin 1874; R. Hertwig 1912; Pelseneer 1926; Mayr 1939, 1963; McIlheny 1940; Bellrose, Scott, Hawkins, and Low 1961.)

The reason why there are about as many males as females was first explained by R. A. Fisher (1930), whose exposition of it leaves something to be desired, but whose idea seems basically correct (additional references: Crew 1937; Bodmer and Edwards 1960; A. W. F. Edwards 1960; Kolman 1960; Willson and Pianka 1963; Verner 1965; Campos Rosado and Robertson 1966; Leigh 1970b). He reasoned from the fact that half of the parents of any zygote are male, half female; a virtual truism. In isogamous protistans with two mating types, where there are more "males" than "females" some of the males would be unable to mate. Hence a gene that produced a smaller proportion of "males" would be selected for up to the point at which equality was reached. A gene for the "minority sex" is always favored, because it leads to the production of more offspring. Now consider the anisogamous protistans. The parent cells, before giving rise to gametes, would each have a certain amount of energy, with which (in an extreme case) they could do one of two things: produce a single egg, *or* produce many sperm. In either case, one parent would have, on the average, one offspring, to which it contributed one half of the genome. It would not matter that (in theory) many sperm could produce many offspring, for (in practice) every male individual would have the same potentiality. Although competition would adjust the number of sperm so as to maximize male fecundity, all males would tend to be equally fecund. It follows that in a protozoan species in which the males each gave rise to eight gametes and the females to one, the probability of any male gamete producing a zygote can be no more than one in eight. The sex ratio is controlled by a strictly reproductive competition and bears no relation to the mythical "fitness" of populations, and in this respect its mode of evolution resembles sexual selection. What matters is not the number of conjugants, but rather the most effective way for the parents to use their resources, for selection does not necessarily favor the "interests" of the entire genome.

It is not strictly true that each parent contributes half of the zygote's genes: as Fisher (1930) puts it, each contributes half of the "ancestry." In man, the father's Y chromosome has fewer genes than does the mother's X chromosome, and as a result the sons derive somewhat more DNA from their mothers than from their fathers. A more extreme case is to be found in *Escherichia coli*, which, as Wollman, Jacob, and Hayes (1956) have stressed, gives a fine example of the development of male and female individuals in a very simple creature. As in eucaryotic organisms, the male expends the energy used in contacting the female and transferring the DNA, while the female remains quite passive (K. W. Fisher 1957). Yet only a small proportion of the genome is donated, to the "detriment" of what gets left behind. A number of evolutionary theories, some of which are discussed later on, go astray on just this crucial point. Selection favors not just genes, chromosomes, organisms, or social groups, but rather it favors whatever happens to gain some competitive advantage.

It has been maintained that anisogamy has adaptive value in maximizing the number of unions of gametes within populations (Kalmus 1932; Scudo 1967a). Reasoning that the union of gametes is analogous to the kinetics of gaseous molecules, it has been calculated that "the maximum number of unions occurs when the microgametes constitute rather less than ½ of the bulk gametic mass" (Scudo 1967a:291). This implies that there should be a numerical preponderance of females, and a higher output of female gametes than male ones. But since it is clearly advantageous to the individual male, in competing with other males, to maximize his output of sperm, the theoretical argument is dubious; the same consideration would apply to the production of male-producing individuals. Furthermore, the sex ratio does not deviate as expected from such considerations. In some taxa, males are less common than females, but in others the reverse relationship obtains, and a closer look shows that this has nothing to do with group advantages. Some animals, such as starfish, produce large quantities of sperm; others, such as birds, produce only a little. However, the difference is readily explained as a consequence of the fact that males can expend energy on such activities as hunting for a mate or caring for their offspring as an alternative to merely generating sperm.

A sample of organisms collected from the field will often be found to deviate from the expected numerical equality of the sexes. Frequently this reflects only what is called the "apparent" sex ratio, and the one-to-one relationship actually still obtains. Some of these divergences result from an imperfect sampling technique, as when the males live in one area, the females in another, and the entire range is not collected, or merely because the sample is too small. The apparent sex ratio may also deviate when the sexes mature at different ages. If the males remain in a juvenile condition for some time after the females have started to breed, adult females will outnumber adult males. Conversely, the males can mature earlier, so that they appear to be more numerous than the females. Similar effects result from differential mortality (Geiser 1923), as often happens when the males die after fertilizing the females, who then survive long enough to lay the eggs and perhaps to care for the young.

When sexuality is labile, different rules may apply (for sex determination see Bacci 1965; Crew 1965). If sex is phenotypically determined (as it is in some marine invertebrates), and the individual is male or female according to environmental influences, we should anticipate variation in the sex ratio in different parts of the habitat. On the average one would expect a balance leading to the usual equality in the entire population, but local differences in density and other influences could produce considerable deviation in a given sample. More important, an individual may have certain alternatives to providing half the ancestry of its offspring. It may provide not half, but all of it, as in self-fertilizing hermaphrodites and the like. In the case of such hermaphrodites, or of parthenogenetic females, any males in the same population should tend to be less numerous than individuals with female reproductive organs. Systems of this type should be expected to evolve where low population densities restrict mating or where recombination is not particularly advantageous. Some nematodes, for example, tend to have an excess of parthenogenetic females or to include self-fertilizing hermaphrodites. These are mostly fresh-water or terrestrial forms. In some parasitic nematodes, the proportion of males increases with more worms in the host (Maupas 1900; Honda 1925; Nigon 1949a, 1949b, 1965). Similar examples are available from other taxa (see Vandel 1925, 1941; Arcangeli 1932; Reinhard 1949; Delavault 1958; Andersen 1961). In all these cases it would seem that the sex ratio deviates from unity when some version of asexual reproduction, or its functional equivalent, has produced special conditions where Fisher's model applies only when modified.

According to Fisher's model (see also Leigh 1970b), differential mortality of the offspring combined with parental care implies deviations from the usual ratio. Fisher cast his argument in terms of the effort expended upon offspring of each sex. He posited that equal amounts of energy should be devoted to male and female offspring. In other words, the return depends upon the amount invested, and not upon the absolute quantity of the commodity itself. If it takes equal amounts of resources to produce sons and daughters, but 90% of the sons and none of the daughters die before reproducing, then the amount invested in sons and daughters will remain equal, but the return on the individual son who does reproduce will be ten times that of a daughter. We would have a balance between a high-risk, high-return and a low-risk, low-return strategical alternative. Parental care would alter this situation to some extent, for should a number of the sons die while still being cared for, the parents would have fewer sons in which to invest their resources. Were the resources apportioned equally between all offspring at a given moment, the consequence of a one-to-one sex ratio would be that less than half the effort would be expended on sons, shifting the balance in favor of producing more sons. Investing proportionately more in younger daughters and older sons would shift the sex ratio back toward unity.

Fisher's model, at least as originally formulated, leaves out a great deal. The notion of parental expenditure was vague, and no consideration was given to changes in resource allocation. Also, conditions about the modes of competition and of inheritance were left implicit; but, as we shall see, such conditions are not invariably met. Nonetheless, it stands to Fisher's credit that he brought out two very fundamental points. First, sex ratios are indeed determined by reproductive competition between individuals. Second, the model rightly attributed the adaptive significance to the needs of the parents rather than to the needs of the offspring.

As things now stand, Fisher's model must be amended when certain implicit conditions of genetic structure and competitive interactions are not met. In certain cases, the sex-determining mechanism itself raises problems. Thus, where sex is determined through a balance between homozygotes and heterozygotes, inbreeding produces a temporary deviation, followed by return to equality (Scudo 1967c). Other situations have been proposed in which the sex-determination mechanism could produce a permanent shift (W. D. Hamilton 1967). We need not go into these here, since the focus is on environmental factors.

Fisher's model also presupposes an unlimited competition for mates, but, as W. D. Hamilton (1967; see also Hartl 1971) points out, such need not always occur. He gives the example of certain insects, in which mating occurs largely between brother and sister. In this case, it is to the mother's advantage that she produce many daughters and few sons, for it is the output of the entire family that determines her reproductive success. One son can service the entire family, giving rise to a phenomenon that we may view as a form of self-fertilization on an extended scale. Hamilton suggests that intermediate stages exist, for example the wasp *Melittobia acasta*, in which a considerable amount of fighting between males occurs (Browne 1922). Hamilton notes that such male combat would not be expected in a strictly inbreeding system, but his inference presupposes, of course, his own ideas on kin selection.

One naturally wonders if perhaps a different mode of competition between the males could force the sex ratio to deviate in the other direction. It is interesting that an excess of males has been recorded among certain birds in which competition for females seems to be intense, but we have only a correlation to go on here. Maynard Smith and Ridpath (1972) describe a curious situation in the Tasmanian native hen, *Tribonyx mortierii*. Here, some of the competitive units are ordinarily "trios" composed of two brothers and an unrelated female. Such families have a higher reproductive output than those with a single male. Since the males get less than twice the output when in trios than in pairs, Maynard Smith and Ridpath conclude that the individual male loses fitness, which is compensated for because he aids his brother, who has genes like his own. Their argument, however, is not conclusive, for they do not treat all the possibilities (e.g. that the system evolved under different conditions). More importantly, they maintain that the polygamy had to come after the change in sex ratio. This is a purely theoretical inference. Not all the possible influences of competition have been worked out, and the alternatives should be examined in the light of additional data.

HERMAPHRODITISM

A division of labor, with concomitant energetical advantages, would thus seem adequate to explain the differentiation of sexes at the cellular level. One might dispute whether we should speak of hermaphrodites among unicellular organisms. An hermaphrodite is generally understood to have the reproductive apparatus of both genders, and some people feel that the difference should be a morphological one. Among metazoans and higher plants the distinction between male and female is on the whole clear and unequivocal. Even in hermaphrodites, eggs and sperm are easily recognized. In the more complicated organisms, the advantages of a division of labor are, if anything, more obvious. Yet hermaphroditism is very widespread, and one would suspect that under certain conditions combining labor is better than dividing it. Indeed, there would seem to be many such conditions, for we may distinguish several kinds of hermaphroditism. At this point it seems reasonable to classify these, and to reform the ambiguous terminology.

The nomenclature of hermaphroditism. In zoological terminology, which will be followed here, an hermaphrodite is any individual organism with both male and female reproductive organs. The term "bisexual" should be avoided, because it may refer to a "bisexual species," i.e. one having both genders present rather than asexual or parthenogenetic individuals, or on the other hand it may mean having both sexes in one individual. The botanical terms (see Knuth 1906) *monoecious* and *dioecious* should be applied only to plants. Some of the botanical terms will be defined here; but we will avoid using them as far as possible since they are hard to remember. An *hermaphroditic* plant is one with flowers having both male and female parts. A *dioecious* individual would have all its flowers male or all female. A *monoecious* one would have both male and female flowers. The situation gets more complicated where there are mixtures of male, female, and hermaphrodite flowers on the same plant. Thus, *andromonoecious* and *gynomonoecious* individuals would have male and female flowers respectively, in addition to the hermaphrodite ones. An individual with all three is said to be *coenomonoecious*. Matters are further confused by the fact that a population can include both hermaphrodite and gonochoristic individuals, the latter of either sex or both. In an *androdioecious* species some have all male, some all hermaphrodite flowers. A *gynodioecious* one has some purely female plants. A *trioecious* condition is one in which all three occur. Other possibilities exist, but botanists have largely refrained from adopting names for all the combinations and pervertations. Our section on "positions" in flowers would be incomplete, however, were we not to mention that the hermaphrodite flowers may exist in two or more morphs within a species, with pollen transfer occurring from one morph to another. In *heterostyly*, the stamens and the style are both of different length on the two or three forms. But other possibilities are realized, such as depositing pollen on one side of an insect and receiving it on the other.

Among hermaphrodites, we may distinguish two kinds: *sequential*, or changing from one sex to the other, and *simultaneous*, or having both sexes functional at the same time. Sequential hermaphrodites are either *alternating*, meaning that they change sex more than once, or else they are either *protandrous*, beginning the period of maturity as males, and then later becoming female, or *protogynous*, beginning as female, then becoming male. Often animals switch from a functionally gonochoric state to become hermaphroditic, or undergo the opposite change. As no adequate and unequivocal terminology exists for these phenomena, the following is proposed. In *adolescent hermaphroditism* the organisms attain sexual maturity as simultaneous hermaphrodites, and later become gonochorists. In *adolescent gonochorism* the organisms mature as functional gonochorists and later become simultaneous hermaphrodites. *Adolescent protandry* is the form of adolescent gonochorism in which the gonochoric stage is male, *adolescent protogyny* the form in which the gonochoric stage is female. This terminology avoids the ambiguity that so often results from calling animals protandrous simply because the sperm begin to mature first. These organisms are often functionally hermaphroditic from the time they mature, for in the so-called male phase they can receive and store sperm. Likewise, "protogynous" tapeworms simply begin to mature the ovaries in the younger part of the body: the mature portion of the worm is simultaneously hermaphroditic (Johri 1963).

The functions of hermaphroditism. In an earlier work (M. T. Ghiselin 1969b) three hypotheses were evaluated which seemed to explain most instances of hermaphroditism among animals. For factual details and a general phylogenetic survey, the reader is referred to this paper. The opportunity will now be taken to review some old notions in more detail, and to broaden and revise the earlier hypotheses. Considering hermaphroditism in the context of general sexual biology has naturally led to some new ideas.

Consider first the possibility that *hermaphroditism is primitive.* As was mentioned above, the notion of the sexes being united in "lower" organisms and becoming separated in "higher" ones goes back to antiquity. In modern biology, the same idea developed connections with the notion of trends toward specialization, differentiation of parts, and division of labor. The failure of such early morphologists as Gegenbaur and Haeckel to see that hermaphroditism is not necessarily primitive reflects the difficulty they had in overcoming their pre-evolutionary heritage. Nonetheless, the problem was adequately solved by F. Müller (1885). But even to this day, the hermaphroditic state is often mistaken for a remnant of primitive conditions.

Perhaps all living multicellular organisms have in fact descended from hermaphroditic forebears. If so, we must infer that the hermaphroditic period occurred in the remote past, for all the evidence now available either conflicts with the hypothesis that hermaphroditism is primitive in metazoans and metaphytes, or at least implies that it was not a feature of the ancestral stocks. To be sure, the sexes have in some cases been separated secondarily; yet in others they have united. We need only trace the distribution of hermaphroditism along the major branches of the phylogenetic tree to see what has happened. Invariably, the hermaphrodites differ from their gonochoristic relatives in having features, such as parasitism or a sedentary existence, that are likely to affect reproduction. And it may often be shown that the hermaphroditism evolved after the shift in ecology. The problem is complicated somewhat by the fact that, contrary to popular thought, "primitive" organisms are specialized. They survive by virtue of their adaptive superiority under certain conditions of existence, such as parasitism. As the topic is widely misconstrued, it seems worth the effort to survey the evidence.

A detailed consideration of plants will not be included in this review, for the task is difficult even for a trained botanical anatomist. We nonetheless should observe that in flowering plants, Lewis (1942) says that hermaphroditism (individuals with hermaphrodite flowers) is the general rule. Of the British flora, 2.25% are dioecious (i.e., the individuals are gonochorists); the condition being widespread and scattered taxonomically, it is therefore secondary. It is present in advanced orders, and is thought to be an outbreeding mechanism, but this notion is unsubstantiated. Monoecy (hermaphrodite organisms, but gonochoric flowers) is more common (5.4%) yet still rare. Since the flowers here bear vestiges of structures proper to the opposite sex, it is thought that their ancestors passed through an hermaphroditic stage. However, the example of analogous rudiments in man, in structures known to have evolved in gonochoric organisms, shows that this argument is not reliable. For whatever reason, hermaphroditism correlates positively with reduced motility and with difficulty of gametic union between individuals; higher plants exemplify this. That some advanced forms have lost it must be balanced against the fact that many other advanced forms have retained it. At any rate, there is no necessary connection between anagenetic level and hermaphroditism in flowering plants.

Beginning now a review of the animal kingdom, we may note that some of the Porifera, or sponges, are gonochorists, while others are simultaneous or sequential hermaphrodites (Hyman 1940; Sarà 1961). These organisms are sessile as adults, hence specialized in certain respects; they represent a very early offshoot of the metazoan lineage at best. And although "primitive" with respect to their histology, their sexual biology is unique among organisms, and nothing like it could have given rise to that of other Metazoa (see Gatenby 1920; Lévi 1956). They fertilize their eggs internally, with the sperm being transported to the egg by an amoebocyte. The main lineages of metazoans originally fertilized outside of the body, and the shift from external to internal fertilization has never once been reversed, so far as we know. Many sponges brood their young during the early stages of development, and brooders in general tend to be hermaphrodites. Some evidence, admittedly weak, suggests that the gonochoric sponges are forms that have secondarily ceased to brood. Various other reproductive peculiarities obscure the historical picture, but suggest that from the point of view of evolutionary mechanisms, sponges present some important opportunities for new research. Sex itself appears not to be very common, for the group relies extensively upon various forms of asexual reproduction (Burton 1928, 1949a). Larval behavior (Bergquist and Sinclair 1968) and the repression of sexuality among intertidal forms (Bergquist, Sinclair, and Hogg 1970) suggest that many sponges are comparable to "weeds" (see the previous chapter). Burton (1948) found that certain sponges settle gregariously, later fusing together to form a single colony. This phenomenon would render the notion of an hermaphroditic individual meaningless in the present context. It also gives us an intriguing possibility that some multicellular organisms have passed through an evolutionary stage analogous to one known from certain insect societies, with "multiple founders" as it were (see chapter 8). As if this were not enough, evidence has been given that the mother contributes cells to the developing embryo (Warburton 1961; T. L. Simpson 1968). The facts ought to be checked, of course, but at least the sponges are sufficiently unconventional that they should not be treated as ancestral metazoans.

The radiate phyla, or Coelenterata, are probably derived from a gonochoric stock. The Cnidaria (jellyfish, sea-anemones, etc.) have only occasional hermaphrodites. Among these the hermaphroditism often correlates with viviparity, as in sponges, but in this case the care of the young has demonstrably arisen secondarily (Carlgren 1901;

Gravier 1916; T. A. Stephenson 1929; Thorson 1946). Likewise hermaphroditism seems to go hand in hand with the invasion of fresh water (Föyn 1927). The Ctenophora, or "comb jellies" are all hermaphroditic (Willey 1897; Komai 1922; Krumbach 1925; Riverberi and Ortolani 1965). This implies (weakly) that hermaphroditism evolved early in the phylogenetic history of the group, but not that they have never been gonochorists, for the ctenophores are descended from the Cnidaria. They converge with a number of pelagic groups in their hermaphroditism, which is retained in the aberrant benthic members of the phylum.

The flatworms (Platyhelminthes) would seem to be basically an hermaphroditic phylum. They are generally considered to be ancestral to the related, and structurally more advanced, Rhynchocoela, which are for the most part gonochoric. This fact might lead one to infer that hermaphroditism is primitive in the platyhelminth superphylum, but the facts do not bear out the inference. The true flatworms fertilize their eggs internally, and this means that the reproductive biology of extant forms is in some respects more advanced than that of the rhynchocoels, which principally fertilize the eggs outside the body. A very few of the flatworms are gonochorists, but these are unrelated to each other and have close affinities to hermaphrodites. A parasitic turbellarian *Kronbergia* is one of these (Christensen and Kanneworff 1964, 1965). The parasitic classes include two families of flukes (Trematoda), the Didymozoonidae and Schistosomatidae, and a cestode, *Dioecocestus*, in which the males and females live together in pairs, and a loss of hermaphroditism is demonstrable by rudimentary organs and intermediate forms (Fuhrmann 1904; Dogiel 1966). Gonochoric tapeworms in general are unusual, and taxonomically isolated (Schmidt 1969); vestigial organs (Mettrick 1962) and intermediate forms (Czaplinski 1965) show that the sexes have separated. When hermaphroditism does occur in the Rhynchocoela, its secondary nature is clear from the fact that it is more widespread, and more common, in nonmarine and in commensal species (Coe 1939; Hyman 1951a; Gontcharoff 1961). We thus see that the animals can evolve in either direction.

The "aschelminth complex," or pseudocoelomate phyla (Hyman 1951b), would all seem to be basically gonochorists. Their relationships are debatable; but where hermaphroditism occurs it seems to be related to special habitats and ways of life. Exclusively gonochoric are the Acanthocephala (A. Meyer 1932–1933), Rotifera (Remane 1929–1933; de Beauchamp 1965), Kinorhynchia, Priapuloidea, and Nematomorpha (Dorier 1965). The Gastrotricha constitute an exception, being hermaphroditic or parthenogenetic with hermaphroditic ancestry. Hermaphroditism is only occasional among the Endoprocta and Nematoda (see above).

The phyla with a good trochophore larva are all related to annelids, and thus form a natural assemblage, even though the properties of the common ancestor are hotly contested. The phylum Sipunculida has but one known hermaphroditic species, noteworthy also for extreme reduction in size (Åkesson 1958). In annelids themselves, the class Polychaeta is generally considered ancestral to the others. Polychaetes are basically gonochorists, but family after family includes some hermaphrodites, and these in general have some peculiarity in their mode of reproduction (for details, see below and M. T. Ghiselin 1969b). The archiannelids are small creatures which may, as their name suggests, include some very "primitive" forms; but in this group hermaphroditism is exceptional. The class Myzostomida is hermaphrodite, but specialized for commensalism or parasitism on echinoderms (Fedotov 1938; Prenant 1959). The Echiurida are gonochorists (Baltzer 1931). The classes Oligochaeta (earthworms) (Stephenson 1930; Avel 1959) and Hirudinea (leeches) (Mann 1962) are often treated as a single group (Clitellata). They have lost the free-swimming trochophore larva, and their hermaphroditism is thus an advanced character.

The mollusks have often been treated as examples of a group in which hermaphroditism is primitive, in spite of overwhelming evidence to the contrary. It is hard to say which group of mollusks is the most "primitive," since all are modified in one way or another. A good possibility is the chitons (class Polyplacophora): only one hermaphrodite species has thus far been reported from this group (Heath 1908; A. G. Smith 1966). It broods, and this is a most unusual, and obviously secondary, feature in chitons. The solenogastres (Aplacophora) could be modified chitons, or they could be a very early offshoot. In either case it doesn't matter, for the more primitive, free-living forms are gonochorists, while the commensal or parasitic members of the groups are hermaphrodites. The only extant monoplacophoran, *Neopilina*, has separate sexes (Lemche and Wingstrand 1959), as do all known members of the classes Scaphopoda and Cephalopoda. The Bivalvia include a few, taxonomically scattered, hermaphrodites, but these for the most part are dwarf forms, live in the deep sea, or brood their young (Pelseneer 1912; see also below). The Gastropoda (M. T. Ghiselin 1966c) are primitively gonochoric, although their sexual lability has led some workers (Bacci 1947a, 1947b, 1947c, 1951a, 1965) to view them as basically hermaphrodite. Yet the hermaphroditism in the ancestral subclass, Prosobranchia, is sparsely distributed, and by no means occurs in the forms most likely to represent the ancestral condition. In the two subclasses probably derived from it, and often treated as a single subclass, hermaphroditism is simultaneous, or more rarely adolescent protandrous or adolescent protogynous. The marine Opisthobranchia are highly specialized, yet only one minute form has (to judge from its close relationships to specialized forms) reverted to gonochorism (M. T. Ghiselin 1966c). The other subclass, Pulmonata, includes the highly successful snails and slugs of land and fresh water, as well as the marine stocks that polyphyletically gave rise to them. It would be indefensible to treat either opisthobranchs or pulmonates as anything but highly evolved and dominant organisms.

That remarkably diverse and abundant phylum, Arthropoda, has a singularly low proportion of hermaphrodites. The explanation lies in locomotion, for the phylum owes much of its success to the functioning of limbs and wings, and the exceptional hermaphrodites on the whole may be shown to have restricted motility at the time of mating. Among groups transitional between the annelids and arthropods, hermaphrodites are unknown: Tardigrada (Marcus 1929), Onychophora (Peripatus and related genera) (Bouvier 1905, 1907), and an aberrant group of parasites, the Pentastomida (Heymons 1935). Among trilobites, pycnogonids and arachnids, normal hermaphroditism seems entirely wanting, although it would be surprising if a few cases are not ultimately discovered. The mandibulate complex includes the crustaceans on the one hand, and the myriapod-insect lineage on the other; it is problematic whether their common features are due to convergence. In the latter group, hermaphroditism is exceedingly rare, occurring only in some of the most highly modified insects. One of these is the wingless fly Termitoxenia, which lives as a commensal in termite nests (Wasmann 1900, 1902; Assmuth 1923; Megelsberg 1935). Another is a coccid "bug" in which the hermaphroditism is linked to parthenogenesis and a sedentary way of life (Hughes-Schrader 1930). In crustaceans, hermaphroditism is scattered, and the exception, but not unusual. One might think that since what experts consider the most "primitive" subclass, Cephalocarida (Hessler 1969; A. Y. Hessler, Hessler, and Sanders 1970), is made up of simultaneous hermaphrodites, gonochorism has been secondarily acquired in the rest. However, the extant forms of Cephalocarida have been modified in some respects, and this may very well include their reproductive biology. They are quite small, they do not swim, and the eyes have been lost. The other hermaphrodites either live in unusual habitats, such as fresh water or the deep sea, or else are parasitic and sessile, and therefore suffer from locomotory difficulties. In certain instances such as barnacles (for details see chapter 7), hermaphroditism has been both gained and lost.

Three minor phyla are often considered intermediate between the two main branches of Bilateria, the Protostomia (just treated) and the Deuterostomia. Some workers have grouped the transitional phyla together as the Tentaculata. All are attached, or at least sedentary, organisms, and the hypothesized relationship between them is debatable. The Brachiopoda have only two known genera of hermaphrodites, and these brood (Senn 1934; Atkins 1958; de Beauchamp 1960; Rudwick 1970). Hermaphroditism is the rule in both the Phoronida (Cori 1939) and the Ectoprocta (Polyzoa or Bryozoa *sensu stricto*), but there are exceptions. Here again, a correlation between anagenetic level and the separation of the sexes is without foundation.

Among the "lower" Deuterostomia, gonochorism would seem to represent the ancestral condition, in spite of the fact that a sedentary way of life may have characterized the ancestral stock. The rather aberrant, wormlike Pogonophora (which are increasingly being treated as annelids, not deuterostomes) have separate sexes (Ivanov 1963). In the phylum Chaetognatha (Hyman 1959; Alvariño 1965), hermaphroditism is said to be universal, but these creatures are so highly modified that even their relationship to other deuterostomes is debatable. Their reproductive biology includes sperm storage and perhaps self-fertilization. Active predators, so numerous that they must be considered dominant organisms in the plankton, they can hardly be dismissed as either archaic or degenerate. The phylum Echinodermata is generally linked to the Chordata via the Hemichordata on the basis of embryology. In echinoderms the most "primitive" extant class, Crinoidea, is wholly gonochoric (Hyman 1955), as is the class Echinoidea (Harvey 1956). In the three remaining classes, Asteroidea, Ophiuroidea, and Holothuroidea, hermaphroditism is taxonomically scattered, and largely restricted to forms with brood protection. Since care for the young is demonstrably polyphyletic, and since hermaphroditism evolved subsequent to brood protection, the facts clearly show that all three classes were primitively gonochoric (for detailed argument, see M. T. Ghiselin 1969b).

The Hemichordata ([van der] Horst 1927–1939; Hyman 1959; Barrington 1965) are often treated as an independent, annectent phylum, close to the Chordata. However, the only apparent reason is that many systematists habitually trade an historical relationship for a convenient diagnosis, and a good evolutionary classification might well unite the two phyla. The burrowing Enteropneusta ([van der] Horst 1932–1956; Burdon-Jones 1951, 1956) are all gonochoric, and so are most, but not all, of the stalked Pterobranchia (Sato 1936; [van der] Horst and Helmcke 1956a; Stebbing 1970). Passing to the unequivocally chordate animals, it is inescapable that the Urochordata (tunicates, larvaceans, and salps) are basically hermaphrodite. A few species have secondarily become gonochorists (Berrill 1950). Yet urochordates are highly specialized organisms, adapted to sessile or pelagic life. It has been argued (Garstang 1928; Romer 1958) that the tunicates gave rise to the vertebrate lineage by neoteny; were such the case an hermaphroditic ancestry of vertebrates would follow, but this hypothesis is hard to reconcile with the data of invertebrate comparative anatomy. It demands that many features of the enteropneusts shall have become vestigial in the pterobranchs, then returned to the original condition in the urochordates and acranians. The view of Burdon-Jones (1952) that the pterobranchs are neotenous enteropneusts seems plausible, as does the classical view that true chordates are modified enteropneusts and the urochordates a branch in which the adults became sessile. Be this as it may, the issue is debatable enough that we cannot on this basis infer that the vertebrates are descended from hermaphrodites. The Acrania ("amphioxus") are gonochorists.

Especially in the Vertebrata, there is no real evidence that hermaphroditism is a primitive trait. Although reported from the Cyclostomata (hagfishes and lampreys), it has turned out to be nonfunctional in this group. The same phenomenon of poorly differentiated gender in the early stages of maturation has been recorded in various other vertebrates. And many, including some birds, can be made to change their sexual phenotype quite easily. This being so, one would expect that a normal change in gender would easily evolve in the simpler vertebrates, leading to sequential, or even simultaneous, hermaphroditism. But this change has occurred only in the more advanced bony fishes, being unknown among sharks and restricted to the higher Osteichthyes (Atz 1964; Breder and Rosen 1966). Its absence in amphibians, reptiles, birds, and mammals, where hermaphroditism is invariably an abnormal condition, could mean that it has never been advantageous, or perhaps that it was physiologically or developmentally impossible. Even Darwin (1874) accepted Gegenbaur's view that the vertebrates are descended from hermaphrodites. Darwin pointed out, however, that they must have been gonochoric for some time, and that the presence of rudimentary female organs in the males, and of rudimentary male organs in the females, does not of itself necessarily mean that these structures originated in an hermaphroditic ancestor. These parts, such as the clitoris, have evolved in strictly gonochoric lineages. And the cybernetical reason for their existence is hardly unclear. The embryonic development of the sexual apparatus in both sexes is controlled by a single morphogenetic mechanism that evolves as a unit. A change in the anatomy of one sex therefore tends to influence the properties of the other. An analogous process goes on in the evolution of serial homologues: in the generality of vertebrates, hands resemble feet and left feet resemble right feet; but there never was an ancestral vertebrate with only one limb. Here as elsewhere, then, the notion that hermaphroditism is "primitive" results from invalid reasoning and a lack of acquaintance with the pertinent facts.

A second traditional interpretation has been that *hermaphroditism allows organisms to fertilize themselves*. This view would seem to have been especially popular among those who failed to appreciate the difference between reproduction and sex. Especially in plants, hermaphroditism does seem to be important in allowing facultative self-fertilization. Even if crossing is advantageous, it may be better to fertilize in this manner than not to be fertilized at all. But if the advantage lies in crossing, an obligate selfer would be wasting effort. On this basis Darwin (1858) propounded the well-known Knight-Darwin law, which states that hermaphrodites do not self-fertilize through a perpetuity of generations. He maintained (Darwin 1859) that hermaphrodites, whether plant or animal, all have a structure that allows them to cross. To the thesis that crossing is advantageous and that plants are adapted such that they can or must at least occasionally cross with another individual, he devoted parts of several important and highly influential works (Darwin 1876, 1877a, 1877b and other editions). Some facts appear to contradict the hypothesis, for evidently some organisms are indeed obligate selfers. But we forget that forever is a very long time; fertilization in such forms has become vestigial and is on the way out. The common occurrence of physiological and structural adaptations that delay or prevent self-fertilization and further outbreeding substantiates the basic generalization (Mather 1940). Physiological blocks to selfing are best understood in plants, because so much work has been done on them, but comparable mechanisms are known in such simultaneously hermaphroditic animals as tunicates (Castle 1896; T. H. Morgan 1942a, 1942b). Early in this century many geneticists maintained that crossing is not particularly advantageous, since inbred lines are purged, through selection, of deleterious recessive genes (Muller 1932). This crude way of looking at things was characteristic of their essentialism in general, which led them to ignore both diversity and population dynamics. We now realize that inbreeding species and outbreeding species have their own "genetical environments" (Mayr 1954), and that each kind has adaptive advantages in various habitats. Outbreeding may not be advantageous to an inbred population, but it is advantageous to the members of an outbreeding one. Selfing has, to be sure, the advantage of saving time and energy, and it is often combined with occasional crossing as in cleistogamous plants (forms with structural blocks to crossing) (Darwin 1877b; Uphof 1938). An ability to get along without a mate would also be useful wherever it is hard to find one, or where variability is of secondary importance.

Many hermaphrodites do fertilize themselves, either obligatorily or facultatively, but the exact details are often unclear. Some of the animals that are thought to be wholly parthenogenetic may produce occasional males. In some apparent selfers, the sperm are necessary for initiating development, but the sperm do not contribute to the genome of the new individual (pseudogamy). Hermaphroditism probably is linked to self-fertilization in a variety of invertebrates, particularly slow-moving ones, parasites, and inhabitants of the land. A few examples are certain Trematoda and Cestoda, a fresh-water triclad flatworm (Anderson 1952), some Nematoda (Nigon 1949a, 1949b, 1965; Nigon and Dougherty 1949), a fresh-water annelid (R. I. Smith 1950, 1958), (dubiously) some barnacles (Barnes and Crisp 1956), and occasional marine (E. Clark 1959) and fresh-water fishes (Spurway 1957; Haskins, Haskins and Hewitt 1960; Harrington 1961, 1967; Schultz 1961, 1969; Kallman 1962a, 1962b; Harrington and Kallman 1968). On the whole these organisms must be considered unusual, and they constitute but a small fraction of the hermaphroditic animals. Even in plants, the analogous autogamy and related phenomena have been overemphasized. Nonetheless, an ability to self-fertilize must be accepted as one valid explanation for the existence of hermaphroditism. The others will now be discussed in turn. These, using the terminology proposed earlier (M. T. Ghiselin 1969b) are the *low-density, size-advantage,* and *gene-dispersal* models.

The *low-density model* is the traditional explanation in terms of selection theory. Traces of it may be found in the nineteenth-century literature (Meyer 1888), but it is lacking in some obvious places (Steenstrup 1846; A. Lang 1888). Tomlinson (1966) and Scudo (1969) have treated versions of the low-density model mathematically. (Also of some peripheral interest are: Altenberg 1934; Mosimann 1958; Smouse 1971.) The present account is mainly a refinement of older ideas, with an attempt to place them in a larger frame of reference. A gonochoric worm that only encountered one other member of its own species during the period when it was ready to mate would have, half the time, the same sex as the other one, and it could not reproduce. Even if it met up with two other worms, the probability that neither would be a suitable mate would be one in four. As hermaphrodites have no such problem, they should enjoy a distinct selective advantage wherever it is unlikely that one individual will encounter another. We should emphasize that this model applies only to simultaneous hermaphrodites. Forms that switch gender derive no such advantage. The term "low-density model" is taken from the nearly obvious principle that as there are fewer and fewer individuals in a given area, the selection pressure for simultaneous hermaphroditism ought to increase. However, what really matters is not absolute numbers nor density as such, but relative numbers and density as related to the organisms' way of life. A population density that would provide numerous potential mates for a highly motile animal such as a bird would be grossly inadequate for a sessile creature such as a barnacle. Once again we need to understand the functional problems of each group of organisms before we can apply the model to individual cases.

If the low-density model gives a valid picture of the real world, we should be able to demonstrate that there exists a correlation between a high frequency of simultaneous hermaphroditism and those habitats and ways of life that are characterized by a low-population density or by its equivalent. A number of lines of evidence suggest that such low-density conditions are realized in the deep sea. A fairly high species diversity (hence few individuals per species), coupled with a sparse food supply (making dispersal expensive), provide theoretical grounds for thinking that this is the case. More compelling is the existence of strategical alternatives having the same effect. Among these, we shall later discuss the modification of the males so that they become purely dispersal stages (see chapter 7). Data are available for quite a number of taxa, showing an unusually high frequency of hermaphroditism in the deep-sea representatives of groups that are predominantly gonochoric. These include the Bivalvia (Pelseneer 1912), Crustacea (Wolff 1956, 1962), and fish of the order Iniomi (Mead 1960; J. G. Nielsen 1966). Probably other groups will be found to show the same trend, and some normally hermaphroditic taxa may be considered preadapted to life at great depths.

There would seem to be a moderate preponderance of hermaphrodites in pelagic invertebrates, but these characterize, in many instances, entire groups, and we cannot necessarily invoke it as a post-adaptation that arose in response to a selection pressure in that habitat. Nonethless, it may have contributed to the success of forms such as the two orders of "pteropods" (Gastropoda: Opisthobranchia), the Chaetognatha, salps, and the Ctenophora. Unfortunately we have little information on how hard it is for these creatures to find mates even though they are known to form large populations.

In parasites, the low probabilities of mating are so obvious as to need no substantiation. Many parasitic groups are hermaphrodites, generally simultaneous ones. The few sequential hermaphrodites often live together as male-female pairs, for example the Myzostomida (Wheeler 1894, 1897; Beard 1894; Prenant 1959). A number of major parasitic taxa are largely hermaphrodite: the flukes (Trematoda), tapeworms (Cestoda), leeches (Hirudinea), and the gastropod orders Pyramidellacea and Sacoglossa. Yet those just mentioned show that we need to be careful, for all of these have free-living hermaphrodites as close relatives. More convincing are some hermaphroditic parasites in taxa that are predominantly free-living gonochorists: a nemertean worm (Keferstein 1868; Marion 1873), a few prosobranch Gastropoda, some solenogastres (Mollusca: Aplacophora), and various Crustacea. In the interest of presenting a fairly complete and unbiased picture, we should mention the groups in which parasitism has not resulted in a particularly high frequency of hermaphrodites: Acanthocephala (A. Meyer

1932–1933; Hyman 1951b), Nematomorpha and Nematoda, and various groups of lower rank. Sometimes the reasons for these exceptions are clear, sometimes not. At present it is hard to explain why the Acanthocephala are never hermaphroditic. On the other hand, the Nematomorpha (see Hyman 1951b; Dorier 1965) are parasitic only as juveniles: the adults escape from the host and swim about at the time of mating. Some nematodes have complicated life histories, and a few are hermaphrodite (see for example Christie 1929). A few parasitic groups have secondarily reverted to gonochorism, as we have already mentioned; but since they often live in pairs, it would appear that the density conditions have changed in favor of a strategic alternative. Parasitic copepods and parasitic barnacles are good examples, and the reason for their gonochorism will be explained and substantiated in chapter 7. On balance, therefore, the rule holds quite well, and we need not invoke *ad hoc* hypotheses to cover the exceptions. Indeed, the ease with which we may account for exceptions on the basis of the more general hypothesis that reproductive strategies are determined by dispersal and population density may be taken as an argument in favor of the model.

As a final example of the applicability of the low-density model, we should definitely include sedentary and sessile organisms. Among these, there exists a whole spectrum in the ability to exchange genetic material with distant individuals, and the development of hermaphroditism correlates as one would expect. At one extreme of the scale, we have groups of animals that can only fertilize internally and are permanently attached to the substrate. As a result, they cannot mate unless they live in groups. Sociality does allow them to exist as gonochorists, but hermaphroditism still helps a great deal. In view of this limitation, internal fertilization ought to be strongly selected against in sessile animals; and indeed it ought to go far toward preventing the assumption of that way of life. It would seem the internally fertilizing animals that are sessile have evolved from motile precursors, and have been unable to re-evolve external fertilization. Why evolution in this case does not get reversed is unclear, but no change from internal to external fertilization is known. Such organisms, although successful, are perhaps not so well off as those that transmit their sperm, or both sperm and eggs, via the water. In this situation, however, we must take precautions not to confound cause and effect. It would appear that simply casting the gametes into the open water is effective only when the population density is high and the whole local group sheds them simultaneously. In the less abundant marine invertebrates, a form of pseudocopulation, in which the animals move together before giving off the gametes, is quite common. The situation resembles that of wind-pollinated plants: the stratagem works only at fairly high densities.

In terrestrial plants, symbiosis with pollinating insects allows sessile organisms to exist at low population densities, for it directs the
pollen to the right individuals. If there are many individuals of the
same species, the probability of any pollen grain reaching an appropriate place when moved at random will be high: hence wind pollination will work, especially if pollen output is maximized. At lower
densities, energy is more effectively used in directing the pollen than
in producing it. The low-density effect is accentuated in plants fertilized by symbionts in a manner that, apparently, has not been noticed.
When a bee visits a flower, it may both take up pollen and transmit it.
No crossing will occur when the bee first visits a plant of a given
species, whether that plant be hermaphrodite or not. To the general
principle that half of the second individuals visited in gonochoric
plants will be males, we must add the consideration that half of the
time the first plant to be visited will be female; and in one quarter of
the instances the second will be female as well. A little calculation,
which we may leave to the reader for his amusement, will show that
only one in four of the second visits will result in fertilization, three
in eight at the third visit, seven in sixteen at the fourth. In hermaphrodites, however, the second visit can always result in a cross. It follows
that the symbiosis imposes a particularly high selective advantage for
hermaphroditism upon flowering plants, and this would explain why it
is so common. (See also Baker 1963.)

By applying this same line of reasoning we can see the advantage
of heterostyly, which, we may recollect, is the existence of two or
more morphs in one species, each morph having different lengths of
pistil and stamen (for review see Vuilleumier (1967). Heterostylous
plants are hermaphrodites which deposit pollen at one point and receive it at another. Distylous forms have two morphs (long and short
styled), tristylous forms three (long, medium, and short) (Darwin
1877b). Distyly has evolved convergently from the monomorphic state
and back again (Ernst 1953). Similar reversions to the distylous or
monomorphic (homostylous) state are known from the tristylous
(Ornduff 1966a, 1972). Others have reverted to dioecism (gonochoric
plants) (Ornduff 1966b). In some species homostylous and heterostylous races are known (Ornduff 1970). The arrangement is thought
to reduce the amount of selfing and other inbreeding (see Mulcahy
and Caporello 1970). That inbreeding has something to do with heterostyly is suggested by the annualness and weedy nature of homostylous forms; the heterostylous weeds have extensive vegetative
growth (Ornduff 1964; Ornduff and Perry 1964; F. W. Martin 1967).
It is not obvious what is cause and what is effect here, however. Nonetheless, the advantage of heterostyly over dioecism (gonochorism) is
clear. The first visit of a pollinator is not wasted, even though no

pollen is received. Tristylous flowers have an even greater advantage in this respect.

If this analysis is correct, many well-known features of flowering plants have been seriously misinterpreted. The problem of documentation, however, is particularly difficult, and the subject deserves extensive study by a number of specialists. The brilliant researches of Darwin on pollination mechanisms at once evoked an immense vogue for that subject: this was one of his contributions that definitely was not overlooked. A series of important works, many of which are still worth reading (e.g. H. Müller 1873; Kerner [von Marilaun] 1878), rapidly disposed of the older views, which were strongly tainted with teleology. By the turn of the century Knuth's *Handbuch der Blüten-biologie* (translation 1906, 1908, 1909) presented a vast amount of data. Research along the same lines continues, lately taking an increasingly ecological point of view (a good place to start is Faegri and van der Pijl 1966). The latest contributions stress competition (Mosquin 1971; Levin 1972) and energetics (Heinrich and Raven 1972). It is clear, however, that additional principles would help. For a simple example, consider the widely accepted hypothesis that protandry has evolved as an adaptation that favors outcrossing. Perhaps, but this can hardly be the whole story. Protogyny would work just as well. A much simpler explanation is that pollen has to be deposited on the symbiont before it can be received by another plant. A temporal division of labor could be achieved if the same flower began with a masculine role, then switched to a feminine one; if the two roles interfered with each other, such a division of labor might be favored by selection. Since pollen grains live only a few hours, this arrangement would work only on a short-term basis. Of course, the combination of labor, that is to say, simultaneous hermaphroditism, may also be favored. Protogeny when it occurs is generally of the adolescent type, and evidently allows facultative selfing; the combination of labor here has obvious advantages.

But it seems at least reasonable that many problematic features of plants may be explained in terms of the division of labor, timing, and competition, doing away with many *ad hoc* hypotheses that often invoke outcrossing. The female flower, for instance, subserves two distinct functions: receiving the pollen and forming the seed and fruit. The male, on the other hand, only serves a copulatory role. Hence, combining the male with the female makes the flower a less wasteful structure. It would appear from their different frequencies that the various flowering types are not all equally effective, at least in a single habitat. But it is not so obvious why, for example, certain species have some individuals with only hermaphrodite flowers and other individuals that are strictly female, while no purely male individuals occur. This phenomenon, gynodioecy, has been considered an outbreeding mechanism (Lewis and Crowe 1956). But why would there not be males here too? One might think that a pure male flower could not then be used to form fruit. But some species do include all three kinds of individuals (trioecism or polygamy), and some have hermaphrodites and males, but no females (androdioecy). Perhaps a better explanation for all these arrangements is one that, although suggested in one of the classics (Hildebrand 1867) has received little if any attention. The first flowers in a protogynous species bloom in vain; hence individuals in these forms tend at first to produce only male flowers, so that although the flowers are protogynous hermaphrodites, the plants that bear them are really adolescently protandrous hermaphrodites. Similarly, a protandrous system should tend to have too many females at the end of the season, favoring the production of some male flowers at this time—a condition that recalls one seen in the Tanaidacea. It requires but a simple elaboration of this general body of principles to explain the gonochoric individuals in a species with hermaphrodites. The separate sexed plants are maintained by selection because they concentrate their effort on the period when one sex or the other is in short supply. It is a temporal division of labor.

The assumption of one or another adaptive stratagem in pollination is determined by the density of the population and the availability of time and energy. (Much as in contemporary warfare, where the choice between sniping and machine-gun fire is determined by the massing of enemy troops and the supply of ammunition.) When the population density is low, and much time is available, the effective stratagem is hermaphroditic flowers pollinated by specialized animals. At higher densities, more flowers of the same species are visited, and the advantages of dividing labor may favor separation of the sexes. If the pollination season is brief, then rapid transfer becomes expedient (see Baker 1965). The effect should be a reliance upon nonspecific symbionts or wind pollination. But these two stratagems are effective only at higher population densities. Hence the workability of symbiosis may itself strongly influence diversity. But the relationship is a reciprocal one.

A widely recognized, but hitherto incorrectly explained, phenomenon will help to show how density affects pollination mechanisms. On islands, flowering plants tend to have separate sexes. The usual interpretation has been based upon the value of outbreeding (Carlquist 1966); but arguments have also been given that self-compatibility is equally favored, because it furthers reproduction in general (Baker 1967). Consider, however, the possibility that on an island the overriding factor may be the lessened extent of the market, which, in all economies, reduces the effectiveness of specialists. A smaller number of species in a given area, while the number of organisms remains constant, implies that there would result a larger number of individuals of a given species per unit of area. The higher population density would then render gonochorism more workable. Hence we need not invoke "genetic systems" advantages to explain the facts.

At this point we must digress a moment upon insular economies in general. It is important to do so, for the topic will recur several times in later chapters. Insular biogeography has lately enjoyed a considerable vogue, thanks to the influence of MacArthur and Wilson (1963, 1967), who attempted to apply the population model. The importance of input *versus* output, in the sense of immigration or speciation *versus* extinction, is now far better recognized. But the limitations of the model need to be known too, so that the phenomena it does not cover will not be overlooked (see J. H. Brown 1971; Simberloff 1972). (For general treatments of island biogeography the following are particularly useful: A. R. Wallace 1876; Gulick 1932; Mertens 1934; P. J. Darlington 1957; Carlquist 1965.) Insular faunas and floras show their own peculiar trends in clutch size, body size, sexual dimorphism, and many other phenomena. Mayr (1965) notes that different taxa display different extinction rates, and in fact one has to examine the biology of each taxon before one can understand any insular trend. It is reasonably obvious that some kinds of organisms will tend to be peculiarly successful on islands, others not. A herbivore or an omnivore should be able to get along in a smaller area, owing to the higher standing crop of food. On this basis we may explain the fact already well known to A. R. Wallace (1876) that pigeons, reptiles, and land snails are particularly abundant on islands, while hawks and many others tend to be impoverished. The limited extent of the market favors the local enterprise and reduces the impact from competitors that need a large area in which to operate. Likewise, the dwarf forms of many large mammals on islands can be explained in terms of the same principles. One needs to be careful here, however, for such adaptations are not means of preventing extinction, say through increasing genetic diversity by a greater number of individuals. The error of such views is apparent when we realize that some insular animals and (especially) plants tend to get bigger, thus diminishing the size of the gene pool and endangering their population. There has been much *ad hoc* hypothesizing about genetic drift, loss of evolutionary plasticity, and the like among insular forms. As Soulé (1972) points out, much of it will not bear critical examination.

When the male element is transported by currents of air or water, but the female remains stationary, the difficulty of finding a mate is somewhat greater than it is where both sexes release their gametes into the water. In marine invertebrates, retention of the fertilized eggs in the care of the mother before the assumption of a free-living stage has repeatedly evolved. This puts a premium on a sort of pseudocopulation, for the eggs are not randomly distributed any more. The earlier account of hermaphroditism (M. T. Ghiselin 1969b) needs to be corrected in the light of this train of reasoning, including the energetical ideas treated in the above two paragraphs and the male-dispersal aspect of the size-advantage model. It remains a brute fact that forms with suppressed larval development are often hermaphrodites, but too much was attributed to the changes in population structure that result when the larval dispersal is suppressed. Reconsidering the evidence, it turns out that in many forms with maternal care, but fair to good larval dispersal, simultaneous rather than sequential hermaphroditism is the rule and these should now be explained by the low-density or size-advantage model. The protandrous and protogynous animals conform to expectations all the better. Among important examples of invertebrates with brood care but good larval dispersal and simultaneous hermaphroditism, we may list the following examples, all of which are relatively sedentary organisms: some sponges (Sarà 1961), a few brachiopods (Oehlert and Deniker 1883; Senn 1934; de Beauchamp 1960), some phoronids (Marcus 1949; Rattenbury 1953; Silén 1954), certain species of the polychaete annelids *Sagitella* (Uljanin 1878) and *Spirorbis* (zur Loye 1908; Bergan 1953; Gee and Williams 1965; Potswald 1967), and quite a number of bivalve mollusks (Pelseneer 1912; Coe 1943). A careful survey of the algae along these same lines might well repay the efforts of a systematic botanist.

The *size-advantage model* is of fairly recent origin (M. T. Ghiselin 1969b) although precursors may be found in earlier work. Particularly important is a paper by C. L. Smith (1967), who treated the subject from the point of view of age rather than size, but obviously the two go hand in hand. Originally intended to explain only sequential hermaphroditism, the model may perhaps be extended to some cases of simultaneous hermaphroditism as well. If a small animal (or a young one) can reproduce more effectively as a member of one sex, while a large animal (or an older one) can reproduce more easily as a member of the other sex, then it becomes advantageous for an individual to switch gender as it grows and ages. Where it is advantageous for a male to be large or a female to be small, protogyny should evolve. Where large size in females or small size in males increases individual reproductive success, the result should be protandry. Since these phenomena are a special case of sexual dimorphism, we shall defer many of the details to later chapters, but we may briefly summarize the general conditions. A large size is advantageous to males when they fight with one another for mates; that is to say where there is one form of sexual selection. This has led to protogyny in quite a number of fishes and perhaps a few other organisms. Among the most interesting examples is a fish from the Red Sea studied by Fishelson (1970). The females change into males, but only when no males are already present, and the change is prevented by a chemical signal emanating from the males. Robertson (1972) provides a second example of this phenomenon; he treats it as a means of optimizing the sex ratio for the sake of the population, and does not even consider the possibility of sexual selection. (For more on protogyny in fishes, see Atz 1964; M. T. Ghiselin 1969b.) Large females tend to evolve when their reproductive output is increased by virtue of size, but the same effect does not occur in the males as when the females carry the young. In some cases this has led to the evolution of protandry.

Sometimes males enjoy a definite competitive advantage when they are small. This occurs when the young stages are much more motile than the older ones. Protandry therefore may be expected to evolve in marine invertebrates which become more or less sedentary in later life. This hypothesis effectively covers the protandry of many animals, such as intertidal limpets (Coe 1944, 1948; Bacci 1947a, 1947b, 1947c, 1951a, 1951b; Hendler and Franz 1971). It may likewise apply to the pandalid shrimps, in which members of certain species and local populations display sex reversal (Carlisle 1959; J. A. Allen 1963, 1965; Charniaux-Cotton 1963; Butler 1964; Rasmussen 1967; Hoffmann 1968) and to some other crustaceans (Dohrn 1950; Dohrn and Holthuis 1950; Yaldwyn 1966; more below). The Tanaidacea, an obscure group of crustaceans related to isopods, would also seem to have a mobility problem, but their solutions are evidently rather diverse. In *Heterotanais oerstedi*, the female lives in a tube, where she cares for her young. The males are said to be of two types: small primary males, which do not feed, and secondary ones, derived from females that have finished raising the young (Forsman 1956). Here, protogyny would seem to be linked with a greater motility at an advanced age, but maybe the secondary males have an advantage in fighting over mates. The group deserves further study, for although a variety of sexual arrangements have been described, their adaptive significance is virtually unknown (G. Smith 1903, 1904; K. Lang 1952, 1958; Kudinova-Pasternak 1965).

Extending the size-advantage model to simultaneous hermaphrodites, we need only add a probabilistic analysis of the local situation. In gonochoric species, the individuals are by definition one sex or the other, and the sex ratio under average conditions is unity. However, organisms in nature do not live under average conditions. If the local population is small, we might expect random mortality to produce fluctuations in the proportion of males and females. If selection ordinarily favors large size in males, it may nonetheless occasionally happen that no large males are available. In that case a small animal could gain a considerable advantage by functioning as a male. The possible disadvantage might be offset if, when the opportunity failed to present itself, the animal could function as a female, resorbing its sperm and using the energy to care for the young. Thus hermaphroditism might be favored by virtue of its flexibility. Since the flexibility is most likely to be useful in smaller populations, we would in effect have a combination of the size-advantage and low-density models. Little evidence is available one way or another, but perhaps the simultaneous hermaphroditism of certain marine fishes fits in here.

A final explanation for hermaphroditism is the *gene-dispersal model* (M. T. Ghiselin 1969b). Where motility is low, both the number and the variety of potential mates are limited, and the genetical environment is thereby profoundly affected. By sampling error, only one male within a deme may be available for a group of females, so that the next generation will all be half or full siblings. Even if this does not happen, brothers and sisters, or other close relatives, will be likely to cross in animals that do not wander far from their birthplace. The result will be both inbreeding and access to less genetical variability. Simultaneous hermaphroditism would insure that there were always as many potential fathers as there were individuals (cf. Kimura and Crow 1963; Murray 1964 for related phenomena). Sequential hermaphroditism could prevent crosses between siblings. Were the members of a clutch all males at the time they reached sexual maturity, and were they all to change into females simultaneously, they could never mate with one another. The opposite change would have exactly the same effect. A similar phenomenon, sometimes called *merogony*, sometimes *monogeny*, occurs in certain isopods (Vandel 1941; de Lattin 1952). Here, the clutches consist entirely of males or entirely of females; Howard (1940) noted that the arrangement prevents inbreeding, and that it also allows a deviant sex ratio. It is indeed thought that the sex ratio in these forms departs from unity, although little is known about the selective aspects of the problem (see G. Johnson 1961; W. D. Hamilton 1967). Petersen (1892) interpreted the widespread tendency of male insects to emerge from their pupae before their sisters as a mechanism preventing crosses between siblings. This view has been challenged (Demoll 1908), and we may recollect that we have questioned the idea that protandry in plants has the same kind of function; but irrespective of particular cases, the hypothesis seems useful so long as it is substantiated.

The evidence in favor of the gene-dispersal model is rather tenuous, and it may apply to few, if any, organisms. The argument rests upon the apparently higher incidence of sequential hermaphrodites in forms that have restricted gene flow when these are compared with their more vagile relatives. Simultaneous hermaphroditism tends to be ambiguous, and pains must be taken to exclude the possibility of a size advantage. One series of examples is provided by marine inverte-brates that have lost their larvae. The only known hermaphroditic species in the neogastropod mollusks is protandrous, and no longer has a larva (E. H. Smith 1967). Likewise, the brooders are often sequential hermaphrodites in polychaetes (Mesnil and Caullery 1898a, 1898b), bivalves (Pelseneer 1912; Hansen 1953; Sellmer 1967), and ophiuroids (Mortensen 1936). In the genus *Asterina* (Echinodermata: Asteroidea) suppression of the larva coexists with different patterns of gonochorism and simultaneous, protandrous and protogynous hermaphroditism (Cuénot 1898; Oshima 1929; Mortensen 1933; Bacci 1951c; Cognetti 1954, 1956; Hauenschild 1954; Delavault 1960; Cognetti and Delavault 1962; Bruslé 1968, 1969a, 1969b). It would seem difficult to apply the size advantage or the low-density model in this case, but one can never be sure.

Restricted gene flow is also caused by the invasion of land and of fresh water. A few terrestrial pulmonate gastropods have evolved from simultaneous hermaphroditism to protogyny (Rosenwald 1926; H. Hoffmann 1927); a size advantage is not obvious here, but someone might be able to think of one. Hermaphroditism, usually protandrous, is common in freshwater and terrestrial isopod crustaceans ("pillbugs") (Jackson 1928; Arcangeli 1932; Bocquet 1967; Juchault 1967; Legrand 1967), an assemblage known for the strategical alternative of merogony. The correlation between the alternatives is one of the more compelling arguments for the model. Very likely too, some instances of sequential hermaphroditism in fishes may be due to restricted gene flow. In this connection it is well to mention a few genecological studies that de-serve more attention than they have thus far received. Hofmeister (1939) and Lillelund (1965) compared populations of smelt (*Osme-rus*) from the open sea with those inhabiting the estuary of the Elbe river. The invasion of a restricted body of water has somehow led to a decrease in the proportion of males (they have a high mortality rate) and an abundance of hermaphrodites. Along similar lines, Weisensee (1916; see also Le Fevre and Curtis 1912; Bloomer 1939; van der Schalie 1966) studied populations of freshwater clams that had been isolated for differing periods of time in ponds cut off from the Rhine. The more recently separated populations had an excess of females. Later on, hermaphrodites appeared, and these gradually replaced the gono-chorists. We have no compelling evidence that restriction of gene flow actually caused the changes in breeding systems, but no other explana-tion has been proposed. Cause and effect here is hard to establish. One wonders if the inbreeding might not simply have disturbed the sex-determination mechanism.

Thus it would seem that in spite of many ambiguities, the hypoth-eses cover the facts; and on the whole it is clear which hypothesis should be invoked. It is equally apparent that many problems remain unsolved or even uninvestigated. Particularly intriguing is the question of how hermaphroditism is combined with other features so as to produce an effective constellation of adaptive strategies. A number of creatures that move slowly about, exploiting sedentary food items of rather patchy distribution are simultaneous hermaphrodites. These include many flatworms, occasional terrestrial nemerteans (Coe 1939), and sea slugs. It may be that combining the search for food with the search for a mate saves resources. The principle of division of labor has its limitations. Combining a pair of activities gives a high return if the two complement each other. In chess, a good move often involves attacking two pieces at the same time; and people often shop on the way home from work.

GONOCHORISM AND SEXUAL SELECTION

Masculine modes of competition. Having accounted for the hermaphrodites we are now in a better position to deal with the gonochorists. Let us therefore return to the functional differences between males and females. Under the simplest conditions, the females specialize in caring for the young, while the males concentrate upon union with the females. As a consequence of their relationship to the next generation, the females have a more narrow range of workable stratagems able to increase their Darwinian fitness than do the males. For the female, the appropriate maneuver is to put as much energy into increasing the number and quality of offspring as is possible, for any alternative would divert energy from the offspring themselves. Only the manner in which the energy is used in obtaining this result will provide workable alternatives: reproduction or maintenance, few large or many small eggs, etc. A male, however, has a broader variety of effective ways to use a given quantity of resources in getting into the gene pool, for his output is not so greatly limited by the amount of sperm he can generate as is that of the female by her capacity to produce eggs. Beyond the small amount of energy that goes into elaborating the sperm sufficient for fertilization, he has an excess that can be exploited in various ways. He can, indeed, simply put out a maximum quantity of sperm, and many externally fertilizing marine invertebrates would seem able to do little else. Here the number of offspring is proportional to the number of sperm, and the amount of energy put into gametes by each sex should tend to be equal. This would occur in spite of the fact that far more spermatozoa were generated than were needed to fertilize the eggs. The facts seem to bear this out, although differences in standing crop may give an impression that the sexes differ somewhat in productivity. The data of Giese, Greenfield, Huang, Farmanfarmaian, Boolootian, and Lasker (1959) on the purple sea urchin, *Strongylocen-*

trotus purpuratus, show the gonad volumes of males and females to be about equal. Yet a form with copulation, or even pseudocopulation, would be able to compete in various ways. He could use some of the energy otherwise lavished on sperm in hunting for a mate. Or he could help care for the young, becoming, in a functional sense, rather like an hermaphrodite. Alternatively, he might direct his resources against other males, preventing them from breeding, and monopolizing the females in his own interests.

Sexual selection. Of crucial importance for understanding the situation with respect to the males is the sort of intraspecific competition that goes on between them. A male can increase his Darwinian fitness by maneuvers that have no effect on reproductive output, other than monopolizing opportunities for mating to his own benefit and to the detriment of other males. The special kind of selection that then occurs Darwin called *sexual selection,* to distinguish it from the other two: *artificial selection* and *natural selection.* In artificial selection, the selective agent (that which does the selecting) is man; in natural selection it is the forces active in the "environment"; in sexual selection it is a mate or a rival. A failure to understand the differentia underlying Darwin's terminology is perhaps one of the main reasons why many have treated sexual selection as if it were just a special case of natural selection. Very few contemporary biologists use the terms as he did. Worse, his interest in it was long misrepresented as an effort to explain away difficulties in his theory. Darwin reasoned that evidence for sexual selection would serve as a powerful empirical argument favoring his theory in general (see M. T. Ghiselin 1969a). Sexual selection provides instances of non-adaptive or even maladaptive characters evolving through purely reproductive competition. Even some of Darwin's more enthusiastic supporters have failed to go along with him on this point. Some of the opposition has arisen through misunderstandings, deliberate or otherwise, of what Darwin said. The term "sexual selection" has often been used to cover any selection for an ability to mate (e.g., T. H. Morgan 1914). A gene that increased the frequency of successful copulations by changing the structure of the genital organs would have an obvious selective advantage to the males. But the selection favoring it would be natural, not sexual. The males would not only increase their own reproductive success, but that of the females as well. The males would have more offspring, not just a larger proportion, and the females would have more offspring too. In purely sexual selection, the females receive no benefit from mating with one male rather than another. Consider another economic analogy. Sexual selection is like attracting customers through advertising that makes the competition's product look bad. Advertising that lets the customers know that the product exists would be more like natural selection. Sexual selection is thus a very special kind of interaction between individuals which affects their mating; but Ehrman (1972) makes the term too general when defining it as *any* mechanism causing deviations from panmixia.

Much of the resistance to sexual selection has been a part of the opposition to selection theory in general. However, many selectionists, of whom Wallace was most noteworthy, have argued against it, and upon grounds that upon closer examination, turn out to be metaphysical. Wallace, whose attitudes reflect a strong dose of teleology, maintained that natural selection would prevent the evolution of genuinely maladaptive traits. To a certain degree this train of reasoning must be accepted. It seems inescapable that an advantage in reproductive competition will only be selected if concomitant disadvantages do not reduce overall Darwinian fitness. And where maladaptations evolve, natural selection will tend to produce adaptations that compensate for them. But nothing implicit in these considerations should lead us to deny that an individual can obtain an advantage in sexual competition at the price of an increased probability of being killed. Neither does the theoretical possibility that selection may produce a given change imply that selection must produce that change. A. R. Wallace (1912: 295) argues that:

Now this extremely rigid action of natural selection must render any attempt to select mere ornament utterly nugatory, unless the most ornamented always coincide with "the fittest" in every other respect; while, if they do so coincide, then any selection of ornament is altogether superfluous. If the most highly coloured and fullest plumaged males are *not* the most healthy and vigorous, have *not* the best instincts for the proper construction and concealment of the nest, and for the care and protection of the young, they are certainly not the fittest, and will not survive, or be the parents of survivors. If, on the other hand, there *is* generally this correlation—if, as has been here argued, ornament is the natural product and direct outcome of superabundant health and vigour, then no other mode of selection is needed to account for the presence of such ornament.

Wallace's reasoning derives from false premises. He treats the world as if that which reproduces successfully were a unit character, rather than an organism. He might just as well have concluded that a firm can drive another out of business by deceptive advertising only when its product costs less than the other's too. Treating parts as if they evolved separately from their organismal context makes it very easy to believe that this is the best of all logically possible worlds, instead of a compromise between biological alternatives.

Many other theoretical arguments against sexual selection are equally metaphysical. It was early maintained that sexual selection will not occur in monogamous species. Darwin countered with the following model (Darwin 1874:229):

Let us take any species, a bird for instance, and divide the females inhabiting a district into two equal bodies, the one consisting of the more vigorous and better-nourished individuals, and the other of the less vigorous and healthy. The former, there can be little doubt, would be ready to breed in the spring before the others. . . . There can also be no doubt that the most vigorous, best-nourished, and earliest breeders would on an average succeed in rearing the largest number of fine offspring. The males, as we have seen, are generally ready to breed before the females; the strongest, and with some species the best armed, of the males drive away the weaker; and the former would then unite with the more vigorous and better-nourished females, because they are the first to breed. Such vigorous pairs would surely rear a larger number of offspring. . . .

This model has been questioned (Bastock 1967 for example) on the grounds that the most successful males would not mate more, but only first. Yet the basic point of Darwin's argument is that the early breeders have access to more and better resources (see also O'Donald 1972). Furthermore, nobody seems to have noticed that the one-to-one sex ratio is an abstraction, valid as a theoretical concept, but essentially inapplicable to organisms in nature. The sex ratio of a species will average around unity, but in a local population, especially a small one, at a given time and place one or the other sex will be more numerous. In fact, given an odd number of individuals, a sex ratio of exact unity is mathematically impossible. Geneticists often think typologically about populations: "the male," "the female," "the sex ratio," and above all else "the environment" are metaphysical entities.

Sexual dimorphism and sexual selection. Darwin enumerated a host of secondary sexual characteristics that appeared to have been produced by sexual selection. The subsequent controversy has largely involved efforts to explain such features as the result of some other process, generally natural selection, but by no means always. Vitalist, often "Lamarckian," interpretations have been repeatedly proposed (Cunningham 1900; Geddes and Thomson 1901; A. R. Wallace 1912; Pycraft 1913; Portmann 1961). We need not review these here, since evolutionary biologists no longer take them seriously. Other, more current, alternatives will be discussed in later chapters. However, we should at this time mention those aspects of sexual dimorphism that only indirectly relate to reproductive competition.

Different reproductive strategies, combined with other biological features, may result in sexual differences in the ecological niche. Rand (1952) and Selander (1966, 1969, 1972) argue that competition between the sexes is reduced by their divergence. This seems improbable when we realize that any increase in the ability to avoid competing with members of the opposite sex would do nothing about competition with members of the same sex. The population would expand, a new equilibrium would be reached, and selection would favor the one best-adapted type. However, such a dimorphism would be workable in a family or other social group. Kilham (1964; see also Ligon 1968) suggests that this may explain the difference in feeding habits of the sexes in some woodpeckers and other birds. But the male and his mate act as a single unit, and the selective advantage derives from their cooperative division of labor, not their individual efficacy. Hence we are dealing with what Darwin treated as a difference in the way of life of the two sexes, but one that has nothing to do with the reduction of competition between them. Likewise, Rand (1954) provides many suggestions as to how flocks of birds, of either one species or several, may cooperate in feeding, avoiding predators, and other activities. Having a diversity of members within these groups could be attributed to interactions that increase the amount of food available to the group, thereby allowing it to be larger and more effective at predator avoidance. Or it could simply mean that cooperation favored feeding or predator avoidance. But any reduction of competition here is not between the species or the sexes, but rather between the members of the families or the flocks. On the same basic grounds it seems reasonable to reject the other purported examples of ecological differences arising as a way of reducing competition. Male ducks in certain species live apart from the females and the young (Rand 1952), but the obvious explanation is that food is scarce in the nesting area, and the males are better off elsewhere. A fish with a big mouth eats larger food (Baird 1965), but the males bite one another, and sexual selection is therefore probable. As the competition reduction idea is frequently invoked *ad hoc*, we shall have occasion to deal with it repeatedly in later chapters.

How differences in sexual role may perhaps influence adaptive strategies in general is evident when we reconsider some problematical phenomena of mimicry. It often happens that the males and females of a given species mimic different models. Or, only the females are mimics, or where both sexes mimic the same model, the female's resemblance to it is closer. Sometimes the females may themselves be polymorphic, resembling more than one model, or else some do not mimic at all. A fairly straightforward case occurs in beetles of the genus *Anacolus*. The males resemble wasps, the females other species of beetles that are distasteful (Linsley 1959). Here the more motile sex resembles the more motile animals in its environment. To see what happens in sex-linked mimetic polymorphism among Lepidoptera (Ford 1953; Magnus 1958, 1963; Stride 1958; Brower 1963; Rettenmeyer 1970; Emmel 1972; Williamson and Nelson 1972) we may simplify the discussion by considering the parental advantage to having diverse offspring, although other ways of treating the subject could work equally well. It seems that the polymorphism somehow reflects the greater effectiveness of mimicry when the mimics are rare. It occurs only in mimics of the Batesian type, which are palatable, and which are eaten more often when common because predators are then less effectively trained to avoid them. In Müllerian mimics, which are distasteful, several uncommon species assume the same pattern because the predators are more often led to avoid it; and for the same reason both sexes are mimics. If it were advantageous for the mimics to be rare, a parent could increase the safety of its offspring as a group by having them adopt different strategies: mimicry and some other form of protection, or a mimetic resemblance to different models. A simple way to produce such an effect would be to link color pattern with gender. But why should there be a very general tendency, extending over large taxonomic groups, for it to be the females that mimic? And why is polymorphism so often restricted to one sex or the other? In view of the differences in competitive relationship between members of each sex, all this makes a certain amount of sense: the females are adapted to survive, the males to reproduce more than other males. Mortality among daughters will necessarily cut down on the parent's number of grandchildren; but the loss of a son's reproductive output may to some extent be compensated for by the success of his brothers. If so, it would be more effective to protect the daughters than the sons. It may well be that female butterflies spend less time flying from one place to another than do the males; this should influence their frequency of encounter by predators, and therefore affect the workability of the protective adaptation. Since concealment in some groups of insects depends on remaining stationary, the greater motility in the males might decrease the selection pressure for resemblance to their backgrounds. These hypotheses are proposed because they would seem natural elaborations of the general scheme: new research will be necessary before we know if they are true. Nonetheless it would seem that some early speculations on such matters are due for a revival (for history see Brower 1963). The only seriously competing hypothesis explaining mimetic sexual dimorphism is that the protective patterns inhibit recognition of the male by the female (Darwin 1874). This hypothesis, although probably applicable in some instances, will hardly cover all of them, especially where, as the behavioral evidence would seem to indicate, the females do not exercise a choice of mates. The notion of sex-limited mutations happening to be those upon which selection has acted is even more dubious. One would not expect such a consistent tendency for the female to be the mimic. Similarly, the greater development of camouflage and predator intimidation in female mantids described by Crane (1952) suggests the need for an ecological, rather than a genetical, hypothesis.

Any discussion of polymorphism within a single sex would be incomplete if it did not briefly mention a phenomenon called "high-low dimorphism" (Bateson and Brindley 1892; G. Smith 1904; Diakonov

1925; Gadgil 1972). In a number of animals with a highly modified male, there may also be males in which the differences are not so great. Although such variation may be continuous, sometimes it has a bimodal distribution. It would appear that an individual adopts one or the other of two adaptive strategies, and that compromises between them do not work. It might happen that both strategies have their advantages, and selection could maintain them within the population. In salmon (see Briggs 1953; Shapovalov and Taft 1954) the males fight, and larger individuals fertilize most of the eggs. Yet a minority of smaller individuals ("grilse") exist in the population, and these have been shown to enjoy some reproductive success, as, for example, when the larger males are somehow distracted. Since the grilse are younger, relatively fewer offspring are necessary to maintain their genes in the species. In the same way, it seems likely that the "horns" of lamellicorn beetles, known to function in sexual combat, cut down on their possessors' motility. These too display high-low dimorphism.

THE VARIETIES OF REPRODUCTIVE COMPETITION

Such is the approximate state of confusion over the theory of sexual selection at the present moment. The whole subject needs to be reconsidered from a more comprehensive point of view so that it is seen as only one aspect of a broader range of problems related to reproductive competition within species. If we review the various possible relationships, it should become apparent not only that sexual selection ought to be more widespread than has generally been recognized, but also that comparable phenomena have been overlooked or misinterpreted. Let us therefore briefly outline the main categories of interaction between and within the sexes, leaving the details for future chapters.

Male-male competition would include both of the classical Darwinian forms of sexual selection. *Male combat* involves actual fighting between the males. Being the reason for the evolution of horns and tusks in males, it has on the whole been too evident to have been denied outright. *Female choice* is a competition between males by maneuvers that "attract" the females. It has led to the evolution of sexual ornaments, but exactly which ones remains a controversial topic. Less obvious means of competition between males have largely been ignored, and new terms will now be proposed. *Male sequestering*, so called because the male sequesters the female, will here refer to instances in which the male takes possession of the female, often by brute force, and thereby prevents other males from mating with her. It is a sort of Rape of the Sabines ploy, one common among crustaceans. A less violent sort of competition than actual fighting, it nonetheless requires physical strength, and is easily combined with male combat— and just as easily confused with it. Or the male may use a version of female choice, in which he attracts the female and keeps doing so because it keeps her out of the control of other males. Some birds do this. The boundaries between seduction and rape are often far from clear. By *male dispersal* we shall mean that the males compete with one another as if in a race to be first in finding and impregnating the females. First come, first service. A scramble for genital union, putting a premium on speed and early maturation rather than on size and strength, is an alternative to male combat and to male sequestering which leads to quite different consequences. In many cases the necessary adaptations are hard to combine, so that the males must do one or the other. On the other hand male dispersal often is combined with a form of sequestering in which males act upon females so as to prevent subsequent matings, as when they use "chastity belts."

Female-female competition is far less common than that which occurs between males. The females engaging in combat with other females would divert resources from their own offspring, and the males generally act in a way that insures the impregnation of all the females. However, such is not the case in forms in which males care for the young, especially when the animals are monogamous. *Female combat*, in which the females fight so as to monopolize the males, does occur, especially, but not exclusively, in forms in which the male cares for the young. *Male choice* obviously exists in man, and Darwin (1871, 1874) discussed the phenomenon at great length. Males acting as selective agents that opt for alternative females may occur in other species as well. There is but limited evidence for what we would call, by analogy, *female sequestering*, still less for *female dispersal*.

Cooperation between individuals provides an alternative way of using the resources that so many animals expend in competing for mates. It involves care of the young or some functional equivalent. With respect to the females, the alternative is but rarely significant, because the care of the young is their normal occupation. Males, however, can readily adopt a variety of stratagems, including care of the young, and the conditions that lead to the many kinds of paternal care are most enlightening. A given amount of energy has to be used one way or the other, and the result will be a compromise between the effects of natural selection favoring care for the young on the one hand, and sexual selection favoring competitive interaction with male rivals on the other. In general, selection favoring cooperation between male and female will lead to adaptations that allow both parents to supply the needs of the young. It is possible that a family unit having both parents acting to raise the young will be less productive than one with but a single parent. This should happen when the father's contribution is offset by his consumption of resources. The situation might be compared to that in an overcrowded factory. In such cases the family might do better if one parent were to emigrate, die, or not grow to a large bodily size. Small or short-lived males are rather common; but as we shall show in chapter 7, a paternal self-sacrifice is here probably not the best explanation.

Parental exploitation is a special kind of relationship within the family that may superficially appear to be a sort of altruistic cooperation between siblings. In effect, the parent acts upon the offspring in his own interest, and in some ways to their detriment. A simple case has already been mentioned, of a parent cell giving rise to many small, male gametes, which, albeit collectively better off, are individually less likely to mate. Likewise, social insects give rise to neuter individuals that increase the reproductive output of the parent. To be sure, not all of these phenomena are generally explained in terms of exploitation by the parents. Darwin (1859) invoked a selective advantage to the family as a unit. Most recently the idea of "kin selection" has been widely advocated. It is maintained that an animal will gain a selective advantage by helping a close relative, since the organism aided is likely to have genes similar to its own. Nobody seems to question the premise that this explains a parent's care for its direct descendants. But it would seem to be more debatable whether we should conceive of a neuter social insect acting so to speak "in order to help her sisters." Something appears fallacious about many of the explanations cast in such language. Would we say that a sperm cell benefits from being small, since it thereby allows the existence of more individuals like it? If a worker bee sacrifices herself in behalf of bees with similar DNA, should we not argue that cells that die in forming hair are doing the same thing? Where a society or an organism constitutes an integrated whole, which reproduces as a unit, such reasoning is both superfluous and misleading. The real matter to be understood is how the system is controlled. Once this point is clear, it becomes obvious that the supposedly altruistic individuals may act by compulsion. They have no choice in the matter, for everything, so to speak, was arranged by their parents.

Hence the problems of sexual diversity impinge upon those of social evolution. The latter will, therefore, be treated in the closing chapters of this book. For insect societies, it will be fairly easy to show that at least some of what appear to be "altruistic" traits may be explained in terms of natural selection through parental exploitation, and the same idea has wide applicability. Yet other kinds of societies exist, and these require additional principles. We shall consider some of the many instances in which it has been overlooked how a cooperative individual receives a benefit in return. And most importantly we will see how the existence of a society brings new kinds of selective mechanisms into play.

5

THE COPULATORY
IMPERATIVE, OR,
MALE COMBAT

*Every living being is a sort of
imperialist, seeking to transform as
much as possible of the environment
into itself and its seed.*

Bertrand Russell

WHY ANIMALS FIGHT

Is aggression good? The problems of male combat can hardly be
dissociated from those surrounding aggression or agonistic behavior in
general (Scott 1958). Students of animal psychology have accumulated
a large body of theory and descriptive material upon this sort of topic.
Yet the literature suffers much from the teleological approach, and
from misunderstandings about evolutionary mechanisms. The defects
extend to both interpretation and raw data. The "ethologists" in particular have developed a curious epistemology, as may be seen particularly well in the writings of Lorenz (1939, 1950, 1964, 1966a), who
combines a Baconian induction with his own brand of Neo-Kantianism.
On the one hand, Lorenz professes that the study of behavior must begin without hypotheses, in order that it shall be objective. On the other
hand, he supports a notion of *a priori* knowledge, implying that we
possess an innate knowledge of the good and the true. His conception

of the *a priori*, however, differs radically from that of either Plato or Kant (see Piaget 1967); rather, it is virtually identical to that of Herbert Spencer (1862). Natural selection in this way of thinking would program organisms to see the world as it really is, and to do the right thing! Yet it seems reasonably evident that success in reproductive competition may accrue to those who misapperceive the world, and who act to the detriment of others. Our suspicions are aroused when we find Lorenz (1966b) writing a popular book designed to show that aggression is really desirable. Indeed, we are amused to find that the original title, *Das sogennante Böse*, is replaced by the more palatable *On Aggression* for the American trade. Whether or not Lorenz's views began in unprejudiced observation, they surely are not derived from it.

Lorenz has not lacked critics. Lehrman (1953:353) and Crook (1970a:xxii) explicitly draw attention to the ideological content and ulterior political motivation of his works. Lehrman (1953) suggests that Lorenz's views are teleological; Lorenz (1966a) rebuts this charge on the grounds that organisms obviously are adapted. Lehrman has been associated with the "American Museum" school, which has tended to overreact against teleology, so that the critique was somewhat misdirected (see Lehrman 1970). Tinbergen (1968) and others have rejected the notion that aggression is spontaneous, an idea that would follow from the sort of hydraulic model for drive supported by Lorenz, but which has found little acceptance (J. P. Scott 1967). If the impulse to attack another animal occurs inevitably and must be satisfied —like thirst—then we must deal with it in one way. If, on the other hand, it has to be evoked from without—as in flight from a predator— we have another problem. Future historians will probably treat much of Lorenz's writings as an effort to revive late nineteenth-century ideas about "social Darwinism." But they are probably better treated as the result of "convergent evolution" than as "atavism."

We are told by Lorenz that aggression is "good" because it helps to keep a certain amount of space between individuals. But it is said to be good for the *species* (Lorenz 1964), ensuring a better use of resources. Yet if being spaced out is so good, why can the animals not accomplish it with less bother? The Darwinian answer is that they can and do thus distribute themselves, but only when being spaced out is expedient for the individuals. When food runs out, or is about to do so, it is obviously advantageous for any individual to migrate (Lidicker 1962). On the other hand, when some food is available, but more individuals are present than the local area can support, it is advantageous to have others migrate. This simple principle suffices to explain the leading features of aggression if we but extend it to include resources other than food, above all else sex.

The relationships between aggression and territory. The idea of an advantage in "spacing out" organisms is a traditional part of "territory" theory (Howard 1920; Mayr 1939). That the defense of a given area assures control over resources is generally accepted, but what these resources might be, and what the aggression actually does, are confused topics at best. Upon theoretical grounds one might predict that these would be anything in short supply: food, nesting sites, mates, etc. Yet in practice it may not be easy to infer what objects are being competed for from the behavior of the territorial individuals. As a general rule the male is more active than the female in defending the territory. But this could mean a division of labor, a difference of interests, or both. In some cases the males will repel other males, but ignore females, while the females will only drive off female interlopers. In this case it is clear that the combat is sexual, insofar as it is a matter of indifference to the individual whether its mate is replaced by some other animal. However, the pair do gain an additional advantage in that they prevent encroachment upon the resources used in support of their young. We must not go so far as to infer that the only advantage to such territories is in competition for a mate. The males obtain a certain amount of control over local resources by holding a territory, even when they do not prevent females from entering it. The notion of Altum (see Mayr 1935) that the males fight, not for the females, but for the territory, is true, but only in a certain, very restricted, sense. They may act in a fashion that is determined by territorial defense, and this is a very good point. But nonetheless, one of the major benefits they may obtain by virtue of holding a territory is obtaining control of the females. Mayr (1972b) argues that since the males are thus competing for the territory, rather than for the females, the advantage must lie in care of the family or in obtaining food. Although the *non sequitur* should be obvious, the empirical evidence he gives in support of this position is even more decisive on the other side. He concludes that the territorial combat between male pinnipeds (sea lions and the like) is not due to sexual competition, but rather it provides for the needs of the family. But, as will be shown in a more extended discussion below, the males do not feed, and they do not protect the young.

Why aggression has become less sanguinary. Another series of problems surrounding aggression have to do with "ritualization" (Huxley 1966; also other articles in the same symposium). Combat, it is maintained, is really a sham. The animals threaten each other; they may struggle for a time but as a rule no injury results. Some workers have asserted that this apparent chivalry has resulted as a group adaptation, even as a means of protecting the species. Yet the notion that damage is minimal has been somewhat exaggerated; it is difficult to say how much. Records of animals being seriously hurt during sexual combat are abundant for ungulates (Hediger 1952; Fraser 1968:85–92). Even

Darling (1937:175), who minimizes the significance of antlers in deer as weapons, nonetheless does mention a young stag being killed by an older one. Damage, even death, is recorded among sparrows (Summers-Smith 1958), various beetles (Hubbard 1897; Webster 1904; Funke 1957; Linsley 1961), dragonflies (Jacobs 1955), spiders (Bristowe 1929), and crabs (Hughes 1966), to give but a few examples. Perhaps the least edifying case among nonhuman primates has been described by Sugiyama (1967) for a langur, *Presbytis entellus*. A male, after obtaining control over a troop of females, killed the infants, which were not his offspring. The dominant male in such troops ordinarily does not harm the juveniles, and experimental removal of the male from another troop led to the death of infants at the hands of other males. Evidently the males compete by killing one another's children. Such practices, of course, are very common among human beings, and no doubt less sentimental approaches to the study of animal behavior would show that our own species is far less exceptional than has been claimed.

Lorenz's (1952) stock example of a curb upon aggression in canines turns out not to be so neat as he tells us. A submissive dog is said to turn his head away from his assailant, presenting a vulnerable surface for attack but evoking a species-preserving restraint that is instinctive. The meaning of the "signal" has been misinterpreted, as one might expect from what appears to be a risky way of expressing submission. One possible reason for turning away (Morris 1956) is that it produces a movement opposite to that which implies an attack. This is an instance of the principle of antithesis (Darwin 1872). Moynihan (1955; see also Chance 1962; Vine 1970) points out the "appeasement" function of such movements. Anyone can observe the phenomenon for himself. According to my own observations, the submissive animal does not provide his assailant with a chance to do damage but, on the contrary, his actions are directed toward preventing it. The beaten dog flees, turning a moving target away from attack and, at the same time, avoiding any action that might provoke his assailant. The most likely place to be bitten during this maneuver is the curved surface of the neck muscles, not blood vessels or nerves, as Lorenz's hypothesis suggests. The point here is not that appeasement or protection or both should be considered the "true" explanation, but rather that the idealized canine has been placed on a pedestal where he does not belong. We should also draw attention to a study on wolves (Jordan, Shelton, and Allen 1967:248) that describes the case of a dominant wolf who stood his ground and was killed rather than submit or retreat. Although this does not affect the competing hypotheses, it does help present the less romantic side of the tale.

The anatomy of animals that engage in sexual combat incorporates many provisions for reducing damage (Geist 1966), such as heavy padding on the shoulders of boars. Yet, as Schaffer (1968) observes for sheep and goats, much damage occurs in spite of these features. A stag's antlers are as well adapted to blocking his opponent's thrust as they are to inflicting wounds. Furthermore, ill-matched opponents seem not to fight: a show of strength proves adequate. These defensive provisions and hesitancy to give battle are viewed by some as evidence that the species is protected from harm by the limitation on damage. Yet protective features are surely advantageous to the individual combatants. Medieval knights did not carry shields to reduce the amount of bloodshed in general, but to keep from being killed. In deer (Darling 1937; Harrison-Matthews 1964), certain individuals have single-spiked horns with which they can do great damage. But like knights without shields, they will not last very long, in spite of certain temporary advantages and the saving of energy (Gadgil 1972). It would furthermore be disastrous for an animal to hazard battle against a stronger opponent, while on the other hand nothing would be gained by continuing the contest once it had been decided. Hence, it is only to be expected that tests of strength might come to replace more sanguinary modes of combat.

The effects of size and age. In the struggle between males, victory, as a rule, but by no means inevitably, goes to individuals of large bodily size. Greater magnitude generally requires a longer period of growth, hence later maturation. In animals that live in family groups or in herds, the younger males either are excluded from the reproductive unit, or else take a subordinate position and tend to be denied access to the females. It has been noticed that the period of active reproduction is often delayed beyond what seems adequate time for the attainment of physiological maturity. In many species, the younger males are fully capable of fathering offspring and will do so if given a chance, yet their sexual weapons and ornaments remain underdeveloped. Such a condition is well documented for the Tule elk, for instance (McCullough 1969). We might also mention certain birds (Ashmole 1963; Selander 1965), and the pinnipeds discussed at some length below. The problem of why such organisms do not immediately enter into sexual competition has perplexed many biologists. One explanation has been that the younger animals profit from a period of learning and exploration during a protracted adolescence (Ashmole 1963; Crook 1965; for critique: Skutch 1967). Similar ideas have been proposed by Lack (1966) for a delay in breeding by such birds as storks. The data presented by Lack do suggest that the older birds raise more young, and experience may indeed be useful in various reproductive activities. Either raising offspring or indulgence in sexual combat will result in physical damage or depletion of energy reserves. Yet such considerations do not wholly explain the phenomenon. The young animals ought to begin reproducing right away, for this would provide the hypothesized experience. They would only adjust the number of offspring or the intensity of sexual combat to compensate for lesser ability. Wynne-Edwards (1962) very ingeniously fits the phenomenon of deferred maturation into his group-selection system. Under his hypothesis, it would help the species by preventing overpopulation. Yet here, as with so many of his explanations, organisms do not behave as the hypothesis would seem to predict. In the first place it is generally the wrong sex, namely the male, that defers reproduction. In polygynous species this arrangement obviously leaves the reproductive output of the population unchanged. Furthermore, since the younger males are kept from reproducing only by the actions of the older ones, and will mate if given a chance, nothing like altruistic behavior occurs here.

Let us consider a simpler explanation. An animal capable of expending resources may have the option of either using them right away or saving them for use at another time. However, he cannot wait forever, especially if doomed to perish of old age. A young animal can afford to exercise a certain option: reproducing when, and only when, the conditions happen to be optimal. Given occasional fluctuations in the availability of food or the presence of sexual competitors, opportunities may present themselves for reproduction with minimal effort. For older animals about to pass their prime, no such alternative exists. They put the full weight of their resources into whatever struggle presents itself. To be sure, the death rate can be adjusted within certain limits by natural selection, and death is itself but one aspect of a general system of adaptive strategies. But it often happens that the animal cannot wait, and has no option but to go all out, succumbing in the effort. Where reproduction terminates in death, the final stages of life are often accompanied by an extreme metamorphosis, and nothing like a period of "training" occurs. The best example is the Pacific salmon, *Onchorhynchus*, in which the male develops a hooked jaw, used in fighting over the females. The animals enter fresh water, undergo a rapid structural change (Davidson 1935), spawn and die (Briggs 1953; Shapovalov and Taft 1954). Other salmonids can reverse the metamorphosis and reproduce more than once, but the degree of sexual dimorphism is here far less pronounced (Tchernavin 1938). Indeed, among fishes that breed in schools, it does not even occur (Noble 1939; cf. Parr 1931). Therefore it seems reasonable to infer that a period of reproductive flexibility in younger individuals is an adaptation to environmental unpredictability, and not a mere preparation for later activities.

Wynne-Edwards has supported yet another alternative hypothesis seeking to explain male combat: it gives rise to a social hierarchy that produces a more efficient use of food, a notion that he explicitly relates to the spacing-out theory. He maintains that the hierarchy thus developed helps to identify the surplus individuals (Wynne-Edwards 1962). But a surplus is a property of a population, not an individual. Not only are the wrong individuals excluded from breeding, again, but the hypotheses to be presented later on will show that we need not invoke such hypotheses to explain the facts.

Another species-advantage explanation is the "best-man" theory of Lorenz (1966b), who attributes it to Darwin. Yet although Darwin seems to have entertained a notion of this sort, he (unlike his grandfather Erasmus Darwin) did not maintain that male combat exists because it benefits the species. He only said that more vigorous males might win the battles; his explanation for fighting was quite different. Crudely satirized, Lorenz's theory is that stags battle in order that the more Aryan individuals will propagate the race. A somewhat similar notion, that the female benefits the species by selecting the more healthy males, has been maintained by Lenz (1917) and Lebedinsky (1932). Others (Verner 1964; McLaren 1967) suggest that the female would derive an individual advantage through associating herself with a superior male. Stoltzmann (1885) even viewed dimorphism as a means of getting rid of the males, hence benefitting the species by giving a larger proportion of the more useful sex!

The "best-man" hypothesis has material implications that readily show it to be false. If it were true, then we ought to find sexual combat more or less randomly distributed among organisms, for all should derive the same advantage. In actuality, as will be shown, the males fight only when environmental conditions bring them into circumstances where combat is advantageous to individuals. Furthermore, the males refrain from combat so long as they can get into the gene pool by other means—hence such phenomena as youthful disinclination to do battle. The physiological and structural dysteleology that result from adaptations favoring purely reproductive competition are reflected in lack of paternal care or even in infanticide, features that can scarcely be reconciled with a general prevalence of adapted species. As to the possibility of females exercising a choice of more vigorous mates, the plain facts of the matter are that in only a minority of instances are they able to do so (but see Trivers 1972). Aronson (1957) suggests that sexual combat may be sexually stimulating. If so, one could treat it as a special case of natural selection, or of sexual selection by female choice. But the natural history data seem to imply that this role is, at most, minor.

THE ENVIRONMENTAL REASONS FOR COMBAT

The effect of high population density. Let us now consider the ecological determinants of intraspecific sexual fighting. Male combat is only likely to result where the effective population density is high; for if the males but rarely encounter one another, their battles will have little effect upon gene frequencies. The same principle naturally applies to female choice and male sequestering, but not to male dispersal. We should, therefore, expect large males and sexual weapons and ornaments to evolve under high-density conditions. There is no good evidence of male combat among deep-sea animals, although it is common in those of shallow water. Nor would it appear to exist in parasites, although some exceptions doubtless could be found. (An approximation is a blood fluke which lives at high population densities; the males sequester the females and have been observed to take them away from other males by force [J. C. Armstrong 1965].) In both these cases, the usually low population density favors male dispersal or no sexual selection at all. But rather than dwell upon negative cases, let us examine a series of animals in which an unusually high population density has accentuated male combat.

Fiddler crabs (*Uca* and related genera) have long been classical materials for the study of male combat. Treated by Darwin, they were later discussed from this point of view by Pearse (1914). More recently they have become favorite materials for students of animal behavior. One can derive much useful information about sexual selection from this literature, especially from the works of Crane (1941a, 1943, 1957, 1958, 1966), Burkenroad (1947), Altevogt (1955, 1957a, 1957b, 1959), Peters (1955), von Hagen (1962), Salmon (1965), Salmon and Atsaides (1968), and Aspey (1971). The main reason for so much study is that fiddler crabs are very common, diurnal organisms, and one can easily watch their activities. The big claw of the male is used for both combat and courtship display. Both sexes are rather pugnacious —the males, which are somewhat larger, especially so. The animals burrow. Their territoriality has been debated, and perhaps differences between species and local populations have confused such issues. Only the males stridulate; in some other crustaceans both sexes produce sound (see Guinot-Dumortier and Dumortier 1960). Verwey (1930) noted that courtship display is more common where population density is high, and this fact suggests that both male combat and female choice have been accentuated by the increased frequency of competitive interactions. Takahasi (1935) says that population density is determined by the feeding habits. Obviously, a common animal must have an abundant source of food. Hence the kind of dimorphism will be strongly influenced by trophic resources. However, the relationship, although it may be observed in one group of animals after another, is by no means a simple one. Thus it is not clear why the "ghost crabs" (*Ocypode*), which belong to the same family but are larger and have different habits, are quite pugnacious but have less size dimorphism (Crane 1941b).

Another very useful group of organisms for the study of male combat are the dragonflies and their close relatives the damselflies (Odonata). The males are, loosely speaking, territorial, gathering near the sites where the females lay their eggs. Observers seem to agree that the males fight at this time, but their interpretations differ somewhat. Moore (1952, 1957) invoked a kind of "spacing out" hypothesis; Pajunen (1963a, 1963b, 1964, 1966a, 1966b) also leans toward this position. Sexual selection has been preferred by Jacobs (1955) and C. Johnson (1961, 1962a, 1962b, 1962c). All these authors have noted that the male keeps other males from interfering with the female's egg-laying activities. This is probably a case of natural selection, but perhaps male sequestering is involved too. Pajunen (1966b) notes that some male individuals are nonterritorial. They may succeed in breeding, however. The existence of individuals evidently analogous to the grilse of salmon means that the stratagem of combat is not inevitably the most effective. The polymorphism also makes it rather difficult to apply some of the population-advantage hypotheses. Of even greater interest is the observation that at the highest population densities, fighting ceases (Pajunen 1966a). If the reason for territoriality here is "spacing out," then one would expect quite the opposite. Alexander (1961) has recorded precisely the same phenomenon in field crickets (Gryllidae). Increased density here leads to more toleration between adults, and to more dispersal. At first sight one might think that a cessation of hostilities at the higher population densities would not be expected, since there should be more opportunities for competition between the males. But here we need to consider the law of variable proportions: the more competitors present, the sooner male combat will experience diminishing returns. So long as only two males are struggling for a single female, the victor will impregnate her. But a third male party may be able to mate with her while the other two are contesting the issue. If so, then some other maneuver becomes more effective.

In other cases, a high population density at the time of reproduction characterizes a given group of animals. Salmon, again, are a fine example. The females lay their eggs in limited areas of stream beds, where the males then fertilize them. The males congregate in these areas, arriving before the females, and outnumbering them at any given time. Such a situation would appear to increase the probability of encounter between males, and their extreme dimorphism may be attributed to this cause. Yet concentration by itself is not sufficient to make combat evolve. A number of fishes restrict their spawning to a limited area, but do not fight. The California grunion, *Leuresthes tenuis*, for example, lays its eggs during high tides on sandy beaches, and large assemblages of them will congregate at one time (Thompson and Thompson 1919;

F. N. Clark 1925). It is not easy to fight when being swept around by the waves.

Perhaps the most useful classical materials for the effects of population density on male combat are the Pinnipedia, or seals, sea lions, and walruses (see Scheffer 1958 for classification). The relationship between polygamy and sexual dimorphism in this group was clearly recognized by Darwin (1874:234). Later workers have elaborated the same basic idea, showing how the gregariousness of some species accentuates sexual selection (Nutting 1891; Bertram 1940; Bartholomew 1952, 1970; Peterson and Bartholomew 1967). The species vary considerably in both dimorphism and gregariousness, and both of these features have evolved independently in separate lineages. The males may become much larger than the females, and often become highly territorial and aggressive during the breeding season. In monomorphic species such as the bearded seal, *Erignathus barbatus*, the two sexes mature at about the same age (McLaren 1958). The highly dimorphic species *Halichoerus grypus*, or grey seal, on the other hand, has females becoming mature at 4 or 5 years of age and beginning to reproduce right away, while the bulls do not start breeding until about 10 years of age; male mortality in this species is high, for the males live only about 25 years while their sisters live about 35 (Hewer 1964). In another highly dimorphic species, *Callorhinus ursinus*, or northern fur seal, the females begin to breed at 3 or 4, the males at 7 to 10 years of age (Kenyon and Wilke 1953). Data on a number of species show that the younger males in dimorphic species are perfectly capable of breeding. Thus, the male southern elephant seal, *Mirounga leonina*, matures at 4 years, but he does not obtain a "harem" until about 15 (Carrick and Ingham 1960). The younger subordinate males can evidently mate upon certain occasions.

Intense male combat results indirectly as a consequence of some peculiar features of pinniped reproduction. Although most of their life is spent in the water, the females must emerge onto the land or ice to deliver the pups. They nurse them for a variable period of time. In some pinnipeds copulation, particularly the first impregnation, may occur in the water (R. W. Rand 1955; Carrick and Ingham 1960; Hewer 1960), but as a rule it occurs on land, during the time the pup is being nursed. Hence the males obtain their reproductive advantage by monopolizing the nurseries. They show no clear evidence of caring for the young, which is only to be expected, since their mates' offspring are not their own. Indeed, one author after another stresses the fact that the bulls often hurt or even kill the pups, especially when hastening to engage another bull in combat. The only exception thus far recorded in the literature is some observations by Eibl-Eibesfeldt

(1955b) on *Zalophus wollebaeki* from the Galapagos Islands. He saw bulls keeping the young from predators. If preserving the pup kept the females from deserting his territory, it might have been to a bull's advantage to aid them.

Population densities become very high, and sexual dimorphism extreme, in the northern elephant seal *Mirounga angustirostris* (Bartholomew 1952; Bartholomew and Hubbs 1960) and its relative from colder waters in the southern hemisphere, *M. leonina* (Harrison Matthews 1929; Carrick and Ingham 1960). In this genus, the "harems" may reach about forty females. The size of the harem depends upon the amount of crowding, and upon the ease with which subordinate bulls can infringe upon the territories of the dominant ones. The male does not attempt to sequester the females in *Mirounga*, but this stratagem does occur in another extremely polygynous seal, *Callorhinus ursinus* (Kenyon 1960; Peterson 1968). In this species, the male may be quite brutal in keeping the female from wandering, grasping her in his jaws and throwing her bodily about. Similar behavior has been observed in the southern sea lion, *Otaria byronia* (J. E. Hamilton 1934). The somewhat less polygynous California sea lion, *Zalophus*, sequesters the females, but to a much lesser degree (Peterson and Bartholomew 1967). Sequestering has also been noted by Cameron (1967) in the grey seal, *Halichoerus grypus*. The cape fur seal, *Arctocephalus pusillus*, occasionally restricts females to his territory; in this species the females display preference for certain males (R. W. Rand 1955).

One may attribute, at least to some extent, the high density that results in male combat to the gregarious habits. But this leaves unexplained the cause of the habits themselves. The high mortality of pinniped pups that results from social breeding was thought by Coulson and Hickling (1964) to be a mechanism for regulating the population in the fashion suggested by Wynne-Edwards. McLaren (1967), Peterson (1968), and Bartholomew (1970) oppose this view with sexual selection models, but one might question some of the details of their interpretations. They maintain that the females select the dominant males because these will have the more successful male progeny. They invoke a kind of positive feedback loop, providing a version of the best-man hypothesis. No empirical tests are given for this notion, which introduces unnecessary complications. All one needs to explain the population structure is a paucity of optimal breeding sites and a trophic base sufficient to overpopulate them. Seals can hardly be expected to know, through abstract reasoning, when to seek a better place for giving birth. Rather they ought to go where they were born, where everyone else goes, or both. That mother succeeded in a given place is clear and sufficient evidence that daughter may. Only when overpopulation becomes extreme, or failure is experienced, should reproductive success reward the experimenter.

Saturation of all available sites for raising the young would necessarily increase social interaction, although other influences might have the same effect. It would appear that pinnipeds often select isolated beaches and islands free from terrestrial carnivores as places for nursing their young. If such sites are limited in number, and if there is sufficient food to populate them heavily, then competition leading to sexual dimorphism should result. Cameron (1970) points out that in *Halichoerus grypus* polygamy is determined by the nature of the terrain: where there is plenty of space, monogamy is the rule. Bertram (1940) thinks that the system in this species is disintegrating, evolving toward loss of dimorphism. He attributes suppression of differences between the sexes to an evolution of copulation in the water, and thinks that where copulation occurs helps to account for the pattern in the entire group. The more dimorphic forms do seem to copulate mainly on land. Many of the monomorphic species would seem to have little opportunity for effective male combat. The tropical monk seal, *Monachus schauinslandi*, for example, is fairly rare, and displays only minor hostility between the males at breeding time (Kenyon and Rice 1959). *Lobodon carcinophagus*, the "crabeater" (which actually feeds mainly on euphausids not crabs), breeds on the ice and is only semigregarious (Bertram 1940). Likewise poor in sexual dimorphism is the sparsely distributed bearded seal, *Erignathus barbatus* (McLaren 1958).

McLaren (1958) has drawn attention to the fact that in pinnipeds the intrinsic rate of natural increase (r) correlates with gregariousness and the formation of harems. That is to say, the faster the animals reproduce, the greater their sociality. He explains this phenomenon in terms of habits. The organisms with a need to "know" their environment take their time in reproducing. One might suspect that the correlation here leads to a misapprehension of cause and effect, for differences in the availability of food suffice to explain the facts. Animals whose food is scarce should reproduce slowly, tending to be conservative and to avoid wasting food. A species in which something prevented the population from expanding to the saturation level would tend to reproduce much faster, since organisms in general reproduce as much as they can. The high reproductive output and gregariousness of certain pinnipeds result from the same cause: limited places for raising the young combined with a surfeit of food. Indeed the ample food supply is one of the reasons for the shortage of places for breeding. The animals keep populating the environment until something limits them. When food does not set the limit, something else does.

The effects of family structure. A social organization that brings several adult males together as members of a herd or extended family should accentuate competition between them. Such assemblages occur in many terrestrial vertebrates, and also in certain whales. Some toothed whales (Gray 1882; Slijper 1958), including the sperm whale (*Physeter*) and the smaller "bottlenose," live in family groups. The males of both species are larger than the females, and the peculiar snout of the latter occurs only in adult males. The groups consist of one breeding male and several females accompanied by their young, while the younger mature males are excluded. The societies probably do not arise from male sequestering, but rather function as cooperative associations of females who aid one another in caring for the young. Both species feed upon cephalopods, and this may help to explain the small size of the herds: there are not enough food animals in any local area to support many of them. In baleen whales the females are the larger sex. They often live in very large herds, which would tend to render combat less effective, for the high population density reasons discussed above with respect to field crickets and dragonflies (see p. 146). Although male combat does work in large herds when it is combined with sequestering, this stratagem is not always workable. The size of the herds of baleen whales may relate to their subsisting mainly on euphausids (which are herbivorous crustaceans). They eat a type of food that has a very high standing crop, and can, therefore, support a very large local population. As a general rule, organisms living on a kind of food that is abundant (and this usually means one low in the trophic pyramid) exist at higher densities than do those in which the food is rare, dispersed, or otherwise limited. Thus, ungulates very often live in large herds, and sexual dimorphism, or sexual combat without much dimorphism, are very common in this group. Terrestrial carnivores, on the other hand, live in much smaller groups. Darwin (1874) points out that the distinct sexual dimorphism of lions correlates with their unusual habit of associating in prides.

Among primates, similar trends are apparent, although the trophic relationships are hard to analyze. Baboons, which live in troops with younger males subordinate, are conspicuous for their strong males armed with enlarged canines and spectacular instruments of intimidation and display. It has frequently been suggested that the large size of the males has evolved as a mechanism for defending the family. There is a problem here, since one would expect the large males, once they had evolved, to exploit their size and strength by driving off predators. For at least some large males, of various species, this hypothesis clearly does not apply: the indifference of pinnipeds to the welfare of the young is notorious. On the other hand, there exists no known fact that can be explained if, and only if, the protection hypothesis is true. Indeed, it is invariably applied *ad hoc*. In the case of the patas monkey,

Hall (1964) could not see how the male would protect the troop, since he has a "slender physique" and "elusive habits," and therefore he interpreted the highly dimorphic male as an adaptation for diverting the attention of predators. Perhaps these animals do resort to such deflective maneuvers, but it is hard to see how we can derive, from this hypothesis, their greater size relative to the females, or their vicious canine teeth. Animals frequently have weapons that have evolved as means of dealing with predators, but ordinarily these are not dimorphic. Likewise, weapons used in sexual combat have secondarily acquired a defensive role. And the horns of many ungulates used by the males in fighting over the females are also directed against predators by both sexes or by one, while in some species either sex or both may use them in agonistic behavior of a nonsexual nature.

The inattention, up until very recently, of primate biologists to the competitive nature of the societies they were studying provides a classical example of failure to see the obvious. Carpenter (1958:242–243) lists thirty-two functions of territoriality, but fails to mention its role in sexual selection. Thompson (1958:307) noted that "in primates, at least, strangers of another species are usually tolerated and allowed to mingle with the group, while strangers of the same species are not. . . ." He did point out that sexual dominance has something to do with this, but failed to draw the obvious conclusion about the adaptive significance of primate societies. De Vore (1962) observed that ground-dwelling primates tend to be more sexually dimorphic than arboreal ones—but he attributed it to defense of the group by the males. Lately it has become apparent that the kind of food available, the mode of predation, and the size of the family are also correlated with this difference in habitat (Crook 1970c, 1972; Itô 1970; Eisenberg, Muckenhirn, and Rudran 1972). Evidently the size of the troop is strongly influenced by the amount of food available, and by the capacity of the troop to increase the extent of the market by moving about and by territorial defense. Protection of the young is also important, especially in open country, and here numbers have obvious advantages. But the effort of the males in larger primate societies is mainly devoted to combat with other members of their own species. The laws governing the size and structure of such societies are economic ones, which apply to all societies: they will be treated at greater length in chapter 8. We should add, however, that human societies are on the whole quite different, being aggregations of monogamous families in which the male plays a major role in providing for his children, although both monogamous and polygamous families are found in nonhuman and human primates alike.

A male can obviously increase his reproductive output by having several mates rather than just one. Yet, as we pointed out with respect to matters of density, he may not always find it feasible to do so. Indeed, this stratagem may even be disadvantageous. If the males care for their progeny, any energy expended in sexual combat must be diverted from the aid for the next generation. Therefore one might expect a complex pattern of sexual dimorphism correlating with how the young are supported. Especially where the females, but not the males, care for the young, we should expect sexual dimorphism. Such relationships are well known; they have been treated from a behavioral point of view by Baerends (1952). Yet a word of caution is needed here. We cannot reverse the train of reasoning and infer that where the males are in a position to care for the young they will do so. Uncertainties about these relationships have evidently led certain biologists to attempt some explanations of polygyny predicated upon the assumption that it benefits the female. We can see that such models are not compelling on the basis of some simple calculations (see Crook 1965). Suppose that a female by herself could raise three young, a male and a female jointly, six. Would this necessarily mean that selection would favor paternal care? Clearly no, for it would be indifferent to the male whether he fathered six offspring by two females or by one. For the female it would make all the difference in the world, but only the crudest sort of teleological reasoning would imply that she must inevitably be able to force a male into a domestic role. One might suspect that a female could select the mate that would be a good father, but equally a male could abandon his family once this had been "demonstrated." Trivers (1972) considers many possibilities of exploitation, "cheating" and the like, which deserve more study, since little evidence is available one way or the other. Be this as it may, marital relationships in a state of nature are purely matters of expediency, and there exist no laws imposing justice or moral responsibility upon them. Whatever really goes on will mirror this truth.

Trying to determine the relative advantages of participating and not participating in various forms of family life can be a very difficult task indeed. When we compare the reproductive output of a male with two mates with that of a male of the same species with only one, and find that the former has twice as many offspring as the latter, does this mean that paternal care, instead of male combat, would not increase the reproductive output of the species? One might think so, but here we are not comparing organisms adapted to monogamy with organisms that are adapted to polygamy. Rather, we are comparing an unsuccessful polygynist with a successful one. Likewise, that families in species with two parents per family are perhaps not twice the size of those with one does not bear upon the issue of which alternative will be selectively advantageous, for the conditions are not necessarily the same. By the same token, when we find a female in a polygynous family outreproducing a female in a monogynous one, does this mean that it is advantageous to the females to share a husband with another female? Not necessarily, and for much the same reason. It could just as well mean that the strongest males are associating with the more desirable females. A valid argument would have to be based on comparing a polygynist before and after experimentally reducing the number of mates.

The main arguments that polygyny benefits the females are based on studies of blackbirds (Nero and Emlen 1951; Verner 1964; Verner and Willson, 1966; Horn 1968; Orians 1969). These birds are of various species, so we shall have to speak in very general terms. Here nesting aggregation seems to have its advantage largely in defense against predators. There are optimal and suboptimal positions within the colony with respect to this function. The females are known to defend their nesting territories from other members of their sex. The males direct their aggression largely toward other males; some help care for the young as well. Under circumstances such as these one would only expect the territories of successful males and successful females to coincide, but this does not necessarily mean that the females gain through polygyny. Additional work will be needed to see if perhaps the females select the males rather than the territory, which would, of course, have the same effect.

Conditions similar to those just described may exist in various polygynous rodents (Healey 1967; Archer 1970; Downhower and Armitage 1971; Ewer 1971). Here, the females and their offspring live in small groups with a single dominant male, who attacks male interlopers but not female ones. The upper size limit of the family is determined through combat between the females, and it seems reasonable to conjecture that it relates to the availability of resources. What makes it advantageous to form such associations could be a variety of functions: such as communal care of the young, defense of the territory, or escape from predators. Thus we discover the selective advantage of polygyny to the female: there is none.

The assistance of a mate may be highly advantageous to both husband and wife. Yet both may be quite indifferent as to who plays this role. Where no biparental family exists, the female is just as well off being fertilized by one male as another, and therefore we should expect no competition for husbands. And feminine jealousy in such forms relates to other resources—such as the egg-laying territories of European cuckoos (*Cuculus canorus*) (K. L. Skinner 1922). Exactly the same fundamental relationship would be expected where only the male cares for the young: the females may fight it out, while the males are indifferent. This kind of female combat is rare, but does occur, for example in the phalarope (*Phalaropus*) (Tinbergen 1935), the American Jaçaná (Jenni and Collier 1972), and tinamous (Beebe 1925; Pearson and Pearson 1955). In a polychaete annelid, *Neanthes caudata*, the male incubates the eggs, often eating his mate after she has laid them. If the male is approached by another female while the first one is still present, the first one will attack the newcomer. A male incubating eggs may fight with either males or females (Reish 1957).

For both sexes within a family to engage in sexual competition is probably not uncommon. In such cases, however, we are faced with a problem. It is difficult to distinguish between competing "for" a mate, and "for" something else. The two may coexist, in fact. The underlying mechanism is neatly brought out when we consider forms in which both sexes contribute resources to the young. It is here, and here alone, that male and female alike engage in sexual combat, and both may be quite indifferent as to who is the mate, albeit neither of these phenomena necessarily accompany care of the young. Female combat of this second kind obviously exists in man, and it has been reported in several species of birds. Examples are the bronze mannikin (*Lonchura cucullata*) (Morris 1957), the flicker (*Colaptes auratus*) (Noble 1936), and the house sparrow (*Passer domesticus*) (Summers-Smith 1958) (see also Tinbergen 1936). Striking analogies occur in certain invertebrates, but it is problematic whether services of a mate, or just plain food, are at issue. Beetles of the genus *Scarabaeus* roll balls of dung, which they use as a store of food for both larvae and adults. At the time of mating a male and a female together roll a ball of dung to an underground bridal chamber. If they are encountered by another beetle, the interloper is attacked by a member of its own sex. Combat occurs only with food, and never between male and a female (Heymons and von Lengerken 1929). The same basic system has evolved in corpse-burying beetles of the genus *Necrophorus*. Each corpse is the territory of one pair of beetles. Males defend it from other males, females from other females, except where they are defending the food supply from beetles of another species; both sexes will attack a competitor, irrespective of its sex, when it is not a suitable mate (Pukowski 1933). Arrow (1951) points out that in all dimorphic scarab beetles, and in some other Coleoptera as well, the males have "horns" only when the males and females do not collaborate in gathering food. It would appear, therefore, that morphological differentiation depends upon an extreme difference between the competitive relationships of the two sexes. But in any case, sexual selection by no means inevitably leads to dimorphism.

Whatever the interests of the female, a male may gain an advantage by caring for his offspring. If he does, then the effective strategy should be monogyny, which entails minimal competition with other males for mates. Whether it will be advantageous for him to care for his offspring will be determined by the conditions of existence, including the opportunities for multiple mating and the return he gets on paternal care. His return will be affected by the availability of food and its mode of utilization, by the effects of predators, and numerous other influences. *A priori*, one might expect sexual combat, female choice, and male sequestering to evolve where the young are not cared for at all, since the males lack the option of acting like females. But the inference only betrays the dangers of *a priori* reasoning, for the exact opposite relationship obtains. Where dimorphism evolves through competition between males in forms that do not care for the young, it is largely a result of male dispersal. The reasons for this apparent paradox are evident when we realize that support of offspring can have indirect effects by accentuating the difference between male and female roles, and by bringing about new modes of competition. The classical form of Darwinian sexual selection occurs largely in forms with parental care. And in these groups we may find dimorphic species with males that leave the burdens of caring for the young to the females, (relatively) monomorphic species in which both parents cooperate, or occasionally the males and females have reversed their roles and dimorphism once again occurs, this time with the females competing and adapted for it. Among cephalopods, the males in some species of *Octopus* are larger than the females, who brood the eggs; while in squid, both sexes are around the same size and die after spawning (Drew 1911; McGowan 1954; Arnold 1965; Fields 1965; see also chapter 7, and, for the cuttlefish, Tinbergen 1937). In the "longicorn" beetles already mentioned, maternal care is highly developed, and so are various kinds of sexual selection. In fish, it is noteworthy that both dimorphism and care of the young are more common in freshwater and in benthic marine forms than they are in those of the open sea. But the cause is debatable, for any of a number of mechanisms might be invoked. Thus, Winn (1958) studied fourteen species of darters (fishes of the family Percidae). He found trends toward movement from lakes into streams and territoriality correlated with increased sexual dimorphism. Itô (1970) suggests that fishes have evolved such reproductive adaptations in response to low levels of food. The case of salmon suggests that although food is obviously important, density of the population needs to be considered as well. Parental care is highly developed in mammals and birds, as is sexual dimorphism, ordinarily of the type with larger males.

Sexual selection as affected by resources. The kind of food eaten influences sexual competition through its effects on parental care in a way that leads to a whole series of relationships not unlike those that we have noted for clutch size, species diversity, and the like. Many of these have been mentioned in the literature, albeit without any real appreciation of the global scheme of things.

It has long been known that the degree of sexual dimorphism and

the development of secondary sex characters in birds correlate with
the type of food they eat and with who in turn eats them (Mottram
1915; Rosenfelds cited in Lebedinsky 1932). Dimorphism is greatest in
seed-eaters, somewhat less in omnivores, and least in carnivores. This,
in part, would seem to be related to the standing crop relationships
already mentioned, but it also has something to do with the ability of
the young to fend for themselves. Hence the juvenile feeding habits,
dimorphism, and clutch sizes all tend to display correlated trends. In
nidifugous species, that is to say, those in which the very young birds
leave the nest to feed, the advantage of having two parents is perhaps
less than it is when the nestlings remain in the nest and the parents
bring food to them (nidicolous species). If the male's help is less important, he may go in for alternative ways of obtaining reproductive
advantages. In hole-nesting ducks, the larger clutches already mentioned are accompanied by greater sexual dimorphism. The idea of
Bergman (1965) that small holes are better utilized by small females will
explain one of these phenomena but not both, and comparable relationships exist where nothing like it would seem to apply.

The trophic situation is analogous in mammals (Orians 1969).
Herbivores are more dimorphic than carnivores, with certain noteworthy exceptions. A male carnivore can bring meat to his offspring,
but a herbivore has no such option. Yet once again, this effect may be
hard to separate from standing crop relationships. Geist (1971) presents
copious data showing that ungulates of higher latitudes and during the
Pleistocene tended to be larger, more social, and more polygamous.
Deer, such as the Irish "elk" are a good example. He relates this to the
dispersal pattern: a better food supply is available to pioneering groups,
which he thinks tend to be evolutionarily progressive. Very likely,
however, a standing crop relationship may be the crucial factor here.
Given abundant rainfall and not many competing herbivores, it should
be possible for the habitat to support more individuals of a given species,
and these could, for the same reason, attain larger body size. Fewer
species implies more individuals of a given species per unit of area as
well. The large numbers of individuals, especially where herds are
formed, would increase the opportunity for male combat, of course.

In hawks, owls, frigate-birds, jaegers, and skuas, the males are
smaller than the females (Hill 1944; Cade 1960; Willoughby and Cade
1964). This seems odd to many ornithologists, but small males are the
rule among metazoans, and the explanations invoked for birds would
seem to rationalize appearances in that aberrant taxon alone. The
phenomenon has been attributed to the fierceness of carnivores: a big
female is necessary to protect the young from their father (Hagen
1942; Amadon 1959; for critique see Storer 1966). Of equally *ad hoc*
nature, obviously, is the notion that a small size is actually better in

combat between males (Schmidt-Bey 1913). Others (Selander 1966, 1969; Earhart and Johnson 1970) think that it reduces competition by partitioning the niche between the two sexes. A division of labor with the males eating smaller food items has been suggested (Brüll 1937; Hagen 1942; Frochot 1967). Birds are about the only group of animals in which the smaller males obviously can have any of these reasons for their size relationship. The correlation between the degree of dimorphism and the kind of food is a general trend, however, as is its parallel with a concomitant trend in the number of eggs per clutch. Therefore we should seek to attribute the whole set of phenomena to one fundamental cause, allowing us to dispense with such hypotheses. Balgooyen (1972) has lately proposed just such a model and has backed it up with much research on the American kestrel, *Falco sparverius*. The female in this species is considerably larger than the male, but field observations and measurements of bills and talons show that they take food of the same size; this excludes the hypothesis that they avoid competition or subdivide the trophic resources by size. The male defends the territory and helps care for the young; like all other hawks he shows no tendency to eat them. The sexes do, however, display a clear-cut division of labor. The female concentrates upon incubating the eggs, during which time she is fed by the male, and occasionally spelled on the nest by him. When the young are hatched, he continues to provide most of the food to the young, but later the female has to join him, for the family needs a great deal of food. The male is smaller because he can locate and carry food less expensively; and he fetches it from a greater distance. Such is true of predators in general, which benefit from being large insofar as it helps to subdue their prey, but which do not achieve this advantage without having to pay for it. Thus the laws governing size dimorphism of hawks are the same as those governing anisogamy. And just as airlines own large and small aircraft better to turn a profit, and better to compete with other firms—and definitely not to avoid competition between the two kinds of aircraft—so it is with aerial enterprise in the natural economy.

It would be most pleasing were we able to explain all phenomena of sexual dimorphism in terms of simple energetical rules of thumb. Unfortunately the complexity of the underlying interactions, with multiple influences at work, makes it very difficult to gain an adequate picture of what goes on. Hummingbirds, for example (Becher 1919; Pitelka 1942; Stiles 1971), are highly dimorphic, and combat between the males is well documented. Only the females care for the young, which are quite incapable of feeding themselves, since they cannot fly. The males should be able to feed their offspring, but they do not. A fairly high population density should indeed favor male combat as an alternative to paternal care. And given the expensive metabolism of hummingbirds, the cost of transporting food over increased distances might give an insufficient return. But the relationships would have to be worked out for concrete instances before we could do much more than speculate. In a number of fruit-eating tropical birds, the same lack of paternal care is also noteworthy. Snow (1963) maintains that the availability of food sufficient for one parent alone to attend the nest in frugivorous and nectar-eating birds freed the males from paternal duties and led, therefore, to sexual selection. Yet it remains unexplained, on his hypothesis, why the pair do not collaborate to raise a larger family; indeed, Snow has evidently reversed cause and effect. It is a general rule that tropical, nidicolous birds tend to have only one parent care for the young (Skutch 1949:439). If, as we have suggested, tropical species live under conditions in which the food supply is inelastic, then the law of variable proportions ought to make paternal care, at least where accompanied by maternal care in the same nest, less advantageous. Given a limited amount of food per unit area, the distance travelled to obtain it would have to be increased if more offspring were to be supported. Transportation being increasingly expensive as distance becomes greater, it follows that doubling the number of providers will not double the supply of resources to the dependents. Nidifugous birds get around the problem of diminishing returns, to some extent, by moving the family from place to place, contributing to their larger clutch size; but this stratagem, too, has its limitations. In a very few species of birds, such as the red-legged partridge, *Alectrois rufa*, the males and females both raise broods with but a single parent accompanying them (Portal 1924). Such an arrangement very effectively overcomes the problem of diminishing returns, but its rarity suggests that the males more often derive a greater benefit from opting for joint parental care or else leaving it to the females.

Geographical patterns in sexual competition. Accepting the general rule that competition gets accentuated toward the equator, one might expect a heightened sexual dimorphism in the tropics. Such a trend would seem to exist, but many qualifications are necessary. It is obscured, for example, when both sexes are subject to sexual selection which, although giving rise to ornaments and weapons in the tropics does not produce differences between males and females. T. H. Hamilton (1961; see also Hamilton and Barth 1962) drew attention to the fact that in warblers and orioles, the tropical species are brilliant but monomorphic, while in more temperate ones, the female becomes dull. He speculates that a difference between the sexes results from a need to tell males from females. As he puts it (T. H. Hamilton 1961:122), "sexual differences in plumage characters might operate for rapid sex-recognition and alleviate in part the hostility between potential mates." It seems more likely that seasonality has created an advantage in early breeding, maintaining the sexual competition between males and at the same time causing the females to concentrate upon a higher output of offspring. In this, as in other seemingly anomalous cases, we must also consider density effects. In our discussion on the loss of hermaphroditism among insular plants, we drew attention to the fact that species diversity affects the number of individuals of each species per unit of area. By the same token, we should expect tropical ecosystems to manifest comparable relationships and have similar effects. Thus, hermaphrodite flowers should be highly successful in the tropics. Now, as the number of species increases there will be two results that should influence sexual competition. First, the variety of competitors of other species will increase, and this, combined with the general stability of tropical ecosystems, will cause each species to operate very near to the saturation level and render the supply of food inelastic: hence alternatives to parental care will become advantageous for the males. Second, the decrease in number of individuals per species will reduce the opportunities for competitive interactions between them. Hence, some kinds of sexual selection will be intensified, others mitigated in the tropics. From general principles, one should be able to predict what will happen with changes in density, but an effective treatment of the subject requires that one study each kind of organism in great detail, which would be outside the scope of this volume. Hence we shall leave this topic upon a theoretical plane. However, the basic thesis has already been tested with referenc to the insular hermaphrodites, and it will be further substantiated in the next chapter, when we deal with reproductive competition on islands.

SOME BROADER IMPLICATIONS

Before leaving the topic of male combat, it seems desirable to treat some peripheral matters. Male combat and its correlated dimorphism form but part of the general evolutionary picture with respect to the size of organisms. Furthermore, the ideas being developed here suggest that we may need to revise our interpretations of many important phenomena. Hence we shall devote this final section of the present chapter to exploring some additional possibilities. It should be clearly understood that this is largely a speculative endeavor, intended to demonstrate that these ideas may prove useful, rather than a serious attempt to test them.

Scale effects and extinction. Buried in the literature we may find numerous references to what may be called "scale effects" (see Rensch 1950, 1960). Sexual weaponry is most highly developed in larger animals. In many groups with horns, a polyphyletic trend has been noted, with increasing body size being accompanied by an even more pronounced enlargement of the horns. Such a trend is well known in rhinoceroses and ceratopsian dinosaurs. Sexual dimorphism of various sorts is likewise thought to be increasingly pronounced in larger organisms (Lameere 1904), although some taxa display a reversed trend.

Scale effects have been variously explained, but none of the hypotheses thus far invoked seem adequate. If we consider only horns and antlers in mammals (see reviews by Anthony 1928–1929; Bruhin 1953; Geist 1966) we might get the impression that a purely mechanical advantage is involved. A large animal weighs more, and his charge produces more damage because of his greater momentum. Hence the return on an investment in horns should be greater in larger forms. That some such relationship enters in is hard to deny, for small deer often have sharp canine teeth instead of antlers. Yet this functional difference explains only the kind of armament, not the degree to which it is developed. Many lineages of beetles display greater dimorphism in larger species (Lameere 1904; Champy 1929; Arrow 1951). However, this trend characterizes groups having small adult body size as well as those that are larger; hence a giant in one genus will have "horns" but be no larger than a "hornless" small member of another genus. Were it simply a matter of absolute size furthering the use of a particular kind of organ in sexual combat, quite different relationships would obtain.

Some workers have considered increased relative size of horns to be an indirect consequence of bodily enlargement (Huxley 1932; see also his references). If growth is allometric, then an increase in size might enlarge the weapons faster than the rest of the body. It does appear that size increase has been produced by extending the period of growth. Lameere (1904) points out that in the giraffe both the male and the female have horns, while the smaller okapi has them only in the male. Thus giraffids are an exception to the rule of more dimorphism in larger forms. Be this as it may, if an original gene for horns were expressed only in animals of a certain size and no smaller, if the ancestral giraffid had horned males somewhat larger than hornless females, and if both sexes became enlarged, then a character originally expressed in only one sex might appear as a useless excrescence in the other. For the giraffe, however, this seems an unlikely hypothesis. The males, but not the females, use the horns in fighting. But horns are present in the foetus of both sexes. (See Spinage 1968.)

Perhaps something of the sort is going on in reindeer and other groups, in which the females have small antlers. However, it seems likely that the females in such cases use them in various sorts of agonistic behavior, although we cannot exclude the possibility that some really lack a function. A general rule has been enunciated that horns occur more often in females among animals that live in large herds (Frechkopf 1946; Hediger 1952). A purely morphogenetic basis would seem to imply a random distribution.

Darwin (1871, 1882) took great interest in the possibility of non-adaptive features evolving through "correlated growth," and hypotheses along such lines are still viable. Yet the question of whether such mechanisms are possible must not be confused with that of whether they are, in fact, responsible for the data. In some of the beetle lineages in question, the females have become enlarged without such putatively allometric effects, and so have some of the males within the same genetical populations. It also helps not to ignore other invertebrates. Sometimes the female has become enlarged as an adaptation to rearing the young, and here too, the larger species are more dimorphic. Likewise, group after group has undergone a reduction in size, rather than an enlargement. Here, the smaller species, not the larger, are usually more dimorphic, and therefore allometry does not apply.

A widespread evolutionary trend toward increasing bodily size (Cope's rule) has been admirably documented by Rensch (1960), following the work of numerous predecessors. However, many of his arguments purporting to show that such a trend is universal are logically unsound. To be sure, we find larger and larger representatives of one taxonomic group after another appearing in the hypothesized sequence. But, as Stanley (1973) points out, the same would be expected if, for one reason or another, the smallest members of a taxon were the first to evolve. An adaptive radiation would very likely ensue, with some groups evolving toward larger size. If so the apparent trend would really be nothing more than the result of cumulative diversification over an extended period of time. If the common ancestor of a group were of more or less average magnitude, then the trend should go both ways. And sometimes it does. The class Crustacea includes not only groups that have become enlarged, such as the crabs, but also others that have become small, such as the copepods. Gastropod mollusks have frequently evolved minute lineages, balancing the many instances of large ones. The extreme reduction of rotifers and gastrotrichs shows that the trend has not been universal for the generality of metazoans, while hummingbirds and shrews belie the rule for higher vertebrates.

The explanations Rensch (1960) gives for the trend show far too much evidence of *ad hoc* hypothesizing. His notion that larger animals are the more vigorous ones is obviously true for instances in which we compare some normal individuals with pathological dwarfs, but is otherwise unsubstantiated. An advantage to the results of allometric growth begs, as we have observed, the adaptational question. Hybrid vigor does, as Rensch suggests, give rise to larger individuals; but these are large because of their vigor, not vigorous because they are large. A longer life too, is the result, not the cause, of bodily enlargement. Although the fairly rapid growth of some big animals and plants shows that factors other than time must be considered, nonetheless it takes longer for an egg to form a whale than a mouse. Small invertebrates, such as snails (Pelseneer 1933) and echinoderms (Buchanan 1967) can have a very long life-span. We tend to assume that extending the duration of life is advantageous, but there is no evidence that animals or plants live any longer than is effective in maximizing their reproductive output. The other explanations for enlarging the size of the organism involve primarily matters of physiological advantages, which clearly do exist, but which cannot be the whole story: more eggs, more neurons, etc. We may dismiss some of these with the remark that body plans in general have undergone an adaptive radiation; in this radiation more and more complicated structures have evolved hand in hand with the acquisition of more components—which implies larger animals.

The most dubious explanation, perhaps, is treating size increase as the Bergmann effect on a geological scale. The idea that large size helps to conserve heat doubtless has some valid applicability. The importance of conserving heat receives independent substantiation from the reduction in surface area through the evolution of smaller appendages in high-latitude animals (Allen's rule). The objections of Scholander (1955, 1956) to the Bergmann effect have been successfully rebutted by Mayr (1956) and T. H. Hamilton (1961). Yet in such rebuttal the hypothesis has been rendered virtually untestable, so that the topic ceases to be a part of objective natural science. It is maintained that the principle at issue applies only to species. We can ignore, for the moment, the species that demonstrably refute it. The question is, if the hypothesis covers some genera but not others, should we not find a better explanation for the phenomenon at higher taxonomic levels? Furthermore, it accomplishes nothing constructive to say that one must invoke additional principles, and then provide no criterion telling us which to invoke. If heat relationships are in fact important, as indeed they may be, we need a system in which heat is but one parameter.

To refute the explanation for size trends in terms of the Bergmann effect, we need but consider some facts. Of course with respect to its sufficiency, the hypothesis is a pushover. Too many poikilotherms, especially aquatic ones, display similar trends for our credulity not to be strained (Ray 1960). And the effort of Rensch (1924) to apply the Bergmann effect to insular animals works for some groups, but has to be rationalized for others. Some poikilotherms (such as reptiles) are said to display, for instance, a reverse trend, becoming smaller where it is colder, while others seem more conventional.

Critical evidence, however, might well be taken from the works of Kurtén (1953, 1965, 1968) who supports the Bergmann effect, but whose data nonetheless refute it—and they refute it for single genetical populations, and for single evolutionary lineages. It does appear that many populations have displayed trends in the direction of reduced bodily magnitude as interglacial periods set in and the climate got warmer (see also W. E. Edwards 1967). And other species of large animals became extinct (good reviews: Thenius 1962; Martin and Guilday 1967; Kurtén 1968). Some extant animals are thought to be dwarfed forms of larger ones—bison for example. However, much dwarfing began at the outset of the ice ages (Kurtén 1953:104). The wooly mammoth, *Mammuthus primigenius,* became smaller, not larger, during colder periods (Kurtén 1968:245). Cave bears (Kurtén 1965:54, 1968) had dwarf montane forms. The reduction in size of some carnivores, such as hyenas and wolves, is mirrored by a simultaneous increase in the size of others, such as foxes (Kurtén 1965:46) and some mustellids (Kurtén 1968:27). The latter (weasels, etc.) are notorious for an inverse gradient: within a species they get bigger, not smaller, at lower latitudes. Efforts to shovel the exceptions under the rug range from the fantastic to the reasonable but still *ad hoc.* Where the animals become dwarfed on islands and mountaintops, it is in order that the population shall preserve its genetical flexibility, avoid inbreeding and the like (Kurtén 1965:62, 1968:156). Yet we are given no explanation for why so many insular organisms tend to become larger—such as reptiles (see Mertens 1934). (For similar hypothetical population advantages, but on a global scale, see Bretsky and Lorenz 1970.) As to the reptiles and birds, which do not always fit, some authors believe that large size favors heat retention in the lizards of warm climates owing to the Bergmann effect; others say that maybe an ability to get rid of *excess* heat sets upper size limits (Kendeigh 1972)!

Yet scattered among such speculation, one occasionally finds a viable counter-hypothesis, albeit one rarely treated as such. Thus it is suggested that maybe southern weasels take a different kind of prey (Kurtén 1953:111; McNab 1971). One might wonder if the extinction of larger carnivores perhaps opens up an opportunity for smaller ones, such as weasels and foxes. A lower standing crop of food materials would favor the smaller predator, which, as was pointed out above, although at a disadvantage in respect to subduing prey, are better off in locating it. Kurtén (1965:62) proposes that dwarfing may have something to do with the food supply and competition. Rosenzweig (1968) suggests that energetical relationships of some kind are involved, and he relates his data for mammals to climatic stability gradients.

Is it unfair to ask that if resources and competition will explain the exceptions to the rules, should we not use the same hypotheses to explain the data that fit the rules as well? It is becoming increasingly obvious that climates have changed in other ways as they have become warmer and cooler. Thus, Barghorn (1953), Axelrod (1967; cf. Axelrod and Bailey 1968), Leopold (1967), and Slaughter (1967) have all stressed the factor of equability. A colder climate may actually be

better than a warmer one, especially when resulting from lower temperatures in the summer. It may be accompanied by greater rainfall, or,
what is much the same thing, higher relative humidity and less evapora-
tion. A longer growing season for plants means more food for the
animals that feed upon them. The importance of desiccation has been
stressed by various authors who have dealt with Pleistocene extinctions
(Osborn 1906; Axelrod 1967; Guilday 1967; Slaughter 1967). The
pertinence of standing crop effects should be obvious here.

Two additional hypotheses about extinction and size need to be
mentioned. The first is the idea that mass extinctions have been brought
about by cosmic rays; this one has been fairly well disposed of (C. J.
Waddington 1967; G. G. Simpson 1968). The other is that man has
played an important role, especially through his hunting activities (*pro:*
P. S. Martin 1966, 1967, 1973; Battisini and Vérin 1967—skeptical:
Hester 1960—*con:* Gill 1955; Leakey 1966; Guilday 1967; Kowalski
1967). There does appear to be somewhat of a correlation here. However, the hypothesis in question will not explain the persistence of some
big animals in Africa. Nor will it explain many extinctions where man
could not have been involved. Even where there is a cause and effect
relationship, it could be an indirect one. In many cases, man has wiped
out other species by introducing rats and other animals. One reason
why this hypothesis has been so attractive has been the tendency for
extinction to affect primarily the largest animals, upon whom man is a
very effective predator. We should examine this point in some detail.

Extinction seems to be unusually common among large animals
(see Newell 1949; for dinosaurs, Colbert 1955). Thus G. G. Simpson
(1931) noted much extinction among the larger fossil mammals of
Florida; but the rodents were little affected, except for the giant beaver
and two large capybaras. The trend toward extinction of large animals
is especially pronounced in those that are "bizarre" in the sense of
having much ornamentation (see G. G. Simpson 1953). That some of
the deviancy in structure is due to, or at least correlated with, the action of sexual selection is apparent from the example of the "Irish elk,"
in which the antlers were most impressive. Morphological aberrancy
likewise characterizes organisms that have become considerably reduced in size, as may be seen in the fauna ("mesopsammon") that lives
between sand grains (Swedmark 1964; Ax 1966). For small animals, no
pronounced tendency to extinction has been documented, perhaps because it does not exist, but perhaps because of inadequate study. Yet
it does seem that we are dealing with a widespread and general
phenomenon, perhaps applying to both extremes of the size scale. The
notion that such features as the spines and the intricate sutures of
fossil mollusks have somehow arisen as developmental byproducts, or
even as manifestations of orthogenesis, have enjoyed at times a certain

vogue. However, analogous features exist in contemporary organisms, and the supposedly non-adaptive parts repeatedly turn out to have definite functions. Rudwick (1964) maintains that we have no way of demonstrating that parts or attributes have no function. Yet valid canons of evidence do exist—such as disorganization and transitory appearance during embryological development. Be this as it may, in the case of such hypertelic structures it is clear that a lack of function has sometimes been invoked on the basis of negative evidence at best, and wishful thinking at worst. Rensch presents no real argument for interpreting them as allometric byproducts. Neither can we go along with the view of Krieg (1937) that they are ways of using up superfluous energy! Such vitalistic notions, which have been invoked with particular frequency for sexual weapons and ornament, have no place in modern biology. We must likewise categorically reject the notion of a phylogenetic aging process that treats the bizarre organisms as degenerate. This orthogenetic view has obvious roots in the analogical comparison of species with organisms, and must not be accepted until we have some evidence that it is true. In point of fact, such "ornaments" have evolved convergently, with similar forms coming into existence repeatedly, at different times, in similar habitats, and in but distantly related lineages; accident displays no such pattern.

A more reasonable idea, one that Rensch (1960) attributes to Haeckel, is that extinction relates somehow to specialization; and, of course, larger organisms are specialized in one sense of that term. Yet expressions of what actually goes on are notoriously vague in such contexts. Stripped to bare essentials, they assert that conditions of existence change, and the species that do not adapt become extinct. If organisms with a "broad niche" are more "generalized" and those with a "narrow" one are more "specialized," and if the former are more adaptable, then it follows which should become extinct. Up until very recently at least, we have not been told how to predict which groups of organisms will become extinct when a particular set of environmental conditions change, beyond such obvious examples as parasites going the way of their hosts. The most recent literature, however, is coming to formulate extinction mechanisms more explicitly, and there seems to be little need for an undefined, or at best vague, notion of specialists going away. It is now becoming clear to marine paleoecologists that the availability of resources suffices to explain not only extinction, but bursts of speciation as well (see especially R. C. Moore 1954, 1955; Rutten 1955; Newell 1956, 1965, 1971, 1972; Bramlette 1965; Bretsky 1968; Cifelli 1969; Lipps 1970a, 1970b; more below). How a particular organism responds will depend upon the situation. In the same environment new conditions will cause one group to flourish, and another to decline, irrespective of whether they are "specialized" in any sense.

But the particulars must be worked out for each individual case. The picture remains for the moment unclear, but nonetheless, for our scale effects, we may state in a very general way what happens.

A general model. Let us hypothesize an ancestral animal with a new kind of bodily organization, occupying, for the first time, a different adaptive zone or a new habitat. Given nothing like it in the same zone or habitat, or only moderate interspecific competition, we would expect speciation and adaptive radiation to ensue. The animals would diversify, forming as it were, variations on a theme, and exploiting the resources in different ways. Such environmental partitioning would involve not only various adaptive strategies, but various combinations of them as well. Given no radical change of environment one would expect such changes to continue, a stable equilibrium being reached only when the environment became saturated with animals of the type in question. The animals ought to differ from the ancestral form in an increasingly large number of features, not only because change is cumulative, but because the ways of making a living most like that of the ancestor will be exploited first and opportunity will be greatest, up to a point, toward the periphery of the adaptive zone. Since one of the ways of becoming different is to change size, the apparent trends toward magnification and minification follow. Occasional reversals would result if innovations were to superimpose additional radiations upon the original one, as happened within the Vertebrata after the origins of birds and mammals. Yet the very ability to become large depends upon a supply of one resource or another, and at some point diminishing returns will lead to the further enlargment of the body having no advantage (just as limits are set to the size of factories). Now, if the supply of the limiting resource should change, those populations that are nearest to the point of diminishing returns may be those that suffer most (like speculators who buy on margin, or businesses that can only turn a profit if they exploit a very large market). If gigantism has been made possible by a large amount of food being present, then a decrease in the supply may wipe out the giants before it will affect their smaller competitors. If gigantism gives an advantage where food is scarce, then a higher standing crop of food will tend to eliminate the giants through competition with forms that operate best when food is more abundant. But, of course, more or less food may cause other groups in the same environment to flourish while one becomes extinct. We might then expect sexually monomorphic forms to be adjusted to more or less mean conditions in a range of possible constellations of adaptive strategies. Dimorphic forms might coexist with them, but here each morph would only realize an advantage under a particular set of circumstances. If the advantage in question ceases to be realized, then extinction is highly probable, for a reversion to the monomorphic state would bring the dimorphic form into competition with its relatives. We may therefore conceive of adaptive radiation as a process in which the organisms exploit environmental opportunities wherever they occur, invading the most easily tenanted situations first, and gradually adjusting to more peripheral habitats and ways of life. If more and more special adjustments are necessary as the radiation occurs, it becomes increasingly less probable that all of these will be functionally workable if conditions should change. Hence the extinctions should occur among the more modified forms, although some of these may be in a position to expand and flourish.

In some ways the above formulation only reasserts what has been accepted all along, but there is at least one fundamental difference. When we cast the explanation in terms of adaptive radiation and constellations of adaptive strategies, we repudiate a most pernicious typology. The extremes in an adaptive radiation are not "specialized" with respect to an idealized standard. Rather, the separate lineages have each, in their own unique ways, undergone a sequence of incommensurate modifications. It is not "the specialist" that becomes extinct, but certain kinds of specialists.

A concrete example. In order to apply and to evaluate a model of this type, and to test it, one needs to compare it with the alternatives. The methodology should be reasonably straightforward. If one thinks that man has produced extinctions, one sees what happens when men are and are not present. If the Bergmann effect is at issue, one compares homeotherms with poikilotherms. If one seeks to determine the effects of rainfall, one should contrast marine organisms with terrestrial ones. But obviously a global analysis is a bit too much to ask, and it would tend to belabor the point. The relationships are exceedingly complicated, and a great deal of knowledge of particulars is needed, both of the habitats and of the organisms. An example, the bivalves, will, however, make the point. Lipps and Mitchell (MS) have shown that a comparable model will explain the adaptive radiation and extinction of whales. Hence it is not just an isolated instance, but further study will be needed before we can evaluate its sufficiency, and treat it as more than provisional.

The molluscan class Bivalvia (=Pelecypoda) may be characterized from a functional point of view as a group of "suspension" and "deposit" feeders. We have already mentioned this general mode of obtaining food in association with intertidal zonation, where we focused mainly upon "filter-feeders." As we have to put up with something of a nomenclatural muddle here, let us take pains to define these terms. A *filter-feeder* is an organism that takes up food by "filtering" it out of the medium—usually water, but sometimes air. A *suspension-feeder* is one that takes up particles that are "suspended" in the medium—such as floating microorganisms. A *deposit-feeder* takes up materials "deposited" on a surface—especially detritus which has settled out on the bottom. Many filter-feeders are also suspension-feeders: one might compare them to certain machines that purify the air by filtering out dust and pollen. But others take up loose bottom materials (deposits) which they themselves have suspended—rather as a vacuum cleaner does. Some suspension-feeders, and many deposit-feeders do not have a filter; they use other arrangements, some of which might be compared to a broom and dustpan.

A filter-feeder's economy may fruitfully be compared to that of a fisherman, some of whom would technically fall under our definition of "filter-feeder." Such a way of earning a living requires two basic implements: a filter and a pump. The former traps the food, the latter causes the medium to move through the filter. A fisherman's net is the filter; a pumping effect is produced either by moving the net or by taking advantage of natural movements (currents or the fish themselves). It should be evident that the cost of running the pump, if any, will have to be treated as overhead. As the standing crop of fish or other food item decreases, and the overhead remains constant, there comes a point where the enterprise no longer turns a profit. And once that point is passed, the fisherman goes out of business, or the clam starves. Conversely, as standing crop goes up, the opposite effect occurs, with lower prices or their analogue putting certain kinds of fishermen or bivalves at a disadvantage.

Yet there are various ways of filter-feeding, suspension-feeding, and deposit-feeding, and the amount and kind of energy used to run the pump are not the same. *First*, the pump may be run on energy from food. Most filter-feeders probably could be relegated to this group. They either move the medium (oysters, sponges) or push themselves through it (whales and salps). This stratagem does not work at the lowest standing crop levels. When food is somewhat more concentrated, it works, but only if the volume filtered is large (implying a large pump and filter) and only if little energy is devoted to activities other than pumping. Hence, the relative proportion of the body given over to organs of locomotion and the like must be small. At high food levels, a fairly small filter and pump will do, and other structures can be supported. *Second*, environmental mechanical energy can have a pumping effect. Some filter-feeders exploit moving currents. They are sessile and hold a net out into the water. Others (such as spiders) take advantage of the active movement of the prey. The latter stratagem is a form of trapping, and involves waiting and expending minimal effort. Fishermen, it should be noted, use both approaches. *Third*, we have what may be called "interface accumulation." Suspended matter settles out and adheres to various surfaces, resulting in a locally higher concentration of food. Deposit-feeders are remotely analogous to bottom-trawlers. *Finally*, we may note that one stratagem may be combined with another in such a manner that the effective cost of running the pump is kept down. Thus, many open water filter-feeders use the same motions in feeding and keeping afloat. And in the unproductive waters of the tropics, many suspension-feeders—such as the giant clam *Tridacna*—have symbiotic algae, so arranged that nutrients are exchanged for energy.

As already mentioned in an earlier chapter, the present-day distribution of many forms depends upon the existence of local conditions where nutrients are plentiful and hence productivity is high. Thus it is not only filter-feeders such as baleen whales, but many others, such as piscivorous birds and pinnipeds, that tend to flourish in places where nutrient-rich water comes up from the depths of the sea. Such regions of mixing and upwelling can be identified by the type of sediment deposited there. It tends to be rich in siliceous skeletons. Silicon is rapidly depleted from seawater by small plants and animals, and where little is available skeletons are more advantageously formed of calcium carbonate. It is noteworthy that the fluctuations in abundance of siliceous microorganisms such as silicoflagellates, diatoms, and Ebriaceae are more or less positively correlated, and somewhat independent of those of calcareous forms. To see the pattern clearly, however, it helps to exclude the tropical radiolarians—unicellular animals that can live in deeper waters, where they form their skeletons. The siliceous protists seem to be high-latitude, high-productivity organisms, and there is some evidence that they show a diversity pattern somewhat like that of whales and penguins.

Evidently, the rate of exchange between the upper and lower levels of the sea has fluctuated, affecting nutrient supply. To understand such matters, one needs to consider the climate at various periods (Emiliani 1958; Durham 1959), which in turn relates to "plate tectonics" (Hallam 1963; Valentine and Moores 1972). At various times, the blocks forming the continents have been separated, at others, united more or less into a single mass. Likewise, the sea has covered a variable proportion of the land. The union of land masses and the elevation of continents have tended to produce cooler, unstable climates—through radiation in continental areas, through monsoon effects, and through the movement of air and water to regions of different temperature in general. The scattered and low-lying land masses have had warmer, more stable climates. At times of high stability and low productivity, marine species diversity on a worldwide basis has tended to be higher than in periods of lower stability and higher productivity. This backs up the picture derived from contemporary organisms, but we should note that some groups are known to show just the opposite pattern.

Valentine (1971, 1972) has attempted to treat this relationship in terms of advantages to the ecosystem; it would seem, however, that the exceptions are better covered when one treats it from an individualistic point of view. He points out that rich resource supplies tend to be inversely correlated with diversity, but more energy could lead to greater diversity if certain other conditions are met. It would be interesting to see how far the remarkable richness of the algal and molluscan fauna of Northern California fits in with the larger scheme of things. One might reasonably predict that if studied with better techniques it would prove to be one area in which diversity and productivity are both high. At any rate, the whales and certain other groups do show that not all taxa obey the rule. Furthermore, Valentine's model (which draws heavily on Margaleff's ideas) gives some different predictions from the one elaborated here (which, we may recollect, derives largely from Rützler's sponge work). Below a certain point, reduced productivity should lead to less diversity, not more. If the laws of thermodynamics are true, then no species shall exist where there is no energy; and the limiting case must occur at somewhat higher food levels. Above a certain point, an increase in either stability or productivity should tend to bring about competitive exclusion, hence to some extent tending to decrease diversity rather than to increase it. Obviously, no matter how much energy is available, and no matter how stable the system may be, it is logically impossible for there to be more species than there are individuals.

Now let us consider how diversity relates to the picture of nutritional levels at different periods in geological history. We may pass over with an expression of skepticism the view that oxygen levels had something to do with diversification and extinction; we deal mainly with a later period (but see Cloud 1968; Towe 1970; compare McAlester 1970, with Schopf, Farmanfarmaian, and Gooch 1971). Summarizing recent opinion, we may say that most of the Paleozoic era appears to have had rather low levels of upwelling, perhaps less during the second half. Major faunal changes occurred at the Cambrian-Ordovician and Devonian-Carboniferous boundaries. Massive extinctions at the end of the Paleozoic era (late Permian) are correlated with rigorous, continental climates, at a time when the land mass was largely united. Primary productivity in the early part of the Mesozoic era (Triassic and early Jurassic) was relatively low, but since then it has been increasing gradually. There were short periods of lowered upwelling in parts of the mid-Cretaceous, at the Cretaceous-Tertiary boundary, and during the Oligocene. Such changes have brought about catastrophic changes in the fauna and flora. Thus, low levels of diatom production correlate with mass extinction of archaic whales in the Oligocene (Lipps and Mitchell MS). Cifelli (1969) points out that the decline in both morphological and species diversity of planktonic Foraminifera and Radiolaria at the end of the Cretaceous occurred at a time of high productivity.

Now let us consider what happened to different kinds of bivalves during the Paleozoic-Mesozoic transition. For our purposes we may consider a bivalve as an animal with three main components: (1) a pair of *siphons*, one of which takes up food-laden water, while the other serves for discharge; (2) a pair of *gills*, provided with cilia, which serve as both filter and pump; (3) *palps*, to either side of the mouth, which help sort and transfer food, and which can also take it up in some forms. Other important organs include the well-known bivalved shell, closed by a pair of strong adductor muscles, and a foot, used in burrowing, crawling, and attachment.

The *protobranchs* probably constitute a natural assemblage (order Protobranchia) members of which were ancestral to the other bivalves. Siphons are but moderately developed, and most of these animals cannot burrow very deeply. Some have extensions of the palps which are used in taking up food from the substrate. Yonge (1939) maintained that deposit-feeding with palps was the primitive condition among bivalves in general, but Stasek (1961, 1965) has challenged this view. Whatever the historical situation, it does seem that the group relies heavily upon interface accumulation. The data compiled by Nicol and Gavenda (1964) show that although the protobranchs are widely represented in marine faunas, the group is most numerous in the deep sea. Here, biomass is low, as is the amount of energy available, but species diversity is high (Hessler and Sanders 1967). A slight preponderance of this group in the Arctic is noteworthy, and this fact is of interest to us. They do become more numerous during the Cretaceous, but so do other bivalves. When other groups were undergoing mass extinction at the end of the Cretaceous, and in following periods when others began to proliferate, they remained about the same in terms of numbers of genera.

Septibranchs are also considered a natural assemblage (order Septibranchia), but one that has not given rise to other groups of bivalves. The class is small and homogeneous, consisting of animals with a modified gill. From time to time they suddenly suck up a large volume of water, taking up particles from the surface. They are scavengers, which employ the interface accumulation stratagem; but they take larger particles, such as dead crustaceans. Again, they are particularly abundant in deeper waters, where their feeding strategy has been evolved convergently in certain other groups of bivalves. Their fossil record being poor, we shall leave them out of this analysis.

We shall treat the rest of the bivalves as functional groups rather than taxonomic ones. *Shallow burrowers* and *deep burrowers* can be considered together. Shallow burrowers are but slightly modified clams, with a gill of moderate size, short siphons, a well-developed foot, and adductor muscles of about equal size. (Most commercial clams, such as the cherrystone, *Venus*, and the soft-shell, *Mya*, are of this type.) Deep burrowers differ in having adapted to life at some distance beneath the surface. They do this by various mechanisms, such as long siphons or extensive burrows. Our considering the two as a unit makes sense when it is realized that one is a more extreme development of the other. The modest gill surface and the active burrowing habit necessitate a rich food supply. As one might expect, these organisms are not particularly common in the deep sea. Along with the other bivalves, they gradually expanded during the Mesozoic era; but while many other groups declined in the high productivity environments of the Cretaceous and Tertiary, these organisms proliferated extensively (Stanley 1968). Also, there occurred a significant change between the earlier and later Tertiary, at the generic level; perhaps the explanation is that these forms were markedly affected by lower productivity levels in the Oligocene, as whales were. As part of this assemblage, we might mention the Lucinacea, a small group of aberrant bivalves which McAlester (1966) has said are independently derived from the class Monoplacophora. They feed through a mucous tube rather than a siphon, and, according to J. A. Allen (1958) and Kauffman (1969) are characteristic of areas with little food. Their fossil record shows them undergoing both expansion and extinction at the Cretaceous-Teritiary boundary.

The *sessile and epifaunal bivalves* include several lineages which display a group of evolutionary trends associated with life either on the surface or only slightly below it. Siphons are rudimentary to absent. The foot becomes reduced and the animal either lies freely on or near the surface, or else it is attached by one of two mechanisms: a *byssus*, a group of threads (mussels) or by *cementation*, with the shell fused to the substrate (oysters). Correlated with these trends, there occurs a complex reorganization of the entire body, involving an enlargement of the gill and the posterior adductor muscle, along with a reduction of the head, foot, and the anterior adductor muscle. It is to Stasek (1963, 1966) that we owe both the morphological and the energetical explanations for these changes. By employing a well-known graphical technique, deformed Cartesian coordinates, he showed how the body as a unit has been altered through differential growth. The posterior, gill-bearing portion of the body has been hypertrophied at the expense of the anterior part. Stasek points out that the enlarged gill has an obvious energetical advantage in waters containing little food. We might add that reducing the foot and adopting a passive mode of defense forms part of the same adaptive strategy. The habits of these creatures reinforce this inference. Their attached mode of life, although it precludes flight, permits them to survive by hanging on in places relatively inaccessible to predators. Some, such as *Pteria*, live on projecting objects. Others live exposed on wave-swept rocks (*Mytilus*). Still others are protected in fissures (*Lima*) or holes (*Arca*). Among the non-attached members of this group, quite a number can swim (e.g., some *Pecten* and *Lima*, see Yonge 1953), effectively avoiding starfish and other predators. A few live partly buried. The adaptation of these forms to low standing crop conditions is particularly evident in their distribution: they are abundant in tropical areas with low primary productivity, especially on coral reefs. Exceptions, to be sure, are many. *Mytilus* and oysters, for example, do well in temperate areas. But these flourish where they must remain closed much of the time, as at higher tidal levels and in estuaries. The fossil record backs up the same thesis. Such bivalves expanded rapidly throughout the Mesozoic era, only to decline at the end of the Cretaceous, remaining depressed while the burrowing forms radiated extensively (Stanley 1968). During the Mesozoic certain groups manifested a pronounced trend toward gigantism: Ostreidae, Megalodontidae, Trigoniidae, Pectinidae, Spondylidae, Limidae, Chamacea, and Rudistacea (Newell, 1949; Nicol 1961, 1964; Coogan, Dechaseaux, and Perkins 1969). At the end of the Cretaceous, again, these groups were decimated or entirely wiped out, especially the larger species. Why large size should correlate with a low standing crop of food organisms is easily understood. The young of these forms resemble the more conventional bivalves; the differences arise from allometric growth. The larger the animal, the greater the relative size of the filter. Hence, the giant forms would be those least suited to life under the new conditions. If we ask why they could not get along where there is more than enough food, the answer is very simple: competition, largely with other bivalves. A clam with less gill and more foot would be able to get away from predators more easily. Perhaps before leaving this assemblage, it is well to contrast its extinction and radiation pattern with that of high standing crop taxa such as the mammals discussed earlier in this chapter. In both, the giant forms tended to suffer when the habitat became less favorable, but the one was affected by greater, the other by less, productivity. Only if one knows the comparative anatomy, can one successfully predict which groups will become extinct or flourish as conditions change.

Secondary interface accumulation has evolved in both burrowing and surface-dwelling bivalves. Among siphonate burrowers, the Tellinacea are particularly noteworthy (Yonge 1949). These animals move the tips of their siphons over the surface of the sand or mud in which they live, and take up small particles of deposited material. (A few can switch feeding mechanism, and some have reverted to suspension feeding—see Reid and Reid 1969.) They are successful in a variety of habitats, including both cold and warm waters. The adaptations of the genus *Abra* to life in the deep sea have been traced by Allen and Sanders (1966). With increasing depth the gill becomes reduced, while the gut is lengthened and more adapted to retaining fecal pellets. According to Knudsen (1970) the epifaunal bivalves of the deep sea tend to utilize larger particles of food. In this they are convergent with deep-sea fishes and many other organisms in the same habitat.

It must be conceded that the foregoing is only a brief sketch of what has happened in one group, and in but part of geological time. And, obviously, an example was chosen in which the hypothesis did appear to succeed, so that it does not constitute an adequate test. More work is needed, before we can say how sound these generalizations are, and how much they need to be revised. But that is just the point. Such work is, in fact, well under way. And the hypotheses, whatever validity they may have, are testable. No longer need we invoke the Bergmann effect, nor claim that multiple causes render such problems insoluble. The future of paleoecology looks bright indeed. And paleontology will become what sentimentalists claim and pray it never can become: a rigorous, predictive, deterministic, and hypothetico-deductive science. (See also Schopf 1972, and other articles in the same symposium volume.)

6

SEDUCTION AND RAPE, OR, FEMALE CHOICE AND MALE SEQUESTERING

To breed or not to breed,
that is the question.
          Lloyd Morgan

Male combat, by definition, always involves a sort of competition in which the males act directly upon other males so as to monopolize the females. In female choice and male sequestering, the end is the same, but the means differ: the males act upon the females rather than upon each other. As has already been mentioned, the various stratagems are not altogether distinct. Indeed, many of the problematical aspects of female choice can be resolved when we invoke a form of sequestering instead.

FEMALE CHOICE

Why female choice has been so controversial. Female choice has long been a controversial subject, partly because theory has been inadequate. But there have been philosophical and methodological problems as well. Hence it seems reasonable to include some remarks on the history of the subject (see also M. T. Ghiselin 1969a). Some of the resistance to Darwin's ideas on sexual selection was motivated by the same impulses that underlay the opposition to selection theory in general. Evolution by orthogenesis, or even by strictly natural selection, left open the possibility of divine providence and foresight, and was therefore not entirely unpalatable. But sexual selection allowed for nothing of the sort. Nature, in producing contraptions rather than contrivances, and acting against the interests of species, gives rise to a spectacle of purposelessness and triviality. Darwin realized this, and much of his work on sex was explicitly directed against teleology.

But not all of the resistance had to do with mystical or ontological ideas. Epistemological arguments were brought forth as well. Darwin's approach exemplifies the modern hypothetico-deductive method. He came up with a theory, derived predictions, and checked out the data of experience to see if the predictions were correct. Sexual selection, including that by female choice, was no exception. Conjectures, such as the notion that females selected the males, were indeed hypotheses; evidence was, in point of fact, presented; and Darwin's conclusions really did follow from his premises. Anyone who consults the documents can see for himself that this was so.

Yet some of Darwin's critics have presupposed a different epistemology. A fine example is Thomas Hunt Morgan, an American embryologist who turned to genetics and later won the Nobel Prize for his research on *Drosophila*. His arguments (Morgan 1903) against natural selection appear so flimsy and irrelevant to us moderns that at first one may be tempted to question the man's rationality. He keeps chiding Darwin and his followers for presupposing questionable premises; yet no such premises occur in Darwin's works themselves. What for Darwin are hypotheses, induced to be true on the basis of experience, Morgan represents as suppositions from which a deductive construct is derived. Morgan is notorious for having opposed the chromosome theory of inheritance, later having had to recant on the basis of his own work. His attitude has perplexed many commentators. G. E. Allen (1968) suggests that Morgan was looking for experimental rigor. He was, but this leaves unanswered the question of why so much unfounded criticism. The reason would seem to be that Morgan believed it possible to attain apodictic certitude in science. If only he could establish a few basic premises, then he could reason as a geometer would, from one certitude to another. Such a position is no longer tenable, yet in those days the vision of *a priori* knowledge was still maintained, particularly by such German Neo-Kantians as Windelband, Rickert, and Dilthey. It is an interesting and by no means irrelevant point that another embryologist, Driesch (see Driesch 1908; Rádl 1930:362–364), embraced the same sort of Kantian desire for certitude.

He, too, opposed Darwinism and he explicitly invoked the views of Kant and the Neo-Kantians (Driesch 1908:I:13–14, II:238). The two make a bizarre pair: Morgan a materialist with a long and brilliant career in the laboratory, Driesch a professed vitalist who abandoned science for philosophy. Both were outstanding experimentalists, and perhaps their early success contributed to their metaphysical follies. Yet another embryologist who turned to genetics, Bateson (1894:v–vi), criticized evolutionary biology on much the same grounds as did Morgan; he, too, attacked the chromosome theory of inheritance. For all these men, however, a word of caution seems in order. Epistemological arguments are frequently invoked to justify conclusions arrived at for other reasons. The motivational aspect of this matter deserves further historical study.

Darwin's evidence and that of his critics. Darwin's arguments supporting the hypothesis of female choice are basically efforts to show that no other hypothesis is consistent with the data of natural history. His approach resembles that used by physiologists in establishing the function of an organ, in which it is reasoned that the parts are intricately adjusted in a way that should produce the right effect at the appropriate time. If the females are to select the males, then certain conditions have to be met. And if this selection has, in fact, occurred, groups in which these conditions have indeed been met will tend to show a higher incidence of sexual characteristics of a certain kind. The females must possess the appropriate sort of physiological equipment to perceive differences between males: eyes for visual stimuli, ears for song. The females must be presented with a choice—a condition best met when several males display together as a group (now called a "lek display"). Whatever instrument is conjectured to attract the females should be arranged in a way that brings about this effect at the right time and place. A dimorphism, with more conspicuous males, is good evidence and when the dimorphism only develops in the breeding season, so much the better. Behavioral evidence also helps. Special structural adaptations may serve to conceal the sexual ornaments except when they are visible to a female at breeding time. When males posture before the females, displaying their secondary sexual characters and doing so in a way that maximizes the visual effect, it is hard to deny that something is going on between the two individuals. Of course, we also need to have ways of telling whether the display is used in intraspecific competition between the males: for this we have to apply the sort of evidence assembled in the last chapter.

Darwin marshalled the evidence with his usual originality and rigor. His critics, however, burdened the subsequent literature with fallacious arguments. Time and again, they invoked negative evidence, arguing that it was not known whether females exercised a choice, or providing anecdotal accounts of females that seemed indifferent to the display. For instance, Seth-Smith (1925) observed that the female argus pheasant, *Argusianus argus,* did "take an inquisitive interest" in the male's display. But Bierens de Haan (1926) failed to see this in one pair in a zoo, and concluded that the display does not stimulate the female. Yet an hypothesis that has survived an experiential test should not be lightly dismissed simply because it might be refuted on the basis of experiments that have not been done. Nor should a null hypothesis be rejected on the basis of negative evidence and experiments without the proper controls. As one might expect, the work of such pioneer field observers as Selous (1906–1907), and a great deal of subsequent research as well, has more than adequately disposed of the negative evidence. And a vast body of additional material shows Darwin's thesis to be substantially correct. At least some secondary sexual characters have indeed been evolved through sexual selection by female choice.

Some theoretical objections to female choice. Yet establishing that some female choice has occurred does not tell us how significant it has been. Its prevalence is a strictly empirical issue, to be determined in the field, laboratory, and museum, or so one might think. Nonetheless, theoretical considerations do enter in as well and some of these are partly metaphysical. Many evolutionary biologists have thought natural selection to be so effective a mechanism of change that it ought to counteract altogether the influence of sexual selection, or at least to render it insignificant. Impressed, perhaps, by the apparent perfection of adaptations, they have been loath to accept an hypothesis that even suggests dysteleology. The contrary proposition, however, that nature is full of biological contraptions, is the only one that can be reconciled with the organisms. Furthermore, the notion of the omnipotence of selection derives from false premises of a purely theoretical nature; and those who support it may do so for metaphysical reasons, not because the facts support their position. We have already mentioned Wallace; his exchange of ideas with Darwin on female choice is one of the more dramatic episodes in the history of biology (M. T. Ghiselin 1969a). Darwin held that adaptation is most imperfect, and invoked not only sexual selection, but also what we would call pleiotropy and related phenomena as sources of maladaptive changes. He cited an impressive body of embryological evidence in support of these notions, including much work on artificial selection. Wallace did not go so far as to reject the evolution of adaptively neutral characters, but held that really harmful ones will not evolve. As we have pointed out, the difference between their views depends in part upon whether one treats selection as acting to change an integrated system, or whether one views organisms as being made up of "atomic" components. The latter attitude is very common among people who think in terms of "bean-bag" genetics (for defense of this approach see Haldane 1964). Obviously one has to take an intermediate position: both autonomous units and interdependent systems are present. But anyone who presupposes that the universe is structured one way or the other will wind up in a state of confusion, especially if he studies models instead of organisms.

For our purposes the most pertinent of Darwin and Wallace's arguments about female choice had to do with the lack of dimorphism in some brightly colored birds. If such colors were useful to the males, then one had to account for their presence in the females. Darwin attributed this phenomenon to the equivalent of pleiotropy: a variation not limited to one sex was selected because of its advantage to the males. It would, therefore, be inherited and expressed in the females, but without the females being selected. Wallace maintained that when brilliant color did evolve, natural selection should counteract its production in the females so long as it was harmful. The two hypotheses are not wholly incompatible. But the overriding issue still has to be resolved: the degree to which, as we would now put it, modifier genes will counteract the effect of pleiotropy. Nothing in the synthetic theory of evolution implies that they inevitably will do so, although some might wish to argue this point. No matter: there is no place for *a priori* reasoning in evolutionary biology, and we must rely, as Darwin did, on the organisms. Wallace argued that a brightly colored female would attract predators. He pointed out that when both sexes are conspicuous, the female tends to be concealed while on the nest. Darwin, on the other hand, examined some embryological data, and thought otherwise. He predicted that if the females had evolved a form of protective coloration used at the time of raising the young, then they ought to develop that color pattern at the time of sexual maturity. They do not, and he rejected the hypothesis that birds had secondarily evolved a camouflage that protects them while on the nest. The inference may have been a bit rash. A secondary loss of sexual coloration could evolve in the females without any sign of the earlier condition by selection of genes that continued the juvenile plumage into the female adult stage. If we accept T. H. Hamilton's (1961) view on the loss of color in female orioles, then something of the sort has actually taken place. Darwin's fundamental position would thus seem to be correct: it is organisms, not genes or traits that are adapted to environmental circumstances, and selection favors the organisms that strike the best balance between conflicting selection pressures. Yet Wallace was also right, in so far as compensation for maladaptive side-effects will evolve when this is both advantageous and possible, and bright coloration should tend to evolve where it does the least harm to the individuals. Thus, it is only to be expected that sexual selection

will operate most strongly where the males have the least opportunity
to aid their offspring; but it will also operate most strongly where they
have the most opportunity to devote their resources to purely sexual
competition.

Alternatives to female choice. Beyond such theoretical issues, we
have the possibility that a better hypothesis than female choice could
be invoked to explain some, if not all, sexual coloration and ornament.
Hypotheses are never established with certitude, and warning coloration or other phenomena are known to play a role in the evolution of
sexually dimorphic patterns. We have already discussed some of the
counter-hypotheses in earlier chapters. That a bright color might result from an overflow of vital energy (A. R. Wallace 1912 and elsewhere; Geddes and Thomson 1901; Krieg 1937), or as a mere byproduct of variation (Kammerer 1912) are no longer taken seriously. The
remaining alternatives, based on selection, deserve further examination.

A number of alternatives to female choice involve some kind of
group selection. The most serious effort along these lines, again, has been
that of Wynne-Edwards (1962), who maintains that many of the
phenomena of display are advantageous to the population. His reasoning usually takes the form: it is hard to imagine x; maybe y; therefore y.
In spite of the fallacy of his arguments, the phenomena he discusses are
well worth considering, for it requires a certain amount of ingenuity
to see how individuals actually benefit. The lek behavior, or communal
nuptial display, of males in quite a number of animal species, especially
among birds, provides a fairly subtle form of interaction which
Wynne-Edwards (1962:206–216) fits into his system. (For critique,
see Downes 1958; Braestrup 1963.) Many individuals will gather together and court as a group. The males definitely compete with each
other, since they defend territories, and the females are stimulated by
the group display. Sexual selection provides a mechanism for an individualistic advantage. As has been pointed out (Armstrong 1942;
Braestrup 1963, 1966; Snow 1963), a group of males may present a
more effective stimulus than would a solitary individual. Not only
would they increase the intensity of the stimulus, but also, through
individal differences, its variety; this is perhaps the reason for the
extreme sexual polymorphism in the male *Philomachus pugnax*, or ruff
(see Hogan-Warburg 1966). Given an equal probability of gaining
a female's favor, the "rivals" would actually enjoy an advantage over
the nonparticipants. It would be like any other exploitation of society
for personal gain. An interesting example of how such communal display works is a weaverbird, *Ploceus cucullatus*, studied by Collias and
Collias (1969, 1970). The males, which are polygynous, attract the
females by constructing nests, doing so in colonies of varying size. It
was found that the large colonies attract more females per male than

do the small ones. Among insects (Downes 1958, 1969; see also Richards 1927; Parker 1970), group display by the males is not uncommon, but the selective mechanisms are not yet adequately understood. For both the lek and the weaverbird system, Lack (1968) proposes that the advantage to the males lies in avoiding predation. His hypothesis may account for origins, but it seems inadequate to cover hypertelic development where it occurs, and the insects with communal displays fit the scheme with difficulty. In birds, at least, the females are known to display preference for certain males, and the rewards in terms of impregnating numbers of females may be high. (See especially Selous 1906–1907; J. R. Simon 1940; Armstrong 1942; J. W. Scott 1942; Robel 1966; Kruijt and Hogan 1967.)

The "best-man" hypothesis discussed in the previous chapter has been applied to female choice as well as male combat. Lenz (1917) and Lebedinsky (1932) view sexual coloration as a mechanism that advertises the male's well-being to the female. The female, by selecting the healthier males, would therefore purportedly benefit the species by a sort of natural eugenics. We can see how an analogous mechanism might be applied to male combat. Verner (1964) along similar lines suggests that female birds may select the more aggressive males; but his is an individualistic explanation: the female benefits from being associated with a better defender of territory.

Indeed, some of the species-advantage models can be recast in strictly individualistic terms. If the more vigorous males were in fact likely to have genetical or other resources that would help to raise the Darwinian fitness of the female, then she would benefit by acting as a selective agent in a manner analogous to the breeder in artificial selection. Lloyd Morgan (1890–1891) and T. H. Huxley (1894 in Huxley and Huxley 1947:44) presented some thoughts along these lines, but they never followed them up. At first sight, there would seem to be a certain amount of empirical evidence in favor of this hypothesis— equally consistent, to be sure, with the group selectionist interpretations. Observers of lek behavior have often noted that the females mate most frequently with the dominant males. An analogous preference occurs in primates (Wickler 1967). Belding (1934) claimed to have seen female salmon aid larger males in driving away smaller ones. Yet the evidence thus far brought forth does not suffice to exclude the possibility that natural or sexual, rather than artificial or group, selection is operating. Male behavior of all sorts may be stimulating to the females. Any aspect of sexual dimorphism may be accentuated by sexual selection, and therefore to invoke a eugenical advantage may lead to serious misconceptions. For instance, male human beings may favor women with large mammary glands, but probably not because these provide a more copious supply of milk.

The possibility of the females gaining a eugenical advantage through associating with males that are effective in sexual combat was mentioned in the previous chapter; but, as we saw, the data on pinnipeds seem to indicate that no such mechanism operates. Furthermore, a preference for a dominant male in primates may be rewarded by economic, not just eugenical favors: sycophants and flatterers have probably been with us for a very long time. So have wife-beaters: one of the most effective forms of male sequestering among primates is punishing the females who mate with a rival. Even if selection did favor the preference of females for those males most successful in competing for mates, through a kind of positive feedback effect, the true explanation would be only a more intricate version of sexual selection. The result would be dysgenic, not eugenic, and not an advantage to the species. If we admit of a selection by the females of better fathers, we still have to find out how effectively the males can opt out of their paternal role once this choice has been exercised (see Trivers 1972).

It does not seem a reasonable hypothesis that extreme sexual dimorphism, such as that which occurs in birds of paradise, exists because it speeds up evolution (Gilliard 1962, 1969). Perhaps one might get that impression from the prevalence of sexual selection among highly modified birds. Even if such a correlation exists, however, this in itself would hardly tell us anything about the actual causal relationships. More important, there are the general connections between dimorphism and ecology, which render such hypotheses superfluous.

Another group of alternatives to female choice invoke some kind of *natural, rather than sexual or group, selection.* For example, a brightly colored male could be useful as a defensive mechanism. O'Donald (1962) attributed this hypothesis to Mottram (1915) and Cott (1954); but evidently these writers only meant that the males can be conspicuous because they are of little use to the species. Be this as it may, by distracting the attention of the predators from his mate, or from his offspring, a conspicuous male could gain a considerable advantage. That deflective adaptations do exist is common knowledge: birds often feign injury when a predator approaches the nest. In this behavioral reaction, the parent's ability to fly obviates any serious problem of a concomitant risk; but a brightly colored father is exposed to the attack of predators at times when he would not protect them at all. Furthermore, the ornaments are so arranged, and the behavior is so adjusted, that other members of the same species are most strongly affected. The deflection hypothesis resembles the idea that large males evolved as a means of defending the family; and it has many of the same difficulties. The more extreme cases of sexual ornamentation occur in situations where the male does not care for the young. He does not even accompany the family and may actually reside at a considerable distance from it. In highly promiscuous birds, in which dimorphism is common, the males would have no way of recognizing their own offspring, and the efficacy of such adaptations would be low. One would expect dimorphisms of this type to evolve only where the male exposes himself to minimal risk, or else where he obtains a large return from it. We have little evidence that they have evolved, even where either of these conditions is met.

That sexual markings and displays function in species-recognition was proposed by A. R. Wallace (1912:217) and is well established. Cutting down on the production of sterile or inviable hybrids could obviously have considerable advantage. G. C. Williams (1966) has argued that promiscuity affects the difference in reproductive behavior of the two sexes; for in this case a female that paired with a male of the wrong species would be at a severe disadvantage, but the male would perhaps waste little more than a few sperm. Males in some groups of organisms are, indeed, exceedingly promiscuous. Certain male hermit-crabs, for instance, evidently exchange spermatophores with anything that might be a female, even other males with whom they are fighting (Hazlett 1968). Some such influence may be responsible for differences in the degree to which one sex or the other exercises care in the selection of a mate, but there is no reason to think that it is responsible for the hypertelic developments. Here, as elsewhere, the excessive development, not the mere existence, of secondary sexual characters needs to be explained.

In this vein, Mayr (1963) has maintained that much sexual ornamentation has to do with isolating mechanisms rather than sexual selection (see also Sibley 1957; Petit 1958, 1967; Bösiger 1960). Yet the available tests clearly show that this hypothesis is not applicable to the materials thus far examined. Spieth (1968) has lately summarized his extensive work on sexual behavior in *Drosophila*. He finds that different organs are used in species recognition on the one hand, and sexual competition on the other. Dilger and Johnsgard (1958; see also Kear 1970) have attacked the species recognition hypothesis for ducks on the grounds that the Mandarin duck, *Aix galericulata*, and the wood duck, *A. sponsa*, are brilliantly colored, yet neither now nor recently have lived near close relatives; but female choice does occur. Mayr's main empirical argument is based upon the loss of dimorphism among insular forms which have no closely related species living in the same area. However, we could easily oppose the counter-example of the Galapagos finches. As Lack (1968; see also Swarth 1934; Lack 1947; Amadon 1966) points out, they are less brilliantly colored than their mainland relatives; but this is a classic example of similar species coexisting. P. R. Grant (1965) extends the rule of insular monomorphism to birds in general. But this view needs to be qualified. A. R. Wallace (1876) provides a lot of material which, albeit rather subjective, does show a pattern. Birds of large, semi-isolated, tropical islands such as New Guinea show a high degree of sexual ornamentation. Those of tropical but smaller islands with a rich fauna are only a little less ornamented. The pigeons and parrots, which Wallace thought to be more diverse on islands than on mainland areas, are most ornamented there as well. We may observe that the rather depauperate faunas of the more peripheral portions of the Malay Archipelago and of the Galapagos would seem to provide evidence that the economy is different. Of particular interest are some data from a small, high-latitude oceanic island, Tristan da Cunha (Hagen 1952). Here the birds lay small clutches, and they have a long period of juvenile immaturity; they seem not to have been greatly affected by sexual selection. A pattern like this suggests conditions under which physical survival of the individual gives a high return, and a large number of offspring cannot be supported.

To deal effectively with such problems, however, we need to understand how insular population structure influences reproductive competition, and, therefore, sexual dimorphism. There is an extensive literature bearing upon this subject (see, for example, Kramer 1951; Crowell 1961, 1962; Ashmole 1963; P. R. Grant 1965, 1968; Selander 1966, 1969; MacArthur and Wilson 1967; E. E. Williams 1969; Soulé and Stuart 1970; Morse 1971), and differences in sexual dimorphism between mainland and island populations are well documented. However, some groups become less dimorphic, or less brilliant, on islands, while others show the opposite tendency. Generally ecologists come up with some kind of *ad hoc* hypothesis involving competition, which they treat as almost an occult influence, rather than trying to find out how the animals compete. If competition is thought to be more intense on islands than on mainland areas, dimorphism has evolved to reduce its effects; if less intense, it allows the exploitation of relatively empty niches. When, on the contrary, we are dealing with lack of dimorphism, the intellect proves equally flexible. If it is believed, again, that competition is more intense on islands, the lack of dimorphism means that the dimorphic forms have been excluded; if competition is relaxed on islands, then obviously monomorphic species mean a lack of intraspecific partitioning of niches.

When we realize that competition is of several kinds, and that one form or another affects all organisms, prudence suggests that we need to take a more careful look at what really happens. Different kinds of organisms will be affected in different ways. As was suggested in the discussion on the loss of hermaphroditism among insular plants, one of the most important differences between island and mainland forms has to do with the extent of the market. The smaller area supports an impoverished biota, but at the same time increases the number of individuals of a given species per unit of area. The result will be an increased probability of aggressive encounters between males, and therefore sexual dimorphism will evolve through male combat. This mechanism seems adequate to explain the greater sexual dimorphism of insular *Anolis* lizards, which Schoener (1967) attributed to lack of competition between species and "a more thorough utilization of environmental resources," whatever that means. The well-known Komodo dragon probably owes its large size in the males to sexual selection, as Carlquist (1965) says (see also E. R. Dunn 1927). Soulé (1972) studied insular populations of the lizard *Uta stansburiana* from the Gulf of California. He found a tendency to increased size, and also that variability is greater on larger islands. Both these points had already been made by Mertens (1934) for insular lizards in general. The trends, as Soulé points out, are readily explained as adaptive responses to new selection pressures. Mertens also observes that lizards are more aggressive on islands, and tend to be omnivores. Snakes, he says, are fewer on islands because they eat largely mammals; they become dwarfed more often than lizards do. We are as yet in no position to erect a comprehensive explanation for such phenomena in terms of resource supply and population density, but the prospects look encouraging.

On the other hand, the lessened number of species may tend to decrease the stability of insular ecosystems, creating a "boom and bust" economy, and otherwise increasing the elasticity of supply in food. This would explain the trends toward lesser sexual ornamentation on smaller islands and on high-latitude ones as well. For some kinds of organisms this would tend to make certain strategical alternatives to sexual competition more effective than actively competing for mates. Thus, for example, a male bird of paradise could either wait until no competing males were present in his locale, or hunt for the same conditions.

One of the most popular alternatives to female choice has been argued with great skill and coherence by Huxley (1914, 1923, 1938a, 1938b), and many still go along with his views. The idea is that display helps to establish the pair bond and stimulates and coordinates the reproductive processes through its influence on the nervous and endocrine systems. Getting the sexes together, and keeping them together as a family unit, would, like more effective copulation, exemplify natural rather than sexual selection. So, too, would having a male evoke the production of eggs by his mate at the right place and time. Yet the existence of such processes, although resolving many difficulties, in no way does away with the hypothesis of sexual selection. If reproductive activities are going to work, then of course they need to be coordinated one way or another. Courtship might play such a role, but this does not tell us what came first or what accounts for the details. Nor does a basic utility in coloration exclude the possibility that an extreme development has resulted from sexual competition: both hypotheses might be true.

Mutual display would indeed seem to be a better explanation for the occurrence of nuptial plumage in birds than Darwin's pleiotropic effects. Both the behavioral and the morphological phenomena predominate in birds that cooperate in raising the young. Courtship does keep the pair together, but still we need to be careful in deciding why. Huxley (1923:253), noting that display continues long after the pair has been formed, concluded that sexual selection could not account for it. Yet as we saw in the case of male combat, sexual rivalry continues after mating. Thus it would appear that the presence of highly developed secondary sexual characters in both male and female may likewise result from a combination of male choice and female choice.

As a final series of alternatives for the utility of sexual ornamentation, we may consider *forms of sexual selection other than female choice*. Male combat naturally comes to mind. Animals in which the males are known to fight with one another are often brilliantly colored. And from the way they behave in such combat, it is clear that the markings are directed toward their rivals. The reason would seem to be an intimidatory effect, perhaps especially useful in driving off males that are not sufficiently ready to reproduce as to hazard battle. On this hypothesis we should explain at least some of the coloration and similar features of fishes (Noble 1938), reptiles (Noble and Bradley 1933; Hecht 1952; Eibl-Eibesfeldt 1966), birds, and mammals, and also some of the songs of insects, amphibians, and birds. Major Hingston (1933) applied a strictly combative interpretation to sexual ornaments in general. Many of his explanations are quite close to the more recent hypotheses about human dimorphism summarized by Guthrie (1970), who unfortunately cites nothing published before 1937. Wolf (1969; see also Pitelka 1942) says that female hummingbirds which defend feeding territories are, like the sexually aggressive males, brilliantly colored. Yet in many such cases a mixed effect may occur. Such fishes as the sticklebacks use their livery in attracting females, in intimidating other males, and in defending the nest (Semler 1971; see also Morris 1952, 1958; Gandolfi 1971).

MALE SEQUESTERING

A final alternative to female choice is male sequestering, a possibility that has scarcely been considered in the previous literature. The use of brute force by the male to obtain and keep a mate is well documented for ungulates and some other vertebrates; we have already mentioned a few examples. The male, in addition to driving off his rivals, often herds the females so as to keep them under his control.

Less widely known, but equally significant, cases are not uncommon among invertebrates. Bateson and Brindley (1892) relate how a dimorphic beetle, *Dynastes*, carries off the females. The species displays a high-low dimorphism, with some males lacking the anatomical structures used in this maneuver. Nonetheless, they possess the behavioral repertoire, and go through the motions—without the least success. Since the smaller males are perfectly capable of reproducing, it would seem that the only function attributable to the structural and behavioral peculiarities of the larger ones is competition with other males.

Sequestering by inhibiting access of other males. A good number of male insects have secretions or other mechanisms that prevent the females from mating again. Their significance was long a problem (see Bryk 1919). Eltringham (1925) attributes its solution to unpublished work by Poulton. (For an extensive review of such phenomena in insects see G. A. Parker 1970.)

Crustaceans often manifest a form of sequestering that appears to have evolved partly because of their moulting (Nouvel and Nouvel 1937). When a crustacean moults, its body is soft; hence for a time it is quite vulnerable to the attacks of predators. In some, but not all, copulation can occur only when the exoskeleton of the female has just been shed, and her body is still pliable. The effect of this arrangement is that copulation is only briefly possible, and the female is, at the same time, in a vulnerable position. The males under such circumstances embrace the females, defending them from other males. At the time of moulting, the male has been seen to aid his mate in getting out of her shell (Unwin 1920; review in Patel and Crisp 1961). Afterward, they copulate, and then may continue the association for perhaps a few days, until the female has hardened. The males in such species are larger than the females, and may mate with several in succession. There are at least two hypotheses that would explain this behavior. The males might be competing by sequestering the females, or they might be protecting their soft-bodied mates from injury. From the comparative evidence, we may reasonably infer that both hypotheses are true. In the freshwater isopod *Asellus aquaticus* the much larger male carries the female about with special legs, and fights with other males; but he releases her as soon as they have copulated (Unwin 1920). Here, we have a case of pure sequestering without postcopulatory protection, but one might wonder if the female received some protection while being carried by her possessive husband. The females in crabs of the family Coristidae can mate when hard, but not at all times: there is precopulatory, but not postcopulatory embrace (Hartnoll 1969). In *Cancer* (see Butler 1960; Edwards 1966; Snow and Neilsen 1966) and many other crabs (Hartnoll 1969), the soft female is protected by the male. Evidently he does not do so as a means of sexual competition, however. The male inserts a temporary "sperm plug," which soon dissolves, into the body of the female (Spalding 1942). In the literature it is said that the function of this structure is not known, but it seems to be a "chastity belt" analogous to similar formations of insects.

Schöne (1968) notes an important correlation with habitat in the sexual behavior of crabs. Semiterrestrial ones go in for visual display, while those that are aquatic emphasize sequestering. In part, the difference perhaps reflects the greater transparency of air, but it seems likely that density effects are at least partly responsible. There is no hard and fast rule predicting whether combat or sequestering will be more effective at higher population densities. Animals that live in herds would seem to use them both. However, the biology of crabs is such that a male can sequester only a single female at a time. The stratagem ought to work best at moderate population densities. The male takes advantage of an opportunity for mating, but may have to wait a long time before the female is ready. At higher densities a greater advantage might be derived from combat, since such transactions would be more frequent and locating the females would be no problem. At low densities, the advantage ought to shift toward finding the females and away from combat: in other words, male dispersal would be favored. Indeed, the system in aquatic crabs would seem to be rather a mixture of combat, sequestering, and dispersal. The high density of many semiterrestrial crab populations was alluded to in the preceding chapter. Sequestering would seem to be common among crustaceans only in shallow water where diversity is high and population density moderate. Male dispersal takes over in the low densities of deeper water.

Sequestering by attracting the females. A milder version of male sequestering would involve attracting and maintaining the favor of the females, rather than seizing and holding them physically. If a male were to attract females before it was really time to breed and to sustain the pair bonds, he would monopolize them as effectively as he would by force. This hypothesis differs from the usual Darwinian female choice, both in the mechanism and the empirical implications. In Darwin's model, the female "prefers" one male to another: in this version, the female is led to establish and maintain her attachment. Both hypotheses are true, but they apply to different cases. The results are more or less the same, namely that the secondary sexual characters are accentuated beyond the degree of development necessary for reproduction. But the conditions favoring them are not the same, and the features that become hypertrophied differ as well.

Darwinian female choice occurs in its purest form in the lek display, where a number of males gather together and court as a group. Here the female is presented with different males, from which she does indeed "select" in an almost anthropomorphic sense of that term. The males gain a considerable advantage by inducing each female to mate as soon as she will. But since the females once impregnated go off and raise a family, there is no functional reason for continuing the relationship, and the males exert no effort whatever to do so. Under the hypothesis here proposed, the males could just as well court in isolation, albeit they should have to be close enough to compete. An unattached female "selects," but not in the sense of exercising an option between two or more males before her at the same time. Rather, she is presented with a stimulus, and either accepts it or rejects it. Once the choice has been made, she retains the attachment to the male, until she is ready to be fertilized, and, since copulation may not occur until some time after the bond has been formed, the male has every reason to go out of his way to maintain it. Otherwise his potential mate might transfer her nuptial relationship to some other male.

That such a process has actually been going on in nature is perhaps most obvious for the celebrated "bower birds," a group that has lately been treated in books by A. J. Marshall (1954) and Gilliard (1969). The males construct elaborate "bowers" of striking appearance, which they decorate with colored objects and even a sort of "paint." That the bowers are useful in competition between the males is clear from their habit of destroying one another's bowers and stealing the ornaments. As predicted from theory, the bowers are constructed at some distance from one another, and a period of months may pass before the females are ready to mate. Marshall rightly observes that such an extreme development of secondary sexual behavior is clearly out of line with the hypothesis of a purely physiological advantage.

What gets chosen and why. From the facts thus far adduced, it would appear that sexual competition often involves one or another form of psychological warfare. This proposition can be extended to a very high level of generality. A great deal of speculation has been directed toward how an animal might exploit the behavioral reactions of other individuals. That such interactions do indeed go on is perhaps best seen in the behavioral adaptations of social parasites, such as cuckoos and termite guests (Kistner 1969). Courtship feeding, in which the prospective mate is offered food (see Bastock 1967), may be analyzed from the point of view of reducing aggressive tendencies. Or it may be that the attractiveness of food determines that it shall be used as a stimulus: the two are not exclusive of each other. Neither should be confused with feeding the male as an indirect way of benefitting the offspring—which does occur and which may be an historical precursor of the others (see Royama 1966). The use of nests by male weaverbirds in communal displays implies that the objects biologically useful to the females are exploited by the males as instruments of attraction. The possibility that the bower birds' constructions are ultimately derived from nests has been considered by a number of authors, but all have missed the point: the structures are analogous, not homologous. Marshall (1954) thinks that bowers originated as a form of displacement activity. The resemblance of bowers to nests he explains as the result of a form of nesting that substitutes for some other activity. The resemblance of their decorations to male plumage is thought to arise from displaced male combat. Painting, again, is derived from courtship feeding. Yet the whole setup may readily be explained in terms of presenting a particularly attractive stimulus to the female; the behavioral precursors are determined by the selective influence of the female, and the male's behavioral repertoire is not the efficient cause of change. In the final chapter, we shall reiterate this fundamental point: it is the context in which selection occurs that determines how organisms behave.

Along the same lines, there is a fairly extensive literature on what might be looked upon as courtship feeding in insects (Richards 1927). In some, especially Orthoptera, the female eats the spermatophore (Boldyrev 1912; Gerhardt 1913–1914, 1921; Withycombe 1922; Gabbut 1954). In *Panorpa* the male has large salivary glands, and the female eats the secretion while they copulate (Stitz 1926). A number of male insects feed their mates. The male in some dipterans of the family Empidae presents the female with a balloon in which a dead fly is embedded (Aldrich and Turley 1899). Originally, it would seem, both sexes captured flies; later, only the males did so; still later, the balloon was added, intensifying the stimulus; finally, the fly was omitted (Wheeler 1924; review in Kessel 1955). Gruhl (1924) notes that the food is not necessary for the eggs to develop, and says that the males may fight over the prey. In many insects, the female eats her mate, while often the sexes are dimorphic, with the females feeding but the males unable to do so. If the females are strongly disposed toward doing anything that leads them to obtain energy, the use of food as an attractant by the males ought to be particularly strongly favored by selection. The overall picture does not favor the alternatives that the food is used in appeasing the cannibalistic female, or in providing her with food for the benefit of the young. But there may be some truth in such hypotheses, especially since the system has evolved.

The effect of imprinting, early learning, and related processes on the choice of a mate has a number of interesting genetical consequences (see, for example, Mainardi, Scudo and Barbieri 1965; Kalmus and Maynard Smith 1966; Scudo 1967b; Scudo and Karlin 1969; Karlin and Scudo 1969; Braestrup 1971). These are not of direct interest to the present study, but they do serve to reinforce the basic thesis: the structure of the society and the organization of social behavior can strongly influence the manner in which sexual selection operates. Similarly,

Wickler (1967) has analyzed certain aspects of mimicry from a sociological point of view. A male, by resembling a female, receives a certain amount of protection from his rivals in primates known to have male combat. Likewise, Guthrie and Petocz (1970) show how organs resembling weapons, including sexual ones, have evolved as a mechanism whereby the threatening effect is accentuated.

Extending these principles to the full extent of their generality, we may see that the institution of a society brings new kinds of selective mechanisms into play. Sexual selection resembles artificial selection, in that an organism affects the gene pool by determining which individuals will reproduce. As Darwin took pains to show, no conscious exercise of choice is necessary: the breeder need only cherish one kind of animal or plant more than another. It is self-evident that where animals live together in societies, whether of one species or several, the opportunities for interactions of this sort are many. Hence their evolution may differ in certain respects from that which characterizes solitary creatures. It would seem, indeed, that selection of this kind casts new light upon the most problematical aspects of the evolution of society itself. To this theme we shall recur in the closing chapters; but first we must treat the one remaining mode of sexual selection.

7

FIRST COME,
FIRST SERVICE, OR,
MALE DISPERSAL

Journeys end in lovers meeting,
Every wise man's son doth know.
Shakespeare

In sexual selection by male dispersal, the males compete with one another by "racing" to get into the gene pool before their rivals do. Dimorphism evolves, with the males becoming smaller than the females, often very much smaller. The size disparity may occur in either of two ways. The male may cease growing and become sexually mature at an early age, or else it may put its energy into locomotion rather than growth. So far as phylogenetic history goes, in some cases it is better to speak of the females as having been enlarged, but the end result is much the same. A similar "female dispersal" evidently rarely or never occurs, at least not in an extreme form; this is just what one would expect from the usual advantage the females have in directing their resources toward the support of the young. One would, however, anticipate a certain amount of female dispersal in those species in which the sexual roles have been reversed to any degree.

We have already discussed, in chapter 4, how male dispersal relates to protandrous hermaphroditism, and the present chapter is basically an extension of the same idea. The definitive arguments for male combat and female choice have already been published (Darwin 1871). Hence it has not been necessary to burden this account with many factual details, and it has seemed reasonable to emphasize new developments and conceptual problems. Male sequestering, too, is a simple extension of the two classical modes of sexual selection. However, the idea of male dispersal is essentially new, at least in so far as it relates to intraspecific sexual competition. Therefore this chapter includes a large number of examples and a considerable body of descriptive material.

DWARF MALES

The prevalence of small males. Evolutionary biologists have paid very little attention to the phenomenon of dwarf males. An important exception is Darwin (1851, 1873) who was particularly intrigued by the minute "complemental males" that live with certain hermaphroditic barnacles; his classic monograph on that taxon incorporates numerous references to evolutionary phenomena. It happens that dwarf males do not occur in the groups that have been most intensively studied from an evolutionary point of view. They are evidently wanting in birds and mammals, and exceedingly rare among flowering plants—although pollen grains have certain analogies with them. One has to stretch a definition to find them in the class Insecta, and the genus *Drosophila* has males and females of about the same size. They occur most often in smaller organisms, and in the sea, especially at greater depths. However, the forms that do display the phenomenon are by no means rare, and they occur scattered about in the various branches of the phylogenetic tree. They are well developed in some members of the following groups: algae, liverworts, all three classes of flatworms, rotifers, nematodes, annelids, echiuroids, onychophorans, several classes of Crustacea, spiders, gastropods, cephalopods, echinoderms, fishes, and amphibians.

We should note, however, that we are dealing here with a spectrum, and what constitutes a "dwarf" may be somewhat subjective and arbitrary. Above have been listed only a few groups in which size dimorphism is carried to an extreme. Anthropomorphism would seem to be the main reason why one gets the impression that small males are not the rule. It is beyond the scope of the present work to enumerate all the taxa in which moderate or even extreme instances of such dimorphism are known. The discussion to follow focuses upon those cases that are pertinent to the theoretical interpretation. However, as the reader may want to consult some additional references, the following are suggested as interesting or useful. For plants, see the reviews by Goebel (1910) and Kniep (1928). For algae, see Setchell (1914), G. A. Dunn (1917), Myers (1925), and Herbst and Johnstone (1937). Caullery (1961) gives an account of dwarf males in the orthonectid mesozoans which should not be overlooked. An interesting paper on a marine, parasitic flatworm is that of Christensen and Kanneworff (1965). Dogiel (1966) reviews dwarf males in his advanced parasitology text. Also of interest is some recent work on dimorphic Copepoda: Bouligand (1961) and Kabata (1964a, 1964b); on Isopoda: Szidat (1964); on mites: Schuster (1962). For various aspects of sexual selection in spiders not cited below consult the works of Montgomery (1903, 1908, 1909, 1910), Petrunkevitch (1910, 1911), and Kaston (1965). A curious example of very small males among insects is a collembolan treated by Lubbock (1873:108–109). To the materials on gastropods treated in our discussion on protandry, we should add the following on dwarf males: Moore (1937), Gallien and de Larambergue (1938), and Thorson (1965). For fossil ostracods, see Shaver (1953) and Guber (1970). Arkell (1957) and Teichert (1964) review the sexual dimorphism of fossil cephalopods. A general survey of sexual dimorphism in fossil metazoans has been edited by Westermann (1969).

How dwarfing is selected. The environmental conditions most likely to favor sexual selection by male dispersal are: (1) a low population density, (2) a restricted motility as the organisms become larger, especially through a sedentary way of life in the adults, and (3) a premium on long life in the female, but not in the male. A low population density, or what has the same effect, a reduced motility, decreases the probability of encounter between individuals of the same sex: hence there is reduced opportunity for male combat and most other forms of masculine competition that might set up a counter-pressure tending to enlarge the males. Similarly, when it is less likely for individuals who are members of the opposite sex to find each other, those males that spend more time and energy hunting for the females are at a considerable advantage.

Exactly how much reduction of the male should be attributed to sexual, how much to natural, selection, in such cases is problematic. Indeed, the distinction tends to become a semantic one in this context, for the various modes of selection, although rightly kept conceptually distinct, merge into one another by insensible degrees. (Where, for example, does artificial selection leave off and natural selection begin?) A more effective union of individuals here would produce more than just an advantage in sexual competition, but this does not mean that both are not there.

A restricted motility in larger individuals would tend to put the smaller males at a functional advantage, for the males are the more motile sex. Furthermore, since the males are likely to be eaten, they may be better off if they reproduce as soon as possible, provided, to be sure, that a delay will not result in a larger return. The ease with which younger individuals are transported by currents thus helps to explain why dwarf males are so common in the sea—and not unusual in lakes. The effect is somewhat accentuated in forms that have a life history involving sedentary adults. This can occur under a variety of circumstances, to be discussed in detail later on. The adult may have anatomical restrictions imposed upon its structure by the way in which it feeds. It may be sessile, as in barnacles, highly degenerate, as in various parasites, or it may wait for its food rather than pursuing it, as in angler fishes.

In certain environments the energy level is so low that a general loss of motility ensues. The effect upon reproduction is accentuated by the fact that in such environments population densities tend to be very low, so much energy would be necessary in hunting for a mate. One would predict that both the tropics and the deep sea should have more dwarf males than comparable high-latitude and shallow-water habitats. In addition, care of the young may render it advantageous for the female to grow to a larger size or to live longer than the male. It is mostly under these conditions that evolution has enlarged the female rather than dwarfing the male. And sometimes, namely where the male cares for the young, selection has enlarged the male. It seems reasonable to predict that we will ultimately find evidence of female dispersal in such animals as sea-horses and pycnogonids. As was mentioned in chapter 4, a good number of cases of protandrous hermaphroditism (conceivably of protogynous too) may be explained on the basis of this kind of size advantage. The smaller individuals capitalize on their motility, the larger on their greater ability to care for the young or to produce them in large numbers.

SEX RATIO AND LONGEVITY

Among organisms in which the males have become smaller or the females larger, the sex ratio often appears to have shifted away from unity. Yet when such cases are investigated with care, it often turns out that the usual one-to-one ratio still holds, but differences in mortality must be compensated for. And sometimes finding out what is going on severely taxes the imagination. Fortunately the requisite logic for understanding this problem is well-developed in certain branches of knowledge, for the same situation exists in numerous dynamic phenomena. What aquatic biologists refer to as the "standing crop" is not equivalent to "productivity." Organisms that grow very slowly can be very numerous, so long as sufficient resources are available to support them and predation is modest. Conversely, a few animals that grow very fast can support a large number of predators. An equivalent situation exists in commerce. When the rate of turnover is high, the firm can afford a lower margin of profit. Thus, an item kept on the shelf for a year will yield a return of 10% if sold for 110% of its cost. But if ten such items are, during the same period, bought and sold in succession, then in theory at least, the same yield will result if the cost to the buyer is only 101% of that to the seller. Let us see how such relationships are manifested in the natural economy as it affects the sex ratio.

A fairly clear-cut example may be extracted from a work by Beebe (1934) on the life history of a fish, *Idiacanthus fasciola*. The larvae in this species are sexually monomorphic, and feed on diatoms and marine crustaceans at a depth of approximately 100 fathoms. At the time of metamorphosis, they sink into deeper waters (below 500 fathoms) and the two sexes diverge radically in their way of life. The females continue to grow, feeding on fish and becoming approximately six times as long as they were when they underwent metamorphosis. The males, on the other hand, immediately become sexually mature, with an atrophied digestive tract and the sort of "larvoid" aspect that suggests precocious maturation (neoteny). The males have a large photophore under each eye, and their evident ability to signal, coupled with their weak powers of locomotion, led Beebe to infer that the females seek out the males. Better evidence on this point is desirable, for a passive role by the males seems unlikely from what is known of other organisms with dwarf males. Bertelsen (1951), in a brilliant monograph on the ceratioid fishes, enumerates many structural adaptations that facilitate the dwarf male's search for a mate. Be this as it may, it seems inevitable that the males must soon perish, and that the females will live on, gradually diminishing in numbers through predation. They eventually mate with males that are much younger than themselves. At the time of metamorphosis (and presumably also at earlier periods) the apparent sex ratio is unity; but not at the time of sexual maturity, for there are fourteen times as many adult males as adult females. Yet if the post-larval mortality seems high, we may note that Beebe found 14,000 fully developed eggs in a single breeding female.

In some instances the mortality and maturation rates may unbalance the apparent sex ratio so that the females seem greatly to outnumber the males. A particularly interesting example is the phylum Onychophora. This group (*Peripatus* and allied forms) is of great phylogenetic significance, because of its transitional position between the Annelida and Arthropoda. But it is equally intriguing for its many reproductive peculiarities. There is a good series, with intermediate forms, in which we may make out stages in the evolution of viviparity. Some are oviparous, others ovoviviparous with yolky eggs, still others have small eggs and a placenta, being remarkably convergent with mammals. Taxonomic works provide copious data, showing a clear tendency for the females to be more numerous in collections, to have more legs than the males, and to attain a larger body size (e.g., Sedgwick 1888; Bouvier 1905, 1907; A. H. Clark 1914; Kemp 1914; Clark and Zetek 1946) but Marcus (1952) notes occasional exceptions to the rule about legs. Evidently the sex ratio at birth or eclosion is around unity. The dimorphism and the divergence in the apparent sex ratio would seem to be most pronounced in the viviparous and ovoviviparous species (see Mosley 1874; Gaffron 1885; Sedgwick 1888; Willey 1898; Manton 1949). On the other hand, the oviparous species display only a very modest difference in size, and collectors usually get approximately equal numbers of the two sexes (Dendy 1902). Evidently, the females have been enlarged and their life-span has been prolonged because of the care they give to the young. The females are able to store sperm, and there is a remarkably long gestation period. Some, but not all, produce young of different ages, rather than single broods, and thereby protract their life-span all the more. But the males are superfluous in reproduction once they have impregnated the females. So they reproduce as soon as they can, then die.

A remarkable convergence with the onychophoran system has evolved in marine shore-fishes of the family Embiotocidae—the celebrated viviparous perch of California (Eigenmann 1894, 1896; Hubbs 1921; Triplett 1960; Wiebe 1968; De Martini 1969; Shaw 1971). Here, too, a kind of placental arrangement has come into being; the females are much the larger sex, and can store sperm, and the breeding population contains about twice as many females as males. Growth-rates are the same until parturition; thereafter the males grow more slowly, becoming fertile in their first year. Likewise, in the uterus the sex ratio is unity. Females of *Amphistichus* are said to defer maturation until their second year, and in successive years produce increasingly larger broods. The young of both sexes are quite large when they are born, and the female can raise only a few (less than 100) per year. On the other hand ovoviviparous fishes from the same area, but of a different family, Scorpaenidae, are not dimorphic in size; they have no placenta, and the brood numbers up to 2,000,000 individuals (Moser 1967).

SOME HYPOTHESES ALTERNATIVE TO MALE DISPERSAL

The hypothesis of male dispersal explains many hitherto problematical aspects of sexual dimorphism. It reveals numerous unexpected correlations between way of life, habitat, and the differences between the sexes. Such relationships are difficult to account for on the basis of any other hypothesis. These we shall examine in due course. However, two other possibilities need to be considered, one of which seems to be true, while the other evidently has little or no validity. Therefore we shall first entertain the possibility that males have been suppressed by *parental exploitation*. That is to say, we shall see how far they have been dwarfed in the interest of their mothers. Second, we shall inquire whether perhaps a *competition-reduction* model might apply, along lines already suggested for the dimorphism of certain vertebrates. It might take the form of a species-advantage hypothesis, or, in more orthodox Neo-Darwinian terms, it might be cast in the form of an advantage that accrues to the family. We will then be in a much better position to deal with those instances of small males to which neither of these hypotheses are applicable.

Parental exploitation. In our discussion of anisogamy in chapter 4, we observed that the energy expended upon an individual by its parents is determined, not by any advantage to the offspring, but by the advantage to the mother or father. The protozoans provide many examples of the unequal distribution of resources in favor of the "female" progeny. Ordinarily, to judge from the size of the parents, the energy bestowed upon "sons" as a group appears to be the same as that supplied to the "daughters." Such an equality is exactly what one would expect from an extension of Fisher's ideas on the sex ratio, although we should bear in mind that this may not be the correct explanation. Yet in some cases, the fission leads to a smaller amount of resources going to the male portion. One might conjecture that the greater efficacy of a smaller male allows a different use of resources. More likely, the female offspring are capable of asexual reproduction; and therefore the energy lavished upon them is used for something other than just contributing one half the ancestry of the individuals produced by amphimixis.

Whatever the selection pressures, we might look for an analogous situation in multicellular organisms, with smaller males being perhaps associated with other peculiarities of reproduction. In addition, it would be necessary for the mother to "know," as it were, the sex of each individual offspring, and to be able to shunt resources from one sex to the other in the adaptively appropriate fashion. It happens that these conditions are met in forms having parthenogenesis, especially those with haploid males. The mother can determine the sex of the offspring by fertilizing or not fertilizing her eggs, giving rise to females and males, respectively. The problem of male haploidy confuses this issue, because it is not at all clear what is cause and what is effect. At the same time, we have to be careful not to misinterpret the data because of the other reasons for dwarf males. If the right facts are available, however, the problems are not insuperable. When the males are suppressed in the interests of their mothers, the difference in size may be seen at an early age, often with smaller eggs giving rise to the males, larger ones to the females. Such data, however, may be hard to come by.

In liverworts (Bryophyta: Hepaticae) a variety of sexual arrangements exist, but all have a dominant haploid stage (gametophyte) upon which lives the diploid (sporophyte). Some are hermaphrodite, in the sense that one gametophyte produces both sperm and eggs. Some, however, are gonochoric, and these include a few in which the males are small and live epiphytically on normal females; sometimes both normal and dwarf males coexist in the same species (Goebel 1910; Woesler 1935; Ernst-Schwarzenbach 1939). The dwarf male may (Ernst-Schwarzenbach 1939) or may not (Rink 1935; Woesler 1935) come from a smaller spore. This size difference (a volume ratio of 1:1.42) was attributed to the presence of sex chromosomes by Rink (1935). Fleischer (1920) noted that the dwarf males are more common in the tropics than in Europe, a point that suggests that an ecological explanation is in order.

Among insects, clear-cut examples of dwarfed males having a congenital basis are rather common in the Hymenoptera. The haploid males have attracted the attention of many biologists, giving rise to much speculation. The topic has particularly intrigued geneticists, but very little ecological interpretation has been attempted. The genetical conditions necessary for its establishment are thought to be rather stringent, because the haploid individuals would suffer from the full expression of lethal genes. Only a few groups have evolved it, and this is taken as evidence of the difficulty. But the problems may have been exaggerated, and Oliver (1971) points out that it has evolved more often than is generally recognized. Hartl and Brown (1970) have suggested intermediate stages in its evolution, involving the suppression of one set of chromosomes. Hence a gradual origin is possible. The peculiar taxonomic distribution (Whiting 1945; White 1954) suggests that it often occurs among forms with a seasonal food surplus: rotifers, some mites, thrips, white-flies, scale insects, a beetle, and the Hymenoptera. The correlation is somewhat inexact, but we must remember that a great deal of evolution has gone on since male haploidy originated. The adaptive significance may have something to do with allowing control over the sexuality of the offspring. At any rate, it functions that way now, in spite of other effects. (For mechanisms of sex-control see Flanders 1939; Brown and De Lotto 1959.) Caring for the young and providing them with food has repeatedly evolved in the Hymenoptera, culminating in the evolution of complex societies. In the next chapter we shall see how the peculiar dominance of the female has influenced the social evolution of this group. Many Hymenoptera are parasitic, on plants or on animals, and the female plays a very important role in seeking out the host. In this respect parasitic insects differ radically from most marine parasites, for in the sea it is generally the juveniles or larvae that find the host. As a rule, but only as a rule, the male hymenopteran provides his offspring with nothing but DNA, and that only for his daughters and his grandchildren. A fascinating exception, however, occurs in certain wasps of the family Thynnidae from Chile (see Goetsch 1953:24). The male is large and winged and carries the wingless female to an appropriate nesting place. Loss of an ability to fly may occur in either sex or in both, depending upon the circumstances. Thus in the wasps that fertilize figs, and whose habitat is therefore temporary and dispersed, the female for obvious reasons retains her wings; but the male is wingless and bizarre (Mayer 1882). In certain wasps that live under water even as adults (Blunck 1914), wings may be lacking in both sexes or only in the males, and the necessity for dispersal obviously fits in here. On the other hand, females are wingless in a number of hymenopteran species; some even have both winged and wingless females.

The variability thus far described in no way contradicts the general tendency toward a suppression of the males, but only shows that it is ecologically determined and by no means inevitable. Wilbert (1969) has lately studied *Aphelinus asychis*, a parasite of aphids in which the males are smaller than the females. He found that the females emerge from larger hosts. The same phenomenon was noted by earlier workers (e.g., Seyrig 1935 for the Ichneumonidae). The saving in time and energy for the mother should be obvious. Chewyreuv (1913) and various other workers have produced larger males by supplying them with extra food. It is worth noting that dwarfing of the males is recorded from other insects in which the males are diploid, for example, aphids (see Hille Ris Lambers 1966).

The phylum Rotifera presents one of the classic examples of sexual dimorphism associated with haploid males. The males are more common than is sometimes believed, but they tend only to be produced at certain times of year, namely at a population maximum (Wesenberg-Lund 1923, 1930), or, so it seems, not at all. We mentioned in chapter 3 how sexuality is itself associated with habitat and way of life in this group. The same trend is apparent in the degree of sexual dimorphism. According to Remane (1929–1933), the Seisonidea, exclusively marine and not extensively studied, develop only from fertilized eggs (i.e., there can be no male haploidy), have an apparent sex ratio of unity, and are essentially monomorphic. The haploid males occur in forms from fresh water. Moderate sexual dimorphism prevails in the benthic species, and in these the gut of the male shows various degrees of vestigiality. The extreme in sexual dimorphism occurs mainly in pelagic forms, but the males are also highly reduced where the female is sessile —and in this case the male is a free-swimming creature. Wesenberg-Lund (1923) repeatedly draws attention to the rapidity of movement that characterizes the males. All this points to an extreme specialization in the males toward sexual union alone, evolved in response to environmental conditions. But how much of it needs to be explained as an adaptation of the mother, how much in terms of benefit to the son, remains uncertain.

The Cladocera are a group of fresh-water crustaceans whose life histories have frequently been compared with those of rotifers and aphids (e.g., by Wesenberg-Lund 1926). They, too, have dwarf males in some species, and parthenogenesis is common. Since in this group both sexes are diploid, we can infer that chromosome number is no inevitable determinant of the reproductive strategy, a possibility that might be raised by certain geneticists. Ostracod Crustacea are another group of small animals in which similar trends might be looked for. There is a moderate dimorphism in size, and aberrant sex ratios with female preponderance (Fowler 1909) suggest a certain amount of parthenogenesis (C. I. Alexander 1932). But in this group there would seem to be both a sexual biology and a way of life more like those of copepods than cladocerans or rotifers.

In establishing whether parental exploitation does in fact obtain, it helps to know whether the mother actually is responsible for the sexual dimorphism. The difference needs to be congenital if it is not to be ambiguous, for a male once out of his mother's influence often will grow less than his sisters because this is in his own interest. Generally such information is not available, or is difficult to obtain. Although it is beyond the scope of the present work to review all the examples, the point needs to be made that this sort of empirical evidence does exist. We mentioned some with respect to the Hymenoptera. Another example that should not escape attention is the archiannelid *Dinophilis apatris*. Within the same protective capsule, larger, female-producing eggs are formed along with small ones destined to give rise to neotenous males closely resembling a trochophore larva. Sex-determination and parthenogenesis in this species have been studied extensively (see Bacci 1965 for review). Different "races" have different sex ratios, but little ecological work has yet been done.

Competition reduction and related alternatives. A second possible significance of dwarf males might be that they further the interests of the family, or perhaps of some larger group. If a failure to grow on the part of the male somehow renders more energy or living space available to his offspring, perhaps indirectly through a benefit to his mate, his overall reproductive output will be increased. It would be a case of not harming the young, with effects similar to those of caring for them. That some kind of competition goes on between male and female barnacles has been suggested by Tomlinson (1969:28–30). He does not make clear the mechanics of such a competition, but one can easily see how it might occur through interference between mates. We need to distinguish, very carefully, this interpretation from one that would be based upon a competition-reduction model invoking a benefit to the entire species. If the general theoretical outlook embraced in the present work is true, then such a competition will never result in the evolution of dimorphism. Furthermore, various conditions would have to be met before the male can become dwarfed if we are to explain the phenomenon in terms of reducing the competition within the family. Of these, two are crucial: first, the competition must occur in such a fashion that not competing will benefit the male's offspring; second, something would have to happen to offset any concomitant decrease in the ability to compete with other males. There are many natural situations in which dwarf males occur and in which both these conditions are met; therefore one gets an erroneous impression that the hypothesis is true. They are met in sessile and sedentary organisms existing under conditions of low population density. Barnacles would seem a good example of a form in which the hypothesis might be invoked. Another is the celebrated echiuroid worm *Bonellia* (Baltzer 1931). This, like other members of its family, gathers food from the substrate with a long proboscis, and hence individuals in close proximity to one another would obviously compete for food. All of the Bonelliidae known at present feed in this fashion, and all would seem to have dwarf males that can be found living parasitically on or in the females. In another echiuroid, *Urechis caupo*, food is extracted from the water by pumping it through a mucous net (W. K. Fisher and MacGinitie 1928). In this species, competition for food between adjacent individuals should be less intense; the sexes are of the same size, but this is hardly significant, for no dwarf males occur in certain echiuroids where competition for food would seem to be a problem: many monomorphic echiuroids feed much as *Bonellia* does. Zenkevitch (1966) notes that the Bonelliidae are largely deep-sea forms. The low-energy conditions should increase the competition for food, but at the same time they will decrease the population density. Certain deep-sea angler fishes have a male that lives parasitically on the female, supported by a direct connection to her circulatory system (Tate Regan 1925). The competition-reduction hypothesis may be invoked here too, but it lacks a critical test, for much the same reason as in the case of the echiuroids.

Indeed, the competition-reduction hypothesis seems to be both unnecessary and insufficient. Dwarf males evolve where only one of the two conditions mentioned above is met: a low density. They have evolved, for example, among internal parasites where although there should be at least some competition for food it should be less intense than elsewhere. They have also evolved among rotifers and other groups in which a trophic surplus is part of the usual habitat—but they appear when the food has run low, of course. And where there is much competition for food, they show no general tendency to evolve. However, it must be admitted that we really do not know in these cases how much of an advantage might accrue through a minor shunting of resources to the mate. And any attempt to determine the effects of competition on such an *a priori* basis is apt to prove misleading. We can say that many dwarf males contact their mate only at the time of copulation. Some die soon after impregnating the female, and one might get the impression that these are sacrificing themselves in the interests of their progeny. Yet examples to the contrary are readily come by. Male spiders will often move on to another female. And the general motility in such creatures as deep-sea copepods would easily result in the supposed competitors being carried too far away to inter-act as the hypothesis demands.

EMPIRICAL EVIDENCE FOR MALE DISPERSAL

The compelling reason for rejecting the competition-reduction hypothesis is that another one will account for the same data and a lot more besides. It is not the whole story, for, of course, parental exploi-tation applies to a number of cases. In addition, a version of the male dispersal hypothesis in which the mother's life is prolonged as a means of providing for the young, as already suggested for onychophorans and viviparous perch, explains a few more. It would appear that the remaining cases can all be explained by something like one of the models for protandrous hermaphroditism invoked in chapter 4. Where a mode of life has been assumed by adults in which motility has been greatly restricted, dwarf males tend to evolve. The selection pressure favoring dwarf males is concomitantly increased by a reduction in population density. Forms with little motility but a high population density can get along fairly well as simultaneous hermaphrodites. By the same token, simultaneous hermaphroditism works when motility is high, but population density low.

We can see how the foregoing hypothesis applies by treating a few examples of habitats and modes of living in which adult motility gets restricted and in which population density decreases. These are: the deep sea, certain kinds of parasites, and sedentary or sessile filter-feeders. We shall observe that the relationship between habitat and feeding strategy is sufficiently complicated to obscure the view of what is going on, but the difficulties are far from insuperable.

We shall restrict our discussion of deep-sea animals to four groups of special interest. These will be two classes of Crustacea, namely Copepoda and Cirripedia (barnacles), and two groups which, if but very distantly related, show some impressive convergences: fishes and cephalopods. Reviewing the energy conditions in the sea, we may recollect that less and less food becomes available with increasing depth. The decreased energy at once lowers the standing crop of each species, and diminishes the effectiveness of active movement whether in pursuit of food or of a mate. Hence both feeding and reproductive strategies develop energy sparing features: notably a division of labor among inactive forms, or its combination in more motile ones.

Free-living copepods in deeper waters. The Copepoda are a group of crustaceans with an ecological significance in the sea rivaled only by that of the insects on land (a fine book is by Marshall and Orr 1955). The free-living forms are the dominant herbivores over most of the globe, owing much of their abundance to their being small enough to graze on small plants in the water. In an earlier chapter we cited them as a noteworthy exception to the supposed rule that as organisms evolve they become larger. It is thought that copepods have become reduced in size through neoteny (Serban 1960), and the males display this trend to a greater extent than do females; however, in a few species the males are the larger sex (Farran 1926). Very often the adult males do not feed and have degenerate mouthparts (Brady 1883; Conover 1965); sometimes, as in *Calanus tonsus*, adult females do not feed either (Campbell 1934). What appears to be an unbalanced sex ratio has been noted by many authors (see especially Wenner 1972). Steuer (1925) found that in 30,000 specimens from an expedition collection the females outnumbered the males by three to one. Environmental sex determination and other genetical peculiarities known for the group make such data particularly hard to interpret. Sex-ratio problems are complicated by the likely shorter life-span of the males, which will bias the sample (Maly 1970). In addition, the sexes are sometimes dimorphic in habits: the males tend to remain in deeper water (Nicholls 1933). Such a dimorphism would be much better known among animals generally had migration not been studied primarily in forms in which both sexes care for the young. The adaptive significance is the same for birds and copepods, namely getting into an environment rich in food when producing offspring, and otherwise living in one with few predators. If it were merely a matter of the two sexes avoiding competition with each other, as one might be led to infer on the basis of insufficient comparative data, the bathymetric trend would not be evident: the sexes would often show reversed depth distributions, and they would sort out geographically.

High-low dimorphism in the males has been recorded from both marine copepods (Sewell 1912, 1929–1932; Haq 1965) and fresh-water ones (Thallwitz 1916; Gurney 1929). The deep-sea forms have the more degenerate males (Brady 1883). Mednikov (1961) notes that species from greater depths tend to be known only as females. He rejects the conjecture that this is a sampling artifact. He found a decrease in the percentage of males with increasing depth. In two species living at the surface, the sex ratio was unity. An upper bathypelagic species was found to have 35% males, while a lower bathypelagic one had only 23%. A second lower bathypelagic form had only 12% males; the proportion of males here was found to be somewhat higher when the immatures were examined: 38% at the fifth copepodid stage. This situation contrasts markedly with that of the deep-sea fishes noted earlier in this chapter. Tschislenko (1964) maintained that common copepods have a sex ratio of one-to-one, while rare ones have fewer males. Moraïtou-Apostolópoulou (1972) found, in surface samples of two species, more males at the time of maximum population size. Both these authors maintain that having a lot of females benefits the species by increasing its reproductive rate. Volkmann-Rocco (1972) maintains that, since a delay in fertilization increases the number of males, the sex ratio is adjusted in the opposite direction at low density. At least one of these explanations must be false.

Depth relationships among barnacles. The data for copepods establish a correlation between suppression of the males and depth, but they do not tell us very much about how the feeding and reproductive strategies interact. Furthermore, we do not know much about the sexual ecology of copepods in general. A more clear-cut picture may be abstracted from what is known of certain barnacles. The class or subclass Cirripedia includes several orders and suborders that differ considerably in both their way of life and their manner of reproducing. That the two aspects of their biology are so intimately linked has been overlooked, so that the interpretation has been confused, to say the least. The order Thoracica includes three suborders, the stalked Lepadomorpha, or "goose barnacles," and two that live more directly cemented to the surface, a small suborder Verrucomorpha and the more abundant Balanomorpha, or "acorn barnacles." The Acrothoracica are a small order of barnacles that bore into snail shells and other calcareous materials; they, like the Thoracica, are filter-feeders, obtaining their food from the water. The orders Rhizocephala and Ascothoracica are parasites. These differences in niche are very significant for understanding the reproductive strategies. The more tightly cemented, stalkless forms would appear to be adapted to spatial competition and to life under more or less crowded conditions. The burrowers and parasites have more of a problem of finding a suitable place to live than of maintaining themselves there. Sexual systems display a remarkable diversity, and Darwin (1851) compared these systems to analogues in plants. The sexes may be separate and the male may be reduced in size. Or the species may consist entirely of hermaphrodites, or of hermaphrodites and males dwarfed to one degree or another. The small males living on hermaphrodites were designated "complemental males" by Darwin (1851). He attributed the separation of the sexes to a "division of physiological labour" (Darwin 1873:432) and pointed out that the male can be reduced because of his ability to fertilize many eggs with but a small "expenditure of organic matter." He also noted that a gonochoric form had always been found occurring in pairs. Why pairing makes it easier to live at low densities in slow-moving or immobile creatures should be more or less obvious: the pair is in effect a self-fertilizing hermaphrodite, but has the ability to cross. And indeed such "marriage" is well known in marine gonochoric gastropods (see Ghiselin and Wilson 1966), and even hermaphrodites occasionally live as pairs. According to my own field observations, the phenomenon of pairing in marine gastropods is much more pronounced in tropical waters than in temperate ones. The reason again should be obvious: lower population densities and less food.

Darwin believed, for very good reasons, that the gonochoric barnacles he studied are derived from hermaphrodites (see M. T. Ghiselin 1966b, 1969a, 1969b). The more recent data have confirmed his view with respect to the Thoracica, but they remain somewhat ambiguous for the other orders. His reasoning, cryptically expressed, would seem to have been based on vestigial structures. The dwarf males display signs of once having been larger, and one hermaphrodite with a complemental male was beginning to show signs of losing the male organs. Also, we have the evidence of taxonomic distribution. Virtually all acorn barnacles are strictly hermaphroditic, but in some species complemental males have lately been discovered (Henry and McLaughlin 1965; McLaughlin and Henry 1972). Change, via an intermediate state with complemental males, clearly goes in the direction of a separation of the sexes.

In the Lepadomorpha, the whole range of variation is seen in the genus *Scalpellum* and its close relatives. Hoek (1883, 1884; see also Gruvel 1900; Stewart 1910) provided the compelling argument that the sexes have indeed been separated, rather than hermaphroditism being secondary. The hermaphrodites without complemental males live in shallow water. With increasing depth the males become ever more degenerate in structure, and gonochoric species become more common. It would be absurd to think that the sequence goes the other way —as ridiculous as taking a kiwi for an animal in the process of evolving wings. Animals do not have useless organs in order to prepare their remote descendants for environmental contingencies to be met countless generations in the future. More importantly for the present discussion, the bathymetric relationships fit in quite nicely with the global pattern of energetics. Shallow-water barnacles generally live in very crowded circumstances and at high population densities; hermaphroditism works quite well for them. As energy resources diminish and population density goes down, they can continue to reproduce by remaining social, but the lower standing crop of food in the water would favor their spreading out more. It seems likely that the difficulty of obtaining food selects for early maturation in the male, rather than there being a competition within the family; but one way or another the basic argument would be the same.

The burrowing barnacles (Acrothoracica) are gonochorists with dwarf males (see Berndt 1907; Tomlinson 1969). If they have a fair degree of substrate specificity, the arrangement may further their success under low-density conditions. But we still would not know whether gonochorism or hermaphroditism had been the original state. We shall deal with the parasitic forms below, when we consider parasites as a group.

Deep-sea fishes and cephalopods. Extreme reduction of the male has been recorded from several lineages of deep-sea fishs (for reproduction in fish, see the compilation by Breder and Rosen 1966; for deep-sea animals see N. B. Marshall 1954). The life histories of these creatures are inadequately known: our earlier account of *Idiacanthus* is one of the few exceptions. Another is the deep-sea angler fishes of the suborder Ceratioidea, which we have mentioned in relation to the parasitism of males. Again, the larvae develop near the surface, and a metamorphosis occurs at the time when the animals take up their adult existence at depths of around 2,000 to 2,500 meters. From the structure of the male it appears that he must seek out the female at once. He has well-developed eyes, and no way of feeding himself; according to Bertelsen (1951) only some species have the well-known connection between the blood streams of the male and the female. The females make their living as adults by trapping their prey. The term angler-fishes is most apt, for they wait, expending a minimum of energy, and lure other animals toward their enormous mouths. The approach is used partly because population densities are so low. The basic reason for the dwarf males is that motion is incompatible with this feeding strategy. Tate Regan (1926:14) relates the dwarf males to the low population densities and poor motility, but does not treat the selective mechanism. It bears repeating that some deep-sea fishes have saved energy by becoming hermaphrodites; but these are motile forms (see chapter 4).

Cephalopods of deeper waters display many convergences with fishes of the same habitat. Watery bodies, degenerate musculature, and luminescence in both groups manifest comparable responses of the populations to the same general group of selective influences. The reproductive strategies likewise show correlated trends. Robson (1932) has discussed convergent evolution among the Octopoda of the deep sea. For the reproductive structures we may note the following trends as depth increases: larger eggs, larger and fewer spermatophores, a larger vagina, and smaller males. In cephalopods, reduction of the male attains its maximum in the paper nautilus, *Argonauta*, which has so large a sperm-transfer organ (hectocotylus) (Müller 1853; Grimpe 1928) that he is generally suspected of living as a symbiont with the female. *Argonauta* is a pelagic creature living near the surface and only penetrating to moderate depths; but we may note that it lives in the fairly unproductive waters of the tropics. Its apparent sex ratio favors the males (Pelseneer 1926:320), and the eggs are fairly small (Robson 1932:25; for more on egg size in cephalopods, see Mangold-Wirz and Fioroni 1970). It seems reasonable to attribute their dimorphism partly to the habit of brooding in the female, who constructs a large egg-case resembling a shell, and partly to the general lack of motility. Such cephalopods as *Ocythoe* and *Tremoctopus* from deeper waters have a less marked sexual dimorphism, but they fit the basic pattern. Of *O. tuberculata*, Naef (1921–1923) says that both sexes carry about the case of a salp as a kind of substitute shell. Here again, we seem to be observing a complex interaction between feeding, reproduction, and predator avoidance, but clearly the whole system is under energetical control.

Dwarf males in certain parasites. The males are smaller than the females, to one degree or another, in a large number of parasitic animals. It would be tedious to enumerate these, and quite unnecessary, for an excellent introductory account has been provided by Dogiel (1966). What has not been done is to demonstrate the connection between locomotory abilities and the kind of dimorphism. Where both sexes are motile as adults, dwarf males do not evolve; and whether the females are motile is linked with the manner in which they feed. A simple example of the beginnings of such dimorphism may be seen in symbiotic crabs of the family Pinnotheridae (Orton 1920; W. W. Wells 1928, 1940; Stauber 1945; Christensen and McDermott 1958; Wells and Wells 1961; Hopkins 1964; Pearce 1966; Seed 1969). They live inside such seawater-filled spaces as the gill-cavities of clams, in which they subsist mainly on the food gathered by the host; hence they are true parasites, not commensals as often supposed. The mature female is soft-shelled and structurally degenerate, with a bloated appearance and often a large number of eggs; when removed from their hosts, they become virtually helpless. The somewhat smaller male is much more like a normal crab, with a hard shell and well-developed legs: he is quite motile. (For a convergent system in another crab, see Castro 1971.)

We discussed earlier the moderate reduction of the males that occurs in free-living copepods. The difference in size between males and females is often much greater in the parasitic forms, although a spectrum may be observed. The adults do not leave their hosts, dispersal being effected by the larvae. However, one group, the Monstrillidae, does exactly the opposite. The juveniles are parasitic, with poor locomotory ability. It is the adults that are free-living, with rudimentary guts and only slightly smaller males. They are unusual, although not extraordinary, for marine animals, but the pattern is quite common among insects: Malaquin (1901) compares monstrillids to the Ephemeroptera, or mayflies, in which the adults do not feed, but even better analogies could be drawn with various parasitoids. Another marine group with a reproductive strategy comparable to that of the monstrillids, is the arthropod subphylum Pycnogonida, which has but a single class, Pantopoda (for general biology see D. W. Thompson 1909; Helfer and Schlottke 1935; Hedgpeth 1955). They are sometimes called sea-spiders, but their relationship to true arachnids is most debatable. The juveniles or larvae are nonmotile and parasitic; the males take the eggs from the females and carry the young, helping them to disperse (L. J. Cole 1901; Oshima 1935; King and Jarvis 1970).

In cirripedes of the orders Ascothoracica and Rhizocephala, dwarf males coincide with a highly modified, internally parasitic female. The Ascothoracica are gonochorists with dwarf males in some species. The Rhizochephala were long thought to be hermaphrodites, and this may be the case for some. G. Smith (1906) thought them to be parthenogenetic, with the complemental males no longer functional. He is partly responsible for the myth that barnacles were primitively gonochoric. Recent work has shown that at least some of the purported hermaphrodites are really gonochorists, with the highly degenerate males having been mistaken for testes (Ichikawa and Yanagimachi 1960; Yanagimachi 1961). Yoshi (1931) notes a rudimentary penis in the female of a strongly dimorphic gonochorist, a feature that may or may not be a vestige of an earlier hermaphroditic state. Until more information is available it seems wise to leave the question open as to whether some cirripedes may have always been gonochorists. All we can say is that some gonochoric species had an hermaphroditic ancestry.

Dwarf males in filter-feeders and deposit-feeders. We have seen how the way of life determines that dwarf males will enjoy a reproductive advantage in two modes of functional organization: parasites and ambushers. The basic principles can easily be expanded to cover other sorts of arrangements in which the adult's motility has become restricted, favoring neoteny and the like in the more motile males. We have been giving examples of one such functional arrangement all along. This is filter-feeding, in which food is extracted from the medium in which the animal lives by straining out small particles. A barnacle is basically a filter-feeder, although some have switched to other mechanisms: it filters food out of the water by extending its legs, which are provided with hairlike processes that form a very effective net, then drawing them in. The snail *Crepidula*, mentioned in chapter 4, is protandrous, and lives in stacks with the females on the bottom; it eats food that gets trapped on its gill. It, too, is a filter-feeder, but its filter has a very different origin from that of a barnacle. A comparable arrangement has evolved in sand crabs of the genus *Emerita*, which live in wave-swept, sandy beaches. MacGinitie (1938) decribed movements of *E. analoga*, noting the small size and greater numbers of the males. At night one may observe the small males pursuing the females over the sand. Barnes and Wenner (1968) provide evidence that this species is protandrous (see also Wenner 1972). Efford (1967) reviews sexual relationships in the genus, drawing attention to male neoteny, insipient parasitism, and early maturation. Such echiuroids with dwarf males as *Bonellia* are deposit-feeders: they gather food from the water indirectly, waiting until it settles, then harvesting it with the long proboscis.

It is not the filter-feeding or deposit-feeding habit as such that determines the sexual dimorphism, but rather the connection between feeding and locomotion. When the adult females are relatively motile, the sexes come together as a consequence of their feeding activities. Hence the free-living copepods, which are, crudely and generally speaking, filter-feeders, display but a modest reduction in the size of the males. In the rotifers, yet another such group, we have noted that the ones that live attached to the substrate have evolved a pronounced dimorphism; but, of course, other factors have the same effect. We have seen, too, that population density influences the global picture. Where the number of individuals of a given species per unit area is high, then simultaneous hermaphroditism, even monomorphic gonochorism, will tend to be favored among totally sessile forms. And the capacity for vegetative growth may lessen the disadvantage of being unable to find a mate. We may conceive of motion as equivalent to an increased density. Hence those filter-feeders that move about in the open ocean, such as pelagic tunicates, may have an effective population density equivalent to that of their sessile relatives. And they tend to be simultaneous hermaphrodites. It seems inevitable that kinetic models will be developed, allowing a very sophisticated treatment of such phenomena.

One reason why dwarf males are much more in evidence among marine animals than terrestrial ones has to do with the preponderance of filter-feeding in the oceans. Seawater can be viewed as a rich soup, with many small animals and plants growing in it; but air does not contain a sufficient standing crop of aerial plankton to support many filter-feeders. It takes energy to build and run a filter system, and the expense is particularly great when the animal pumps the water by muscular or ciliary effort. The cost may be somewhat less when energy from the environment is used to bring the animal in contact with his food. But the basic economic principle remains the same: if the animal cannot extract a profit from the medium, he goes bankrupt.

That it is the energetical situation, and not any unknown property of water, that has led so many marine animals to evolve dwarf males, is confirmed by the situation on land. The only group of terrestrial organisms in which dwarf males are characteristic rather than anomalous are spiders, and these include the only successful terrestrial filter-feeders. The small males and the androphagous habits of the females in some spiders have long been known, and indeed somewhat exaggerated. Darwin (1874) accepted the view of Cambridge that the two are closely related. The small, agile male can more readily elude the female. Similar notions along such lines, for this as well as for courtship, have been supported by a number of subsequent authors. However, Gerhardt (1924; also his other papers and Gerhardt and Kästner 1937) provides sound reasons for thinking otherwise. The female rarely eats the male except after copulation, and the cannibalism may not occur at all, whether the male is small or not. The facts are readily explained if we hypothesize that the female is hungry and eats what she can when it is to her own advantage. It is interesting to note that what is said to be courtship feeding may occur. Gerhardt (1924:535–536) notes that the dwarf males relate to the motility of the sexes and compares spiders to other forms. But he is not clear on the mechanism, and his discussion has an orthogenetic flavor. A few more points should clear this matter up. Sexual dimorphism is greater in forms that spin webs, and as well in those that sit on flowers and ambush prey; these are the respective analogues of barnacles and angler-fishes. Sexual dimorphism is more pronounced in tropical spiders than it is in temperate ones (Gerhardt 1928); compare the deep sea. The large size of tropical arthropods A. R. Wallace (1895) attributed to a copious food supply, but we now can see that exactly the opposite influence has been at work. As if this were not enough, many spiders are dimorphic in longevity (Bonnet 1935; Gerhardt and Kästner 1937): the females in some species outlive males by many years.

Thus the patterns of convergent evolution allow us to treat such energetical relationships from a unitary point of view, and to deal with them upon a global scale. When we see organisms of disparate structure but common functional organization responding in the same manner to the same influence, and doing so in the same fashion whether we vary latitude or depth on the one hand, or density or motion on the other, the picture, however incomplete, is satisfying. It would have been possible to extend these arguments to cover more groups of organisms, and to treat the various strategies and habitats in more detail. However, the point has been made, and the general thesis will be better served if we now turn to a phenomenon that on the face of it appears quite unrelated: the evolution of society. Yet as we shall see, the difference is misleading. One universe, one economy, one economics.

8

THE ANTISOCIAL
CONTRACT, OR, SEX
AND THE ORIGINS
OF SOCIETY

*Le premier qui ayant enclos un terrain
s'avisa de dire:* Ceci est à moi, *et trouva
des gens assez simples pour le croire, fut
le vrai fondateur de la société civile.*
                          Jean-Jacques Rousseau

Man, being at once a reflective and a social organism, has repeatedly
compared his own society with those of other creatures. His efforts
along these lines have ranged in outlook from the more or less scientific, to the humanistic, moral, and even religious. In the Bible we find
ants being praised for their industriousness (*Proverbs*, 30:25). Even in
scientific works, from those of ancient Greece down to the most current ones, the anthropocentric outlook has affected the interpretation
of animal societies. Conversely, animal societies have repeatedly been
taken as ideal patterns for reorganizing our own. If a depressing lack
of objectivity has frequently accompanied these comparisons, nonetheless the very act of looking beyond the limits of a single species has
encouraged the formulation of more general principles, and has provided means for correcting the excesses of speculation. In the present

chapter we shall essay yet another synthesis. The way shall be prepared with a brief historical digression. An awareness of our predecessors' history should help in avoiding their difficulties, and in seeing from where our present confusion has come. We can then examine the more pertinent data, largely but not exclusively drawn from the social insects, and see how much sense can be made of the subject in the light of the ideas developed in earlier chapters.

SOCIAL BIOLOGY BEFORE DARWIN

In our discussion on the history of ecology, we drew attention to the "superorganism" concept. The Platonic analogy between an animal and the world (*Timaeus*) reoccurs in only slightly modified form as the parallel between organism and state (*Republic*). The comparison is again a most obvious one, and it has become so much a part of our intellectual tradition that it is embodied in everyday language, as when we speak of "the body politic." In Aristotle's *Politics* the organismal cast of mind is even more evident: he explicitly compares the occupational classes of society to organs of the body (book IV, chapter 4). In his *History of Animals* (book V, chapters 20 to 24, and book IX, chapters 39 to 43) he provides our first scientific account of the social insects. His treatment, which reflects an indebtedness to a highly developed apiculture, is remarkable for both its empirical content and its *a priori* elements. He seems to have known that bees communicate (book 40). He inferred, upon erroneous and largely theoretical grounds, that the "kings" (queens) produce offspring that are other "kings" and workers, while the workers give rise only to drones (*Generation of Animals,* book III, chapter 9). Yet he thought the drones to be sterile, and the workers to be produced without copulation. The statement at the end of this particular discussion might lead one to infer that Aristotle reasoned in the spirit of a modern empiricist: "But the facts have not been sufficiently ascertained; and if at any future time they are ascertained, then credence must be given to the direct evidence of the senses more than to theories,—and to theories too provided that the results which they show agree with what is observed." He is not arguing here that we ought to test hypotheses, but rather commenting upon the difficulty of applying a metaphysical system when insufficient data are at hand. However, the difference should perhaps not be greatly emphasized, for basic premises go unexamined in contemporary science too.

For Aristotle, as for many who came after him, there was no question as to the issue of society being a product of Nature. It seemed altogether reasonable for men or for insects to live together in a state of mutual dependency. Nature does nothing in vain, and language would be superfluous were it not for the sake of communal living

(*Politics*, book I, chapter 2); and society itself was thought to exist for the sake of the good life (chapter 1). The organicist conception of the state, like that of nature in its entirety, was more than metaphysical speculation. It was motivated by habits of thought derived from the society that gave rise to it.

Mankind has but slowly outgrown the myths of his intellectual childhood. And his coming into maturity has been painfully marked by regression and misadventure. In the eighteenth century we find that science evoked an age of skepticism with respect to supernatural causes, yet no abandonment of the divine order. Science expelled God from matter, and its successes provided much of the justification for rationalism. Yet underlying social thought we may readily make out a recurrent conflict between materialism and providence. A confused version of nominalism was used by such British empiricists as Locke to counter the traditional idea of rule by divine right: society is an abstraction, existing only in the mind, and hence it is something artificial. Hobbes, in his *Leviathan*, contrasted the harmonious societies of ants and bees with the tumultuous ones of our own making. He found it possible, however, to imagine a reasonable commonwealth subject to "natural laws." The confusion between the juridical and the physical order is, of course, as inevitable as that between the world as it is and the world as it should be. The doctrines derived from such confusion, whether they be necessary to libertarian ideals or not, lend a persuasive ring of conviction to the literary monuments of Jefferson and Rousseau.

On the other hand, it was equally possible to reinvoke the superorganism hypothesis and consider the individual citizen an abstraction, while society would be the only reality. Much as Buffon denigrated the organisms and maintained that only species exist, so one might argue for an ontological priority of humanity over men. This confusion could then be exploited in arguments for a higher loyalty, as to class, to the state, or to mankind. During the nineteenth century it was exploited by political theorists of various persuasions, for example by Comte. Particularly in Germany, nationalistic movements that have subordinated the individual to the state have had no trouble finding philosophers and biologists who would champion their views and populate academic chairs.

Eighteenth-century thinkers were inclined to make God rather abstract, and to dispense with Him as an immediate participant in the affairs of this world. Yet they rarely abandoned the idea of Providence. The universe was still conceived of as orderly and purposeful, if perhaps corrupt. Hence organisms and society were viewed teleologically, but some kind of ordering principle had to be invoked, one that was either a part of matter, or which, if more transcendent, somehow guided happenings upon the material plane. The systems of metaphysics that were constructed on this basis in effect denied the ills of man's life by explaining them away. If evil existed, God must be accomplishing some more ultimate good by proximate means. Kant wrote an essay entitled *Idea for a Universal History with Cosmopolitan Intent*, in which he argued from the harmony of animal societies to the teleological conception of society in general. He concluded that war is good, because it draws men together. Hegel and others expressed like notions, and, of course, such philosophy has been exploited for ulterior ends. To be sure, the Enlightenment was not without its critics of the teleological interpretations of human life. Voltaire in *Candide* ridicules Liebniz for the view that "all is for the best in this best of all possible worlds." But in the Romantic period that followed men abandoned not teleology, but rationalism; indeed, they abandoned common sense. Perhaps the spirit of the new age is best expressed in a sarcastic remark by Voltaire, in which he told Rousseau that he now wanted to run around on all fours.

A certain amount of strain began to develop around the turn of the century, among empirically minded students of political economy. Adam Smith (1759) embraced psychological notions of a cooperative impulse in his *Theory of Moral Sentiments*. He later developed hypotheses based upon self-interest in *The Wealth of Nations* (A. Smith 1776). Even so, his advocacy of *laissez-faire* economics, however well justified in view of the administrative follies of his times, betrays an enduring confidence in the immanent goodness of the material world. If only restraint were removed, he argued, and natural events were allowed to run their course, then all would be well. The same fundamental proposition lived on into the nineteenth century as a belief in progress, and was embodied in remarkably diverse systems, notably the Marxist and the Social Darwinian. It endures to this day, even though it was rendered untenable in 1859.

DARWIN'S INTERPRETATION OF SOCIETY

A proper understanding of society was impossible until teleogical thinking could be abandoned. The theory of natural selection provided a mechanism for the origin of adaptations. It could replace teleology at the organismal level, and an extension to higher levels of organization could, at least so it would seem, have the same effect. Yet what one might take for an obvious solution poses a number of difficulties when one tries to reconstruct the events that have occurred in natural situations. The apparent altruism of neuter insects might seem out of line with the individualistic conception of the struggle for existence. Darwin fully realized the importance of this problem, and with his characteristic ingenuity provided a solution which, to this day, remains definitive. In the first edition of *The Origin of Species* (Darwin 1859: 237–238) he writes:

This difficulty, though appearing insuperable, is lessened, or, as I believe, disappears, when it is remembered that selection may be applied to the family, as well as to the individual, and may thus gain the desired end. Thus, a well-flavoured vegetable is cooked, and the individual is destroyed; but the horticulturist sows seeds of the same stock, and confidently expects to get nearly the same variety; breeders of cattle wish the flesh and fat to be well marbled together; the animal has been slaughtered, but the breeder goes with confidence to the same family. I have such faith in the powers of selection, that I do not doubt that a breed of cattle, always yielding oxen with extraordinarily long horns, could be slowly formed by carefully watching which individual bulls and cows, when matched, produced oxen with the longest horns; and yet no one ox could ever have propagated its kind. Thus I believe it has been with social insects: a slight modification of structure, or instinct, correlated with the sterile condition of certain members of the community, has been advantageous to the community: consequently the fertile males and females of the same community flourished, and transmitted to their fertile offspring a tendency to produce sterile members having the same modification. And I believe that this process has been repeated, until that prodigious amount of difference between the fertile and sterile females of the same species has been produced, which we see in many social insects.

Recasting this in somewhat more abstract terms, we may say that the particular organism that manifests a trait need not itself reproduce for the genetical determinant of that trait to have its frequency affected. As a first approximation—no more—we find it expedient to think of natural selection in terms of differential mortality. It is often more appropriate to our habits of thought when we consider what happens to organisms, and therefore we refer to these as being "selected." But what really goes on in competition is a series of *events*, in which we may view the competitors as epiphenomenal. A genetical determinant causes a number of duplicates of itself to be produced, and these give rise to phenotypes which, by their joint action, make copies of that genetical determinant replace alternatives in the population. One might say that the difficulty arises from our having a "thing metaphysics" rather than a "process metaphysics." A "thing" has to be "selected," so we ask nonsensical questions about whether it is the gene, the organism, or the population upon which selection "acts." It does not act at all. Or we ask "What are the units of selection?" There are none. There are no "units of copulation" either. By the same token, we are inclined to ask which "thing" is benefitted—the individual, the population, or perhaps the gene. And, once more only as an approximation that does the job, we speak of properties that further the survival of one such "thing" or another as existing "for" its "benefit." This second manner of speaking occasionally creates difficulties, however. When, in the above quotation, Darwin says that "selection may be applied to the family" he means that families may have genetical determinants that are differentially reproduced. Any activity of a member which furthers the competitive interests of that family could be spoken of as "for the benefit of the group." This is not the same as viewing a family as a superorganism, nor to compound the error by treating this selection as a differential mortality of entire family groups. Yet the language is ambiguous, and the thinking may be adversely affected, especially when we do not go back to fundamentals and ask what has been going on. Darwin's writings on the evolution of society suffer from just this difficulty, and so does the literature in general. Our manner of speaking frequently deludes us.

In spite of these conceptual problems, Darwin fully grasped the ethical implications of his discovery. He drew special attention to certain habits of bees (Darwin 1859:202–203):

It may be difficult, but we ought to admire the savage instinctive hatred of the queen-bee, which urges her instantly to destroy the young queens her daughters as soon as born, or to perish herself in the combat; for undoubtedly this is for the good of the community; and maternal love or maternal hatred, though the latter fortunately is most rare, is all the same to the inexorable principle of natural selection.

In *The Descent of Man*, he once more took up the weapons of maternal hatred and sisterly loathing to attack the traditional conception of the moral order (Darwin 1874:114–115):

If, for instance, to take an extreme case, men were reared under precisely the same conditions as hive-bees, there can hardly be a doubt that our unmarried females would, like the worker-bees, think it a sacred duty to kill their brothers, and mothers would strive to kill their fertile daughters; and no one would think of interfering. Nevertheless, the bee, or any other social animal, would gain in our supposed case, as it appears to me, some feeling of right or wrong, or a conscience.

Such pronouncements, which brought humanity face to face with the ultimate horror of objective reality, evoked cries of pain and outrage from Victorian sentimentalists. What had passed for virtue was shown to be a form of expediency.

Yet another basic change in ethical theory was implicit in his hypothesis. Darwin (1874:135) was quick to point out that moral philosophers had been wont to explain social behavior in terms of psychological impulses—such as the desire for pleasure. He maintained that all animals act more or less impulsively. Sentiments might come into play, but their role would be that of a proximate mechanism, having a more fundamental significance in promoting survival and reproduction. It might be that sentiments are a necessary condition for the existence of a society but, without the selective advantage, the explanation was not sufficient. For this reason, psychology could take but a limited role in the scientific explanation of social phenomena. This point has been most inadequately grasped even down to our own time; indeed, the oversight may be regarded as the basic reason why a truly evolutionary psychology has yet to be founded. The insight that the brain is the slave of the gonads is attributed with good reason to Freud, for it was he who rammed an unpalatable truth down the throats of an indignant humanity. Yet it helps to put matters in better perspective when we realize that the Freudian revolution was implicit in the Darwinian.

THE FAILURE OF ANTHROPOLOGY TO ASSIMILATE DARWINISM

Given an indication of how the Darwinian hypothesis could be applied to the evolution of man and society, one might expect it to prove useful in the hands of anthropologists and sociologists. Nothing of the sort took place, although, to be sure, many writers considered themselves Darwin's successors (e.g., Sutherland 1898). Instead of deriving their evolutionary ideas from Darwin, social scientists relied on such marginal intellects as Comte, Spencer, Durkheim, and the psycho-Lamarckians. In particular, the organicist point of view rose to dominance. To make matters still worse, the social and behavioral sciences went overboard in embracing the reaction against natural selection that set in shortly after Darwin's death. Early anthropologists such as Tylor (1865, 1871), embracing basically the idea of a necessary progression, developed Aristotelian *scalae naturae*, so that even now, when cultural anthropologists speak of the "comparative method" they usually mean a kind of orthogenesis. Equally naive and deluding was Social Darwinism: treating all forms of competition as if they were "good" (see Hofstadter 1955).

This "Darwinism" would perhaps have better been named after Herbert Spencer, but even his advocacy of it has been exaggerated. The greatest excesses, it turns out, were really due to Spencer's followers. Spencer (1871b) himself, in a reply to Thomas Henry Huxley, claimed that he did indeed think that government should restrain individuals—as for example in preventing fraud. His position was, to be sure, far from that of Huxley, who claimed that "Satan is the prince of this world." (See also T. H. Huxley in T. H. Huxley and J. Huxley 1947.) Spencer's evolutionism, and his organicism as well, both antedate *The Origin of Species* (e.g., Spencer 1852, 1857a, 1860a). Spencer was also very much of an individualist in a sense, in spite of the fact that he frequently referred to the "social organism." For this he has been accused of inconsistency (see Carneiro 1967), but this charge, again, was largely undeserved. From an ontological point of view, Spencer was indeed an organicist; but his individualism was largely a moral doctrine: the state exists for the citizens, not the citizens for the state, because composite wholes, although real, have no feelings. He remarks (in Duncan 1908: Vol. II, p. 312) that his was a libertarian individualism, in contrast to that of Kant's authoritarian view (which was also a moral doctrine); and he points out that English *versus* German politics helps to explain the difference. Spencer's ideas also changed, especially in response to the theory of natural selection. In his *Autobiography* (Spencer 1904: Vol. I, p. 500) he says that he did not realize some of the teleological implications of his earlier writings until he had a better grasp of mechanisms. This may be seen in his *Social Statics* (Spencer 1850), wherein he reasons much as Adam Smith did: the Author of the laws of nature had ordained a moral order as well, and one with which man should not interfere. And he specifically takes issue with Comte: society to some degree does create men, "but the original factor is the character of the individuals, and the derived factor is the character of the society" (Spencer 1904: Vol. II, p. 465). What was really the matter with Spencer's system was that it tended to be excessively vague. This is especially true of the first volume of his *Synthetic Philosophy*, the *First Principles* (Spencer 1862 and later editions), which reads like much of modern systems theory (cf. Rappaport 1971). As he moved into biology (Spencer 1884) and psychology (Spencer 1886) he became more of an empiricist. And by the time he got to sociology (Spencer 1874, 1876, 1882, 1896), a subject for which he could draw upon a great file of data (much of it published in his *Descriptive Sociology*), he became quite specific, and revised his views extensively in the light of new data. But Social Darwinism, whether that of Spencer or anyone else has to presuppose a teleological world view, and is, indeed, one of the obvious implications of the old cosmology.

Later, particularly under the influence of Boas, anthropologists rejected biology altogether, and advocated a sort of cultural determinism. Indeed, the very popularity of the notion of culture owes much to the myth that, in man, cultural evolution has supplanted natural selection, a view endorsed even by some good evolutionary biologists (e.g., J. S. Huxley 1958). (Overemphasizing the gulf between man and animals is a rampant vice in our science: see, for example, G. G. Simpson 1966, 1972; Stebbins 1969; also Harlow 1958 for critique of Dobzhansky.) As if that were not bad enough, there evolved an organicist movement called "functionalism," which turns out to be the ascription of survival value, to the social group, of features that could better be thought to benefit individuals (Malanowski, Radcliffe-Brown, and others). As Evans-Pritchard (1962) points out, this movement is both an antithesis to evolutionism, and a manifestation of the crudest sort of teleology (see also G. P. Murdock 1971).

In the past thirty years or so, a growing number of anthropologists have been trying to bring evolutionary ideas back into their science. (For a sample, see Steward 1958; Tiger and Fox 1966; F. L. Dunn 1970.) But at least up until quite recently, their evolutionary theory has rarely been that of Darwin and Wallace; rather it continues the organicist fallacies. The point can be made out quite well through a close examination of L. A. White's (1959) *The Evolution of Culture,* a justly influential book, in spite of its drawbacks. The work is outstanding for the efforts to incorporate energetical relationships and adaptive significance into the explanatory scheme. It makes a very strong case for there being an economic utility to many features of culture that might be dismissed as non-adaptive or arbitrary. But one of the most fundamental premises is false: "The purpose and function of culture are to make life secure and enduring for the human species" (White 1959:8). This mistake strongly influences the theory. For example, White comes up with a sophisticated treatment of the rules of exogamy and endogamy. The old notion that it prevents inbreeding is obviously incorrect: the systems are not set up in a way that would be expected if the hypothesis were true. (But see Lindzey 1967, for arguments that inbreeding may have led to an aversion toward incest; disruption of family life could have had this effect too.) He prefers economic interpretations, which usually make a great deal of sense. Marriage alliances and the like are useful in gaining the support of others and in providing for cooperation. Yet his teleological bias causes White to view the systems as altruistic, and as furthering the interests of society, rather than as examples of enlightened self-interest, or as not so enlightened self-interest. Thus he prefers to think that exogamy favors social solidarity to the hypothesis that it prevents disruption. He even thinks that a man practices his trade for the good of society. We shall see later on where this curious notion came from. And he fails to consider the possible influence of sexual selection on such arrangements (see Fox 1972). For other examples of the same approach, the reader may wish to consult a brief volume by Sahlins and Service (1960). According to M. Mead (1958:480) the rather orthogenetic views of both White and V. Gordon Childe "stemmed from a Marxian determinism."

THE FAILURE OF SOCIOLOGY TO ASSIMILATE DARWINISM

A very significant link between the social and the biological sciences may be found in certain highly influential French sociologists. The transfer of ideas from one field to another is evident in a book entitled *Des Sociétés Animales* by Alfred Espinas (1924). Originally published in 1877, it drew heavily upon philosophical writings, especially those of Comte, Spencer, and Hegel (Bowdlerized editions omit the philosophy). And although Espinas discussed biology at great length, nonetheless what really interested him was the evolution of moral sentiments. He could, had he followed Buffon, have considered species to be "individuals" in the sense of composite wholes, thereby providing substance for his affirmative position on the superorganism issue. Yet he rejects this view, and instead explains society as the result of a common psychological attraction of like for like. In other words, a travesty of philosophical idealism was transformed into a biological mechanism.

The influence of Espinas upon sociology was channeled through Emile Durkheim, one of the founders of that discipline. Durkheim, typical of French students of society, carries the psychological theme to its own *reductio ad absurdum* in a work entitled *La Division du Travail Social* (see the 1933 edition; five were published from 1893 to 1926). He explicitly rejects what to many might seem a manifest truth, that the division of labor exists because it is an effective means of increasing productivity. Instead, he maintains that the division of labor, by increasing the feeling of mutual interdependency among men, draws society together, and reduces competition. Durkheim's reasoning is illogical, and one may wonder why. His putatively empirical arguments simply do not test his hypotheses. We are told that labor is not divided as a means of increasing happiness, because happiness seems not to increase as man progresses. But this is not to say that Durkheim's metaphysics is inconsistent or out of line with his system. His interpretation of society (see Durkheim 1938; Benoit-Smullyan 1948) derives from the organicist model, supported by two main methodological assumptions. As already suggested, he treated psychological impulses as aspects of a collective mind. More important, perhaps, is his rejection of individualism. This is another way of saying that he did not accept the fundamental premise from which the theory of natural selection derives. He explicitly rejects the Darwinian view that the features of organisms and societies have resulted from the competitive interactions of individuals, and affirms the existence of altruism. On the other hand, he does provide many perceptive observations and clever speculations, and his works are very useful (e.g., Durkheim 1968 and earlier editions) so long as one does not confuse originality with rigor, as have his followers.

On the later history of sociology we need say very little, and we shall return to this topic only to show how much harm certain biologists have done to that subject (but see Ambrose 1965; Nisbet 1966, 1969; W. L. Wallace 1969; Bock 1970; Wolf 1970). Suffice it for the moment to say that the prevailing functionalism may be dismissed as a form of naive teleology that must be abandoned if ever sociology is to rise above the level of a pseudoscience. Ginsberg (1970) has lately taken his fellow sociologists to task for their teleological habits of thought, but is not sure what should be done.

THE FAILURE OF BIOLOGY TO ASSIMILATE DARWINISM

The inability to cope with the theory of natural selection must not be dismissed as a peculiarity of the social sciences. The exchange of misconceptions went two ways, and biologists contributed a great deal. The analogies between organisms and states were pressed far beyond the limits of good sense. Physiologists such as Virchow likened cells of the body to citizens, while Haeckel (1869) even compared soldier termites to officers. Often such comparisons were ulteriorly motivated. Ideas, political or religious, as to how the world is organized, or how it should be organized, continuously intruded into ostensibly scientific works. Simplistic labelling of a man's politics as left or right wing, in this instance, would be as unnecessary as it is misleading. That Virchow and Haeckel clashed over the teaching of evolution is a "fact," whatever that means, upon which historians can agree. But to some writers Virchow is the "liberal" and Haeckel the "conservative," while others reverse the labels. Likewise, Eiseley (1972) labels determinists and reductionists "conservative," but they might be considered "liberal" in so far as their rejection of vitalism is opposed to theological orthodoxy.

A great deal of moralizing is to be found in the writings of nineteenth-century students of insect societies such as Lubbock (1874–1882). By no means, however, need we suppose that such reflection was always unreasonable or even out of place. Auguste Forel (1928) in a large work on ants discussed the implications of their social life for the prospects for peace among men. A Swiss pacifist, he opposed the anarchical doctrines of Prince Kropotkin (1904) on the grounds that organisms in a state of nature are ever at each other's throats. He explicitly rejected the view of his countryman Rousseau, that man is born good, yet society corrupts him. Against the horrors of World War I, he opposed the example of peace within such federations as Switzerland and the United States, and argued for international government. We might add that Forel was both a socialist and a democrat. Perhaps his experience as a physician and the director of a mental hospital helped him to view humanity in a more pessimistic fashion than he would have had he confined his attention to pure biology.

A bit less down to earth was William Morton Wheeler, whom we have already encountered as one of the leading Harvard crypto-vitalists. It was he who popularized the superorganism approach to insect societies, although he did not invent it (Wheeler 1911). He claimed that he saw in the efforts to found the League of Nations after World War I an innate tendency to cooperation among men and other organisms

(Wheeler 1923). History has since been kinder to Forel. Wheeler's views, although by no means intemperate or extreme, incorporated a strong resistance to natural selection and considerable sympathy for Lamarck and his followers. Nonetheless, he was quite skeptical and uncommitted on the issues of evolutionary mechanism, and he effectively opposed the attacks upon selection theory put forward by Wasmann, a Jesuit who considered insect symbioses inexplicable upon Darwinian principles. He did, however, attack Darwin's theory, citing Kropotkin's *Mutual Aid* with approval (Wheeler 1923). And although his published views were far from dogmatic, he had at least strong alliances with the other Harvard crypto-vitalists. It is rather hard to say just who had the greatest influence upon whom here. L. J. Henderson (see Parascandola 1971), already mentioned as the author of *The Fitness of the Environment*, was originally a biochemist. He later developed a very strong interest in sociology, and his avowedly teleological approach, buttressed through studies on physiological homeostasis, was transplanted into the inappropriate field of social control and regulation. He was responsible for making the works of the Italian sociologist Pareto well known to sociologists in general, and it is said that he took up Pareto's writings at Wheeler's suggestion. Henderson strongly influenced Talcott Parsons, George Homans, and E. B. Holt. Parsons (1970), a Harvard professor, has written an excellent autobiographical sketch, in which he recounts the influence during his formative years of Durkheim and the Harvard crypto-vitalists. Whitehead, we may remember, was part of the same group, and both Whitehead and Wheeler derived much of their organicist philosophy from the works of Lloyd Morgan. The influence of Whitehead is perhaps reflected in Wheeler's definition of "organism" as a "process" (Wheeler 1911).

At the University of Chicago, the same basic tradition was developed by W. C. Allee, of the "Great AEPPS." Allee was an ardent Quaker, who hoped to find evidence among organisms for a tendency to cooperation. Hedgpeth (1971a) has written a charming essay treating the impact of the holistic philosophy on one of Allee's disciples, Edward F. Ricketts, the author of a classic work on natural history, *Between Pacific Tides*. Ricketts's philosophy has been far more widely publicized than one might suspect, for it strongly influenced the novelist John Steinbeck. Together they wrote *Sea of Cortez*, which includes a philosophical chapter by Ricketts, and Steinbeck immortalized his friend as the hero of *Cannery Row*. Allee (1951) tells us of having been inspired by Espinas, and approvingly invokes the critics of Darwin who no longer deserve to be taken seriously, such as Geddes and Thomson, and Keith.

Allee sought to erect a general theory of society. He and his students (see Allee 1951) performed a series of experiments upon chickens, which many cite but few would seem to have understood. It seems not to have been noticed that their design was faulty, a point which one might attribute to their having shown what everyone wanted to believe. When a group of chickens is assembled, they fight with one another until an order of dominance and subordination is established—the well-known "pecking order." It was found that when the order thus established was continually disrupted, the members fell to fighting among themselves and consumed less food and produced fewer eggs (Guhl and Allee 1944). All well and good: when the citizens of a state are at war with one another, everybody suffers. Yet Allee (1952) maintains that the stability derived from the pecking order "may serve to help build a tolerant social unit better fitted to compete or to cooperate with other flocks at the group level than are socially unorganized groups." Perhaps, but the data do not bear out the hypothesis under consideration. It is well known that amputating the testicles may lead to obesity; but the peculiarities of eunuchs hardly imply that men have gonads to keep them from getting fat.

At Chicago the study of insect societies was zealously prosecuted by Alfred E. Emerson (1939a, 1939b, 1949, 1952, 1958, 1960), who worked on termites. Strongly influenced by Wheeler, Emerson drew many comparisons between societies and organisms. By no means hostile to Darwin, he nonetheless took group selectionist ideas beyond what some of his critics have considered temperate (G. C. Williams 1966). Yet in his mature reflections on the superorganism notion he admits the difficulties, pointing out the dangers of anthropomorphic reasoning and warning against the analogical fallacy (Emerson 1952). If, once again, he perhaps sees too much integration in nature, he at least recognizes its importance and has looked for it. Before leaving the Chicago School we should note that the holistic outlook affected a remarkably broad range of disciplines (see Redfield 1942 and other articles in the same symposium).

The social insect literature in French bears unmistakable marks of opposition to Darwinism. True to form, we find Grassé and Noirot (1951:150) accusing the Neo-Darwinists of finalism. A psycho-Lamarckian point of view seems responsible for efforts to explain society in terms of behavioral mechanisms rather than selective advantages (de Beaumont 1945; Deleurance 1952; Grassé 1952a)—mutual attraction reminiscent of the sort favored by Espinas and Spencer.

French attitudes toward societal "finalism" need to be kept clearly distinct from those of other national and linguistic groups. Germans influenced by Kant should, in theory, maintain that teleology is inadmissible science, but good metaphysics. This gives them two options. They can throw out the baby of function and keep the bathwater of true but irrelevant efficient causes; alternatively, they can leave the baby sitting in the bathwater of metaphysical teleology. If there is any link between Germanic and American attitudes toward teleology, it is probably very indirect, and has little if any immediate relationship to Kant. We may attribute the hesitancy of Schneirla (e.g., 1957) and his school (see Piel 1970; Tobach and Aronson 1970; also other papers in the same *Festschrift*) to discuss the selective advantages of behavior to their strongly physiological outlook. American physiology and physiological psychology have tended to develop largely out of the medical tradition rather than the zoological. The "right" questions have been "how" not "why" questions, and evolution has been of little importance in their paradigms. And, of course, the behaviorist tradition, which traditionally played down "instinct," has been very influential in American academic psychology.

This predilection for explaining behavior in terms of "proximate" factors to the exclusion of "ultimate" ones has been justly criticized by a number of evolutionary biologists (e.g., E. O. Wilson 1971; but see Topoff 1972). But in view of the evident failure by so many to appreciate the crucial differences between "how" and "why" questions, it seems worth a little effort to explicate the philosophical issues in a little more detail. Consider an old, if venerable, joke that involves a "how" question: "How do porcupines copulate?" We may observe from the outset that several answers could be given to both this and a similar "why" question: "Why do porcupines copulate?" But, contrary to what is sometimes claimed, the two are definitely not equivalent. For the "how" question, an appropriate answer is the usual punch line: "Carefully." But obviously, when we want to know why, this response simply will not do. Some answers to "why" questions seem much less incongruous, but nonetheless are not really what we are after. We might give a psychological answer: "For fun." Another alternative would be an Aristotelian final cause: "For the sake of the species." Or it could be recast in slightly less teleological terms: "Because otherwise there would be no more porcupines." Answers of this sort contain a considerable amount of truth, but this does not prevent them from being irrelevant. If there are to be porcupines, it is necessary that they shall copulate; and in turn, it is at least logically possible that their enjoyment of the act is a *sine qua non* of their doing so. If that is all the questioner demands, all well and good, but he may have something else in mind. He may want to know what makes the difference— why porcupines copulate but sea urchins do not. In such cases, we would have to respond by pointing out that air is a poor medium for external fertilization; but that is far from the whole story. Whatever answer we may choose to give, just any necessary condition simply will not do.

SOME RECENT SPECULATION

Of late, the superorganism concept of insect societies has been criticized by E. O. Wilson (1968a; 1971) of Harvard. Although in his earlier works (Wilson 1953; cf. 1963) he spoke favorably of it, he now rejects the idea, not because it is wrong, but because it lacks heuristical utility. Thus his grounds for attacking it are epistemological rather than ontological, and his position is rather akin to pragmatism. Such arguments, notoriously easy to invoke and difficult to answer, nonetheless often fail to withstand a more searching analysis. Granted, "It is not necessary to invoke the concept in order to commence work on animal societies" (E. O. Wilson 1968a:31). Neither, we might reply, need one know about chromosomes to do good work in genetics (e.g., Mendel 1865). But does it help? We are told that "The concept offers no techniques, measurements, or even definitions by which the intricate phenomena in genetics, behavior and physiology can be measured." Yet Emerson (1952) tells us how he, at least, searched for, and found, numerous properties of insect societies previously known only from individual organisms; so the idea cannot be dismissed as altogether destitute of heuristical utility. Wilson objects to Wheeler's having enlarged the concept of "trophallaxis" beyond its original meaning of an exchange of nutriment, to a general exchange of substances and messages between the members of the community. He rightly points out that such phenomena have a homeostatic function; yet he could have come to exactly the opposite conclusion. Integration and its attendant mechanisms are necessary conditions for effective functioning of both organisms and societies; and insect societies are integrated to a degree far greater than, say, ecological communities. (For social homeostasis see: Gregg 1942; Grassé 1952b; Noirot 1952; Bier 1954; M. V. Brian 1957, 1958, 1968; Flanders 1958; Shuel and Dixon 1960; Lüscher 1961.) Indeed, Wilson (1971) himself uses such notions as homeostasis, and endorses cybernetics; thus he continues using the concepts, but spurns his heritage by rejecting the metaphor.

Wilson (1968a, 1971) rightly points out that the holistic reasoning that underlay much early twentieth-century research on social insects is now inappropriate. Yet does this justify, as Wilson maintains, a "reductionist spirit" in the future? It seems odd that Wilson (1971:64, 221) praises Schneirla and Bethe for their reductionist approaches, yet does not see that this very reductionism is an integral part of a philosophy that leads them to reject ultimate factors in favor of proximate ones, for which he takes them to task. "Reductionism" is a rather ambiguous term (see below) but it is reasonably clear what Wilson means by it. He maintains (Wilson 1971:319) that "in time all the piecemeal analyses will permit the reconstruction of the full system in vitro. In this case an in vitro reconstruction would mean the full explanation of social behavior by means of integrative mechanisms experimentally demonstrated and the proof of that explanation by the artificial induction of the complete repertory of social responses on the part of the isolated members of insect colonies." One can, however, learn to put a machine together without understanding how it works; and even that insight is irrelevant for the economic question of why the machine is manufactured, or the historical question of what led to its invention.

As Lin and Michener (1972) have pointed out, Wilson may have overestimated the utility of the kin-selection notions developed by W. D. Hamilton (1964). From an ontological point of view, there is little advantage to some of Hamilton's elaborations over Darwin's concept of breeding from the same stock. All that has been added has been ways of making stocks more nearly alike, and hence of increasing the probability that such selection will occur. And, of course, the possibility that ploidy will affect social evolution adds a great deal. From an epistemological point of view, there does seem to be a definite advantage; but this, too, has been exaggerated. Precise mathematical reasoning ought to allow one to predict what will happen in societal evolution under various natural conditions. However, if the calculations are based upon false premises, they may prove inaccurate, and without experiential study, we have no way of knowing how well theory reflects what goes on in the real world. Recent developments provide instructive examples of what may happen to *a priori* reasoning in such contexts, underscoring our need for empiricism.

Hamilton (1964) originally maintained that one can calculate the selective value to altruism on the basis of the number of genes shared by relatives, giving a numerical index of relationship. Thus when a haploid male has a set of diploid daughters, all are identical in at least half of their chromosomes, since they inherit the same paternal complement. This identity, however, would not exist in half sisters in a polyandrous family. They also share maternal genes, and on the average half of these will be identical in any two daughters. By adding the maternal set (giving ½ of ½ of the genes = ¼) we get ¾ identity. By the same token, drones get all their (haploid) genome from their (diploid) mother, having on the average ½ of their genome identical with that of their brothers, and, supposedly, a ¼ relationship to their sisters.

Actually, as Crozier (1970) pointed out, one can come up with a variety of indices for some of the relationships, depending, so to speak, on arbitrary decisions as to how one slices the metaphysical pie. Thus, Hamilton got a mother-son relationship of ½ by dividing her genome into two equal portions and inserting it into two of her progeny. From this perspective, the sons are half individuals, with genomes identical to half of hers. But one could equally well treat the sons as entire organisms, and view each haploid complement as the fundamental unit. If so, the sons would still have a ½ relationship with each other, but mother and son would stand to each other as 1/1, not ½. Each of the son's genes would be represented by at least one entity in the maternal bean bag. From the son's point of view, he is his mother's identical twin, except that every other gene in his body is not there—to state it in the form of an Irish bull. Again, a pancake made with half as much batter may be half as thick, or half as heavy, but it cannot have only one side.

Thus, Hamilton's theory contains a very interesting and instructive paradox. We often forget that a mathematical model often contains simplifying assumptions and other features that may or may not give misleading results (infinite population size, for example). For some purposes it may be appropriate to treat the brother-sister relationship as ¼; a sister will in fact tend to have ¼ of her genome in common with that of any brother selected at random. Hence the fitness of a gene for a sister "helping her brothers" should indeed be less than that for helping her sisters. Is the converse true? According to Hamilton's calculations, it should be. But wait. Consider a male with a gene "for" helping his sisters. That gene is present in half of his mother's chromosome complement. Half of his sisters will inherit that gene. Hence a gene for a brother helping his sister is as advantageous as one for a mother helping her daughter.

Crozier (1970) developed a modified version of Hamilton's model upon like considerations, and came up with somewhat different values. He concluded that the basic argument remained unshaken, although some changes were in order. This may be true, but if some of the calculations were off by a factor of two, is it not possible that others are off as well?

According to Darwin's hypothesis of "breeding from the same stock" the close relationship between members of a society, as predicted from the foregoing considerations, ought to accentuate the tendency to cooperate. But similar conditions would exist without haplodiploid systems. That highly evolved societies exist among termites, which are diploid, demonstrates conclusively that the necessary and sufficient conditions for societies to originate have in fact been met, and that haplodiploidy is not one of them. This fact raises the possibility that something else was responsible for what happened to the Hymenoptera as well. The alternative of straightforward competition between families could also be hypothesized. In addition, what has appeared to be altruism could really be a form of intraspecific symbiosis, analogous to either mutualism or parasitism (see Lin and Michener 1972). We do not invoke kin selection to explain symbiosis between members of different species, so we should not feel compelled to invoke it for what could be symbiosis within species. Now, given a series of alternative hypotheses, it should be possible to derive predictions from them, giving us an ability to show that certain facts are explicable in terms of one hypothesis but not in terms of another. Yet some of the arguments for Hamilton's theory have been defective from this point of view. The evidence can easily be reconciled with the contrary hypotheses. Furthermore, and this is a point rightly stressed by Lin and Michener (1972), data that actually raise difficulties with the kin-selection hypothesis have been brushed aside. Of course, such criticisms can be carried too far, but let us examine the facts in somewhat greater detail. They should convince anyone that the gravest skepticism is in order.

Frequent sociality in the Hymenoptera has been cited as an argument for the Hamiltonian model. All ants are social, and the same phenomenon has evolved repeatedly in a number of lineages of bees and wasps. Yet the fact remains that termites have evolved a high level of sociality, and these animals are diploid. To the lack of sociality among many diploid organisms, we may oppose its absence in many groups with male haploidy, including a vast swarm of hymenopteran species. Wilson's (1968a:36) assertion that he would doubt Hamilton's hypothesis "if only one beetle or cricket could be found caring for its brother or sister" reflects a curious disregard for the adequacy of one disconfirming example, namely termites, in refuting a universal proposition. We do not accept the phlogiston theory because someone has not done a particular experiment that would refute it; we already have decisive evidence. Since termites exist, the hypothesis is by no means necessary to explain the facts, and its alternatives are perhaps sufficient. In this instance, we would violate the canons of logic were we to turn the argument into a statistical one and maintain that the higher frequency of societies among the Hymenoptera supports the hypothesis. The argument would be valid were it the case that societies have originated *at random* among haplodiploid organisms. They definitely have not: rather they have come into existence in parallel from a few closely allied stocks. Something else must be necessary, and it does not matter that for the moment the imagination proves insufficient to discover what this may be.

Another prediction from Hamilton's model was that workers should produce only male offspring, if any at all. This notion derived from the idea that the neuter females are more closely related to their sons than to their brothers ($\frac{1}{2}$ *vs.* $\frac{1}{4}$); but they should equally well produce daughters ($\frac{1}{2}$, or at any rate the same as a queen). Yet the means of sex determination, as Hamilton points out, renders it far easier to produce males than females anyway. In prosperous societies, egg laying by workers tends to be suppressed, their eggs often being eaten rather than giving rise to more insects. When a queen dies, a worker may herself change into a queen, and produce fertile eggs. Sometimes all the workers can do is produce drones, but some may give rise to female offspring by parthenogenesis. A failure of workers to produce queens could result from specialization of the workers to the degree that they are incapable of receiving sperm, and hence of engendering anything but males. Again, the greater sister-sister ($\frac{3}{4}$) than aunt-niece ($\frac{3}{8}$) relationship explains, one would think, why a battle for succession may occur when the queen dies. Yet some of the workers remain subordinate even when they are caring for what appear to be nonrelatives, or only remote ones. West (1967) attempted to salvage this hypothesis by saying that these females have very low reproductive potential when alone, and hence are better off helping even a distant relative. Yet Lin and Michener (1972) argue that the degree of inbreeding in some such forms is insufficient for West's explanation to apply. We may note, however, that an ability to lay some eggs, even ones that only rarely produce offspring, may be retained by the subordinates; and hazarding battle or trying to found another nest might give less return.

Another aspect of social life that has seemed to demand an explanation in Hamiltonian terms is the hostility of worker bees toward drones at the end of the season. The drones are expelled from the hive or even killed, carrying the women's liberation movement to its logical conclusion. However true it may be that the relationship between brother and sister is $\frac{1}{4}$, does this suppression of males really need to be explained in terms of genetics? In Hymenoptera generally, it is very common for the fertilized females alone to live through the winter. Where males can be produced facultatively, they can be dispensed with; and this may indeed be desirable from the point of view of the females (Darwin in G. J. Romanes 1884). During the winter the males would consume food, and, being idle, they would continue to waste energy during the spring. Similar adaptations have been recorded from primitive man: when the going got rough, they ate their wives or children if their neighbors were not available.

A final argument for the Hamiltonian hypothesis derives from the idleness of the drones. A male hymenopteran is thought to derive little advantage from "helping his sister" since his relationship to her is "$\frac{1}{4}$." It follows logically that drones should never care for the young. One might recalculate the relationship ($\frac{1}{2}$ not $\frac{1}{4}$, as we suggested) or ask why they do not care for their brothers ($\frac{1}{2}$), but for reasons we need not go into, things tend to get out of hand when one does. The haploid males are said to be totally idle, yet even in the older literature, we find a number of distressing exceptions. E. O. Wilson (1971:330) said that he knew no exceptions to the rule of total male idleness in social Hymenoptera save the "willingness of male ants to regurgitate food to nestmates." In meliponine bees, the males secrete a certain amount of wax (von Buttel-Reepen 1903). Male bumblebees aid somewhat in incubating the brood (Free and Butler 1959:35). Male wasps of the genus *Polistes* have been seen to give food to the young (Weyrauch 1928; Steiner 1932). Such reports may be erroneous, or we may be dealing with paternal, rather than fraternal, care, such as that known in a few nonsocial Hymenoptera (Lameere 1920) and a host of diploid male insects as well. E. O. Wilson (1971) reasoned that solitary and social Hymenoptera alike should not engage in paternal care, however, and the reason is clear enough: half of his mate's offspring are not his own. But Wilson admitted that some do. W. G. Eberhard (1972) observed male wasps chasing away ants, and a substantial number of additional examples are enumerated by Lin and Michener (1972) who conclude that at least some male behavior is workerlike, not just paternal.

One might argue, on the contrary, that the drones behave like all males, regardless of ploidy, under the influence of comparable selection pressures. They maximize their own reproductive output, subject, of course, to exploitation by their mothers and other influences. The males are specialized for reproductive competition with other males. It is to the mother's advantage that her sons be idle, for they should husband their resources for that very competition. In a colony with numerous females caring for the young, a male's contribution to the next generation will be largest if he succeeds in monopolizing the crosses; yet any resources used to support the brood will have only a small advantage, for he, at best, could raise the productivity of the group no more than a single female could and probably less, for he would have to combine male and female characters. It is noteworthy that young queens tend to be idle too, but here the Hamiltonian arithmetic obviously becomes trivial. Paternal care is only advantageous where the father can distinguish his own offspring from those of other males, for otherwise much effort will be wasted on nonrelatives. The mating system of Hymenoptera is not conducive to the males being able to discover the paternity of the offspring. Termites, on the other hand, are essentially monogamous, and characteristically a male and a female labor together to found each colony. The efforts of both are rewarded by success in raising offspring. As we have mentioned, paternal care is quite widespread among insects. A form of sexual competition, comparable to that which determines the sex ratio and the degree of anisogamy, thus provides the real reason for the reproductive strategies of male social Hymenoptera. Shocking as it may seem to our reductionist colleagues, the haploid individuals behave more like gametes than organisms.

The close relationship among sisters does not exist if their mother mates with a number of individuals. Wilson (1971) admitted the difficulty, and cited examples of multiple insemination, but suggested that maybe the males tend to be consanguineous. Lin and Michener (1972) argued that the mating habits of hymenopterans strongly militate against such views. Indeed, if anything, the system, with females taking nuptial flights, seems to be a special adaptation that maximizes outbreeding. All this is debatable, to be sure, but the point stands that the hypothesis in question should not be invoked wherever it simply makes sense.

Perhaps the most serious difficulty with the Hamiltonian thesis has to do with the sex ratio (Lin and Michener 1972). Indeed the very foundations of sex-ratio theory seem threatened. Hartl and Brown (1970) conclude that the sex ratio in haplodiploid species ought to be the usual one-to-one. Hamilton (1964, 1971) has reasoned that it should never be biased toward the males. For reasons already explained, it does seem to be biased mainly toward the females, and this is what was expected. But in some species the males are more abundant than the females, and, where one omits the nonreproductive castes, it turns out that among social Hymenoptera a strong male bias is the rule (Lin and Michener 1972). Exactly how energy is invested here, we do not know. Nor do we know how competition affects the returns on different strategies in this kind of system. But we can see that there is a contradiction, and therefore a need for better theory.

Lately Hamilton (1972) has reconsidered the whole question, and has modified his views in certain important respects. In the latest version, the haplodiploid system is considered relatively less important than certain other causes of consanguinity. Thus, the rather sedentary habits of wood-eating roaches are thought to have furthered inbreeding, which would in turn have made it more advantageous to the ancestors of termites to have aided their kin. By like considerations he explains much that had previously seemed anomalous. And yet he not only acknowledges, but underscores, the point that additional principles are necessary. Consanguinity alone will not explain why some inbred organisms have not evolved societies, nor will it explain the existence of associations the members of which belong to different species. It behoves us to consider what such additional principles might be. They may turn out to explain the rules, as well as the exceptions.

THE SELECTIVE ADVANTAGES AND DISADVANTAGES OF SOCIAL LIFE

Let us therefore consider what light may be cast upon the subject by treating the origin of society from an economic point of view (see also Wheeler 1923; Michener 1964). Such an analysis must of necessity be carried out at several levels. The activities of industry depend upon the output of individual workmen, the structure of firms, the interactions between them, and the nature of the economy. So it is with the world of bees and termites. A worker bee, bringing food to the hive, does so in the context of a coordinated enterprise. She may change tasks if some other kind of activity becomes more important for the welfare of the community as a whole. The community itself competes with similar units, each of which might be compared to a firm. At the same time, all such communities will prosper or not according to the state of the economy.

In any such discussion, the idea of a *division of labor* must assume a significant place. Fundamental to classical economics, this idea has repeatedly been treated by students of insect societies (Legewie 1925; Rösch 1925; Ökland 1930; Steiner 1934; A. D. Brian 1952; Lindauer 1952, 1953; Michener 1961; Richards 1961; Free 1965, 1966; Spradbery 1965; Wallis 1965; Topoff 1971; E. O. Wilson 1971). Yet we lack a sophisticated discussion of how and why labor gets divided in the colony. This neglect is all the more intriguing, in view of the fact that an analogous situation long prevailed in economics. From Adam Smith's time until the middle of the present century, it was simply assumed that an ever-increasing specialization of enterprise would raise industrial output. It was only when the work of efficiency experts and systems analysts revealed that something more is necessary, namely an effective coordination of the separate activities of the workers, that the traditional view was effectively challenged. The reason is not far to seek. Early economists were not particularly interested in what we may refer to as the "functional anatomy" of production. They concerned themselves with the market, not the factory. We find Adam Smith (1776) giving three reasons why division of labor is advantageous. *First*, it increases dexterity, because a practiced hand accomplishes the task far better. *Second*, much time gets wasted when one has to pass from one task to another. *Third*, concentration of attention upon a single task makes it easier to invent better machines.

Now these aspects of production are obviously important, but they hardly qualify as the entire story. Most significantly, Smith fails to distinguish between two sorts of division of labor, and even modern economists seem not to have done much better. On the one hand, we may note *a competitive division of labor*, which exists between independent artisans or between firms. Here, whether one or several tasks shall be performed by a single unit is determined solely by the output of the firm or the artisan, regulated by the price the goods will fetch. Each worker or firm strives as it were to maximize profit, and specializes or not accordingly. On the other hand, we have a *cooperative division of labor*, such as exists between the members of a single firm. Here, the object is not to maximize the output of each worker, but rather that of the system as a collective unit. An individual artisan within a firm may be very unproductive, in the sense that he remains idle much of the time, but nonetheless his work may be effective for increasing profit, if he provides a crucial service at just the right time. Shifting to the realm of biology, we would say that the subdivision of ecological niches (as into herbivores, carnivores etc.) may be looked upon as a competitive division of labor; that among organ systems or among the members of an insect society exemplifies the cooperative sort. Whether we consider firms, hives, or organisms, the same principles apply. The organic relations between the components determine the functioning of the whole.

Smith's explanations for the division of labor may easily be applied to insect societies. In "higher" bees, different individuals do indeed engage in different tasks. It is clear that a certain amount of practice is advantageous in some of the activities in which bees engage. Before heading out to forage, "orientation flights" are necessary, but once these are finished the same bee is qualified to continue the task without wasting time and energy in preparation. And one bee tends to specialize in visiting flowers, while another brings water to the hive. Hence we can easily see the validity of Smith's first explanation: increased dexterity through practice. The idea that shifting jobs is wasteful applies too. The bees that care for the young have special glands that secrete materials useful for feeding them (Rösch 1927). Those bees that work in the fields must undergo a sort of metamorphosis in shifting to or from domestic labor. Smith's idea of inventiveness through concentrated attention does not work very well for bees, but perhaps one could imagine some evolutionary analogue that is not too far-fetched.

Yet contrary to what one might have expected, individual bees do change from one task to another. The younger workers care for the brood, while the older ones labor outside the hive (von Frisch 1952). Thus we need to distinguish a *temporal* from an *individual* division of labor. Why the workers change tasks has no obvious connection with any of Adam Smith's explanations, and indeed would seem to contradict one of them. Yet the reason is clear enough. Foraging is a risky endeavor, since the worker is subject to the attacks of predators and the effects of bad weather. In order to get the most use out of each individual, it is best to employ the younger members of the colony at "safe" tasks, and only expose them to danger when they are soon to die anyway. This explanation, which is the same as that presented in chapter 6 for the greater frequency of sexual combat in older animals, implies that senescence has made it advantageous for the individuals most likely to die to take risks. E. O. Wilson (1971) on the contrary suggests that the risks have brought about the senescence. He follows some widely accepted theories about aging which may not be true (see

G. C. Williams 1957; W. D. Hamilton 1966). Pearl and Miner (1935: 67) present a survivorship curve for automobiles, suggesting an analogy that may help to explain the alternatives. One might say that automobiles that last a very long time are prohibitively expensive to manufacture, and that beyond a certain point it is cheaper to let an automobile wear out than to repair it. Further, we take better care of a new car than an old one because we find it advantageous to maintain our capital on the one hand, and to reinvest it on the other. The opposing view would be that since automobiles become less valuable as time passes, they might as well be built in a fashion that results in their deterioration. Whatever may be the "true" explanation, it should be reasonably evident that cause and effect could easily be confused here.

The queen is particularly well defended, and enjoys advantages impossible for solitary forms. In the earlier stages in the evolution of insect societies, the queen gathers food, but this task is taken over by the workers once the first brood is hatched. More "highly evolved" ants feed the first workers on energy stored in the mother's body. Time and again, the foundation of colonies has become the work, not of one individual, but of several. The simplest reason for this is that it protects the queen. On a more subtle plane, it would appear that a larger society operates more effectively than a single individual. The arrangement is advantageous, not simply in providing highly productive occupational specialists, but in economizing upon the capital that the firm invests in each worker.

In many situations, labor is better combined than divided. Such is the case when two occupations may be carried out simultaneously without seriously interfering with each other. A familiar example is the combined role of baby-sitter and student. In some cases two functions may even support each other. This is particularly true where one occupation is better done when its practitioners are aware of what goes on in another. The combination of research and teaching may at least be rationalized on this basis, and it is widely considered good practice for managers to have had varied experience within their firms. In the hive bee, the young workers in the nest engage in a variety of tasks, constructing cells, feeding the larvae, etc. (Lindauer 1952, 1953). They do not specialize upon a single activity exclusively, but wander about "patrolling," and do whatever task happens to present itself. The result is that the workers' efforts are effectively directed to the tasks that are most in need of being done at the time. Perhaps a more effective system of management would be preferable from our anthropocentric point of view, but this one works remarkably well, and it may be the best of which bees are capable. We should add that it coexists with a very elaborate system of communication and control, the energetical advantages of which we shall treat in due course. Genuine direction is not vested in the analogue of a manager, but the system regulates itself automatically, as does an organism.

Another reason for combining labor is that a given occupation may be highly remunerative, yet the opportunity for engaging in it appears only part of the time. In this case the usual stratagem is to combine it individually, yet divide it temporally. In seasonal occupations, such as fishing or the tourist trades, it is common practice to have some secondary means of earning a living. The professional soldier may be very effective on the outskirts of a great empire constantly at war; but small and peaceful republics tend to make at least some use of conscripts drawn from other trades.

Labor is combined in quite a distinct sense, and for quite different reasons from the foregoing ones, when groups of individuals unite their efforts at a single task. Here the advantage comes through joining either mere numbers or else a diversity of talents so as to do the task more effectively. This would seem, indeed, to have been one of the driving forces behind social evolution wherever it has occurred. That this is so may be seen by examining the conditions under which societies now exist, and by tracing the stages of their evolution. A simple example is the latitudinal gradient phenomena treated by Richards and Richards (1951; additional references: von Ihering 1896; Evans 1958; Eberhard 1969) for certain bumblebees and wasps. In higher latitudes they tend to be solitary, while toward the equator the size of the colonies increases. Likewise, there would seem to be a general trend toward colonies being founded by larger numbers of individuals acting together toward the equator. Once again, we seem to be observing a manifestation of the general pattern in competitive relationships set up by the worldwide conditions of resource availability.

Early stages in the origin of society can be seen in bees and wasps, especially those that raise the young in burrows (Evans 1958, 1966; Michener 1969). Some species nest together in groups, perhaps originally for no other reason than their use of the same kind of substrate. Several individuals may share a single complex of burrows, with a common opening. An obvious advantage to this setup is that much energy is saved by digging fewer entrances per capita. It is noteworthy that some wood-boring beetles do much the same: sometimes males will dig the openings and attract groups of females to them (von Lengerken 1939). Students of the social insects have reached a considerable degree of consensus (if anything they overemphasize it) that protecting the nest from predators and parasites is a major advantage of society (E. O. Wilson 1971; Lin and Michener 1972). The advantage of numbers in this connection should be obvious. Of great significance as well is the effect of combining and dividing labor. A solitary wasp must necessarily leave her nest unguarded while obtaining food, but a group with a common nursery has the advantage of being able to take turns.

It seems that the founding of the insect society is a particularly difficult task, one that determines numerous features of the adaptive strategy (Harms 1927; Weyer 1930; Goetsch and Käthner 1937; Lüscher 1951; Harris 1958). The same is true of any enterprise. Setting up a factory requires a certain amount of stock, not only to purchase the equipment, but to sustain activity while the factory is as yet so undeveloped that it cannot turn a profit. Furthermore, the efficacy of an enterprise may be proportional to the size of the firm, either directly or inversely. It takes a great corporation to manufacture automobiles, and these days nobody with any sense hopes to get into that business with the capital barely sufficient to found a general store. Just as there are small businesses founded and run by one man, there are small insect societies; but these are in most imperfect competition with the larger ones. Nonetheless, when a great society, populated by many thousands of insects, reproduces by sending forth single individuals to establish new ones, each of these must progress through stages analogous to building automobiles in a small shop, and doing so in the face of competition like that which would come from General Motors and Ford.

From such considerations it follows that under certain conditions it should be more profitable to keep established societies going for a maximal period of time, and to found new colonies as near to the optimal size as possible. Let us examine these two propositions in turn.

In quite a number of social insects, new colonies are founded annually, so that the stock put into building nests is discounted through the year; virtually no capital is put into maintenance during the unproductive winter season, and the whole firm is liquidated in the fall. But this is by no means universal, especially where the resources do not fluctuate very much. At a very primitive level, certain allodapine bees (Skaife 1953; Sakagami 1960) nest in hollowed-out stalks of plants, and several generations successively occupy the same nest. The young receive some aid from their older sisters, and one might conjecture that this is a case of altruism. Perhaps; but a personal advantage could equally well enter in, namely, that of supplying nurses for the next generation, i.e., their own offspring. Be this as it may, each colony does perpetuate itself for some time.

An increased longevity of the mother has been considered an important precondition for the origin of sociality, for it brings her into position to aid, and to control, her offspring (Plateaux-Quénu 1967). There is no obvious reason for questioning this hypothesis, yet it may not be the whole story. Although workers live only a few months or weeks, queens are known to survive for a number of years. The example of seventeen-year locusts shows that such longevity has no necessary connection with sociality, but it may seem peculiar that individuals of the same species should differ this much. Longer life is generally thought to be purchased at some cost in fecundity, physiological vigor, or other concomitant disadvantage, since maintenance is expensive and perhaps difficult. Very likely the dissolution of the colony when the queen dies wastes resources, and prolonging her life so as to maintain it yields a considerable return. The loss of a queen in many species of Hymenoptera leads to the production of males by workers, prolonging the period of reproduction but not sparing the colony. In many social insects, a worker can become a functional queen, while others can replace the reproductives from immature individuals as yet uncommitted to a sterile caste (Noirot 1956; Plateaux-Quénu 1961). The ability to generate new reproductives by parthenogenesis may be secondarily acquired in some social insects, perhaps as an adaptation extending the life-span of the society. Many large societies of army ants, bees, and termites have achieved a sort of "immortality" like that of protozoans. They grow to a certain size, and split into two or more subunits, each with its own reproductive individuals. This "sociotomy" (Grassé and Noirot 1951) has the obvious dual advantages of sustaining a going concern and founding new ones at optimal size. It is not yet clear which of the two advantages is the more important one.

Whether founding or maintenance be the crucial advantage, it is clear that a sort of partnership between two or more individuals may be more productive than solitary labor. In order for such a partnership to work, and in order for it to be selected for, either the partners must each gain something, or else one of them must be forced into the union. For societies of termites, honey bees, and ants, there is little difficulty, for these are families. The workers are the offspring of the reproductives, born to involuntary servitude under conditions where their reproduction is prevented by special adaptations. The society can here evolve as a single reproductive unit, and the classical interpretation (Wheeler 1923) was to view every society as a family. Yet in certain insects, notably the polistine wasps, societies are founded by groups of fertile females (Pardi 1948, 1952; Eberhard 1969; see also Michener 1958; M. V. Brian 1965). One of these becomes the queen and the others function as workers. The queen is the most aggressive, suppressing the others by force and monopolizing, more or less, the group's reproductive output. Why, one might ask, should a wasp enter as if by choice into such an unequal partnership? It has been suggested that the founders are sisters, who gain an indirect selective advantage by furthering the success of replicas of their own genes, in accordance with the model of W. D. Hamilton (1964). So it may be, but selection of the more orthodox, individualistic sort seems adequate to cover the facts.

As Hamilton (1964) points out, an individual would be better off cooperating than not, so long as it increased her overall probability of reproducing successfully. Thus, if two females together raised more offspring than two separately, the gain would be more than balanced by the loss. And since wasps have no way of knowing who will be queen, they simply fight it out. One might wonder why the vanquished do not merely leave, and go elsewhere. At least part of the reason is that in accepting a subordinate position, they do not lose everything. Their suppression is only a matter of degree. A dominance hierarchy is established and, should the queen perish, her immediate subordinate will take over. And reproduction by workers is not necessarily prevented altogether. A broadly comparable situation exists in beavers (Bradt 1938; Tevis 1950; Hediger 1970). The yearlings remain with their parents and younger siblings but once mature, they are driven off. They are said to participate in supporting the family as a whole, a phenomenon that we may interpret as owing to their having a vested interest in a firm that they may inherit if their parents die. Skutch (1935) has noted some comparable activities by young birds that may be explicable on the same basis. "Helpers" aid a pair of adults in raising young. Skutch himself invoked group selection here—the helpers would be altruists. Lack (1968:73–74) suggests that they are helping their siblings; but he adds that they are largely males, and that they often take over when the original father dies. We may propose, therefore, that helping could be a means of forming an alliance with a widow. J. L. Brown (1970) says that in the Mexican jay, *Aphelocoma ultramarina*, the groups cooperate in ways additional to helping raise the young; therefore the "helpers" may be getting something in return. Also we have seen how young males in quite a range of species do not compete with their elders in sexual combat, but will nonetheless reproduce if given a chance. The notion that subordinate insects derive a greater advantage from helping their sisters than from helping their nieces has already been mentioned. Removal of the queen brings about much the same sort of struggle between workers in bees as that already mentioned for wasps (Sakagami 1954). Maybe in some of these forms the kin-selection hypothesis will work, but we should not overlook the alternatives. We do seem to be observing, in the subordinate individuals, the kind of cooperation that goes on among the henchmen of a despot, all of whom hope to succeed to the throne.

Our hypothesis that insect societies have evolved because of their economic advantages may be tested in another, rather more indirect, manner. This is to show that once societies are established, their degree of elaboration (particularly their size) has its upper limits set by the nature of the enterprise and the condition of the economy. Admittedly it is a somewhat risky induction to extrapolate backward from social to nonsocial animals, and to say that what controls the size of societies is wholly adequate to account for their existence. So perhaps one should view the following discussion as not decisive with respect to origins, but rather as providing a general hypothesis that makes this advantage seem reasonable.

We have mentioned how, on a geographical basis, the size of some insect societies tends to diminish in a gradient toward the poles. The idea that climatic instability increases the availability of resources fits in well. One might think that increased competition in the tropics should be reflected in the evolution of relatively larger and more complicated societies. And it does appear that the largest ones—those of certain termites—occur in warmer climates. Yet merely invoking competition, without finding out what actually controls the size of the colony, tells us little.

The size of an insect society is evidently not limited by the fecundity of the reproductive castes, except in the short term. Of course, in a non-evolutionary sense, such limitations may be very significant. One can easily imagine how an increase in the food available to a colony might lead to a queen's reproductive ability being taxed so much that she could not take advantage of the opportunity. Yet in the long run, the fecundity of social insects seems capable of expansion adequate to transcend the limits encountered thus far. In very large societies, the queen's abdomen is greatly enlarged, and she becomes virtually an egg-laying machine. Records exist of one termite queen laying 86,400 eggs per day (one per second), and the process may continue uninterrupted for years (review in Nutting 1969). When the physiological limitations of a single insect are reached, the colony can still increase its fecundity by adding more fertile individuals, and this does, in fact, occur. Hence we come to the same basic inference with respect to the limitation of reproductive output as did Lack for birds. It is the same basic problem in only slightly different form, and the same principles apply. Food provides the limiting resource and control is affected through the extent of the market, which, in turn, relates to the nature of the enterprise. Of course, we must always consider how predation fits in, but this seems to set more the lower limits than the upper ones; and other resources must have some impact too.

An important relationship between the kind of food and the size of tropical ant colonies has been unearthed by E. O. Wilson (1959). He found that more nearly omnivorous species have larger colonies than those that specialize on a few kinds of prey. This makes a great deal of sense, for if an ant can eat only one kind of food, a large group of them would have to forage at an excessive distance from the nest, and expend a great deal of energy just bringing it home. Hence diminishing returns set the limit. The same principle in somewhat more elaborate form, recalling the situation in nidifugous birds, underlies the feeding mechanisms of army ants (see Schneirla and Brown 1952; Schneirla 1957; E. O. Wilson 1958). These creatures are omnivorous predators, which may live in very large groups. At certain times they wander about, and their voracity at this time is quite remarkable. The motile stage occurs—this is the critical point—when the larvae are developing, and when, therefore, the need for food is greatest. Movement increases the extent of the market. Hence we may observe the application of a principle that has proven useful for context after context in the present work: motion affects the effective density of populations.

Thus one necessary condition for the very existence of any society whatsoever is the presence of food at sufficient density that its members can derive economic advantages from communal life. In general, however, it would seem that a "hunting and gathering" type of economy will not support a very large society. A high standing crop of energy must be available to the group. Furthermore, this energy must not be ephemeral and unreliable. When the supply of resources collapses, the society of necessity is dissolved, and individual enterprise may be in a better position to rebuild when conditions happen to improve. For this reason accelerated development of human societies has occurred under conditions where the food supply was both copious and reliable. (For the origin of the state in man, see Carneiro 1970, who gives some historical background.) In general, this has meant irrigated grains, such as wheat, rice, or corn. However, the exception of migratory fishes from the highly productive waters off the western coasts of both North and South America shows that, although the principle broadly conceived is true, a narrower rule that falls under it admits of exceptions.

Likewise, the larger associations of social insects owe much of their success to having adopted a feeding policy that provides a copious and reliable food supply. Termites, for example, utilize cellulose, a substance rich in energy, but highly inert, so that a lot of it is available in many terrestrial habitats. They exploit it with the aid of a formidable array of symbionts. The inert material can be collected in large quantities, stored, and transformed into food under regulated and protected conditions.

Ants, which are largely carnivores, extend their economic base by such means as migratory colonies and diverse food. Sometimes, as in fungus-growers (see Weber 1958) and forms that cultivate aphids, a sort of agriculture may be said to have evolved. Ants resemble termites, rather than most other Hymenoptera, in so far as the winged individuals serve only for dispersal and the founding of the colony. Unfortunately we do not have the sort of comparative material that would tell us how the first ant societies came into being: all known ants are social, and their living relatives are rather distant ones.

For wasps, we do have a good series of forms bridging the evolutionary gap between individual and group life. These wasps are carnivorous, albeit related to herbivores, but it appears that the social ones have widened their feeding habits somewhat. At least the solitary forms seem to be largely trophic specialists (Evans 1966). As one might suspect, trophically specialized wasp colonies are small.

Bees, whether solitary or social, derive the main part of their sustenance from nectar and pollen. The vast majority of these hymenopterans are solitary, but social lineages have several times evolved within the group. Their food meets the two conditions, quantity and stability, which would provide a good trophic base for communal life. Plants transfer a substantial amount of energy to their pollinating symbionts. Furthermore, pollen and nectar are generated at a fairly constant rate within many communities, and a further advantage is that they can be stored.

Although a particular honeybee will, at any given time, concentrate her effort upon a single kind of flower, nonetheless, the colony as a whole must be viewed as a trophic generalist. The ability to exploit a variety of plants allows bees to maintain a large establishment. They use a substantial number and many kinds of flowers relatively close to the hive. But we need to ask what advantage this large society has over the independent labors of the solitary forms. The answer is that in a sense there is no advantage: the pollinating economy has been divided into sectors that are exploited by two distinct kinds of enterprise. To see what has happened we need to consider the reciprocal relationships that exist between the two members of the symbiosis. From the point of view of the plant, two considerations tell us what leads to effective fertilization. On the one hand, there is successful transfer of pollen to another individual of the same species; on the other hand, there is always the need to keep it from going to a flower of a different species. Both of these needs can be met by one of two strategical alternatives, each involving its own kind of specificity. The first is a highly specific pollinator, one that, ideally, visits but a single species of plant. The second is to reproduce during a specific period in the year. In the former case, morphological and physiological adaptations on the part of insect and flower alike are necessary to ensure the specificity, and the need for avoiding inappropriate transfers would favor those plants that exclude generalists. Such a partnership could work either if the plant had a long breeding season, or if the pollinator could be very closely attuned in the timing of its life cycle to the breeding season of the plant. A stratagem of this sort would not readily give rise to sociality, because the flowers would not be common enough to support a large group, nor would they last long enough to make the enterprise profitable for it. The foraging habits of solitary pollinators ought to have a number of features that can be predicted from such considerations; that such is indeed the case may be seen from a very interesting study by Janzen (1971), but little work of this sort has been done.

The alternative of a short season exploited by generalists would naturally lead to an alliance with the social forms. As a mechanism for avoiding transfer with the wrong plants and for attracting the insects, the pollinating season would tend to become temporally partitioned by the different species of plants. The result would be a series of different kinds of flowers, each successively providing the insects with a rich supply of energy close at hand, adequate to sustain the colony. Because they are generalists by necessity, the social bees would be attracted to such flowers. They would be particularly adept, too, at exploiting them. With large societies, they would be very numerous, and chance of itself would determine that they, rather than the solitary ones, would visit the flowers. They would be more reliable, since their diverse food supply would tend to maintain their populations at a fairly constant level. Finally, their social organization allows them to exploit briefly flowering species more effectively. The ability of social bees to communicate with each other, as through the celebrated "honey dance" (see Lindauer 1965; [von] Frisch 1967), permits them to bring the full weight of their numbers to bear upon the food soon after it appears. A solitary bee must hunt until she herself locates it. The social ones can disperse over the countryside, and the successful scout can inform her sisters where the food is. Here, numbers plus communication have an obvious competitive advantage. A saving in the dimension of time gives a profit in that of energy.

Communication, indeed, would seem to have been one of the driving forces behind the evolution of all societies. A host of calls and pheromones are noteworthy features of social organisms. Von Frisch even speaks metaphorically of a "language" of the bees. Language, however, does not in and of itself explain the existence of society, although it helps. Language is useless for those who have nothing to say, and simple associations can get along with little or no interchange between their members. However, it greatly furthers the economic success of many societies, and economics will probably provide the best clues to its origin and history.

Thus, division of labor, extent of the market, and diminishing returns all help to explain the efficacy of social adaptations. It is, again, a matter of the efficacy rather than the efficiency, because the amount of energy expended bears no direct and simple relationship to success in the struggle for existence. One needs to be aware of the competitive situations that actually occur. An inefficient, but useful, worker can mean a very efficient and productive factory. And there is no reason to assume that output of all sorts must be maximized. Thus the employees in a firm might suffer from the effects of crowding, but this does not mean that management will be better off if it reduces the work force. Likewise we should not consider it paradoxical when we find individual productivity in an insect society less than what it could be. One should study firms, not just employees, if one is not to fall into reductionist fallacies.

Nonetheless, in order to understand the whole, it does help to consider the parts and the relationships between them. A functional anatomy of the firm is essential but it ought to recognize what relationships really count. Consider E. O. Wilson's (1968b, 1971) notions on what determines the number of castes. Much of what he says, although well worth saying, proves to be mainly statements of principle rather than inferences from theory. Thus he makes the good point that colony output, and everything determining it, should be optimized. Now clearly one need only restate the same principles to come up with the inference that if, and only if, the return on having a specialist caste outweighs the adverse effect of not having those individuals do something else, will that specialist caste evolve or endure. So why are there not more castes? Naturally, a "fluctuating environment" makes the nonspecialists less effectual. All well and good, but just how does it do so? Clearly, business cycles do affect economies, and instability should have one effect or another on how labor gets divided. For example, they reduce the extent of the market. We need such concrete effects of instability for a genuine understanding of the phenomenon. Wilson argues that the number of castes is determined by the number of tasks to be done. We may admit that this helps set a bottom limit, for there can be no more castes than individuals. And it makes a certain amount of sense that a soldier will have a particular set of contingencies that determine its specialization. Yet as Adam Smith (1776) so well demonstrated with respect to the manufacture of pins, the extent to which labor may be divided can be carried out to great lengths indeed. Furthermore, as some tasks are carried on all the time, we cannot speak here of contingencies in the sense of tasks to be carried out if certain possibilities eventuate, such as an attack by predators. The notion of a finite number of tasks is one of those simplifying assumptions that may be useful but that may not always be valid.

E. O. Wilson's model leads to some predictions that can be challenged if we oppose different premises. Wilson (1971:347) infers that "If, in the course of evolution, one caste increases in efficiency and the others do not, the proportionate total weight of the improving caste will decrease." This generalization holds if there is in fact a set ratio between the different kinds of work that must be accomplished. Thus, if a colony needs to produce a set number of eggs, it should be better to do the job with one queen than with two. There is no reason to assume that the same principle will not apply to certain other tasks; and yet it need not apply to all tasks. The mix of professions could just as well result from a balance between the advantages to different functions set by diminishing returns. If one function becomes more efficient it may give a larger return if proportionately more, not proportionately less, effort is given to that function. Thus in a factory, the employees may consist of management, production workers, and maintenance men. If the workers become more efficient, the profit per worker will rise, and it may be better to hire more workers and manage and maintain less. Likewise, where the number of insects foraging for food on the one hand and guarding it on the other is set by the returns on fetching that food at increasing distance and preventing its theft, a more efficient transport system ought to shift the equilibrium toward more foragers, not fewer.

Thus the model is valid, but only for "fixed costs," and serious errors result when one ignores variable proportions and considers the amount of effort put into a given task as a constant. One readily falls into conceptual mistakes of this sort when dealing with economic theory, as is obvious when we reflect on how hard it is to manage an economy. One might think that given a captive labor force, the less the workers are paid, the greater the profit; but an unhealthy or a starving employee will not do much work. We assume that a rise in the price of a commodity reduces sales; but occasionally the opposite happens (in what economists call "Giffen goods"). Nor does a drop in prices always decrease the number of workers producing a commodity; consider what printing did for books. Given such difficulties, it is manifest that one must know a great deal if one is to understand what governs the division of labor in any economy, be it a natural or a political one. Nonetheless, it should not be difficult to make out what kinds of explanation should apply.

The efficiency and efficacy of individuals will tend to affect the ratio and presence of castes. Thus, recalling an earlier model, we would expect special castes to develop where a saving is affected by subdivision of food-handling into finding, transporting, and storage subunits. Ants that locate nectar are sometimes smaller than those that transport it; and these, in turn, differ radically from the "repletes" which do nothing but store it.

The magnitude of the enterprise determines the efficacy of caste mix. Thus, the minimal number of soldiers adequate for defense should be one if any. In very young colonies, the investment in a soldier should not repay the expense; likewise there should occur a period during which there will be only one soldier, for two will not give twice the protection as one. In larger and larger termite colonies, one would expect the number of individuals to increase faster than the area exposed to attack by ants; hence, as in larger human societies, a smaller proportion of soldiers should be adequate to defend the frontiers.

In the same way the degree to which labor is divided will depend upon the extent of the market. The larger and denser the society, the greater the number of tasks that can be constantly done by a single individual. And the greater the stability the less time will be lost in shifting from one mix of professions to another. Hence the principles laid down two centuries ago by Adam Smith go a long way toward explaining why societies tend to be more complicated in the tropics, and why the larger ones tend to be populated by a greater diversity of castes. How far supplemental principles need be invoked is a question we might as well leave unanswered.

SOME UNPOPULAR CONCLUSIONS

The evolution of society fits the Darwinian paradigm in its most individualistic form. Nothing in it cries out to be otherwise explained. The economy of nature is competitive from beginning to end. Understand that economy, and how it works, and the underlying reasons for social phenomena are manifest. They are the means by which one organism gains some advantage to the detriment of another. No hint of genuine charity ameliorates our vision of society, once sentimentalism has been laid aside. What passes for cooperation turns out to be a mixture of opportunism and exploitation. The impulses that lead one animal to sacrifice himself for another turn out to have their ultimate rationale in gaining advantage over a third; and acts "for the good" of one society turn out to be performed to the detriment of the rest. Where it is in his own interest, every organism may reasonably be expected to aid his fellows. Where he has no alternative, he submits to the yoke of communal servitude. Yet given a full chance to act in his own interest, nothing but expediency will restrain him from brutalizing, from maiming, from murdering—his brother, his mate, his parent, or his child. Scratch an "altruist," and watch a "hypocrite" bleed.

9

A NEW THEORY OF
MORAL SENTIMENTS, AND
CONCLUDING REMARKS

The reasonings of philosophy, it may be said,
though they may confound and perplex the
understanding, can never break down the
necessary connexion which nature has
established between causes and their effects.
Adam Smith

It follows from what has been said in preceding chapters that what we call virtue and what we call vice have alike emerged as the results of natural processes. It follows too that there can be no appeal to the natural order of things for a decision as to what is right and what is wrong. Success or failure in the sexual contests we have reviewed provides no guide to right conduct, nor does the rest of this purposeless world. The efforts of J. S. Huxley (in T. H. and J. Huxley 1947), Skinner (1971), and others (see Ebling 1969) to derive moral imperatives from evolutionary principles, of course, represent efforts to resurrect the teleological cosmology. More than that, they may be dismissed as thinly disguised efforts to gain adulation for having restored to mankind his vanished self-righteousness, and to take the place of prophets who are no longer listened to. Our moral discourse, it would seem, is tainted with hypocrisy.

It does not follow, however, that biology tells us nothing of utility about our ethical relationships. Man can be predator and prey alike, yet biology tells him nothing about who should eat whom. Once we have determined, however, that we prefer not to be fed upon by mosquitos, the scientific insights become exceedingly useful. Likewise, no matter how much we know about the causes of social behavior, we cannot derive rules for conduct that do not ultimately depend upon value judgments. Nonetheless, given a set of values, it certainly helps to know what is going on in society. We find it easier to understand both our own motives and those of others, and thereby to take effectual action. Most valuable, perhaps, is our ability to withstand oppressive forces. No weapon is more effectual against hypocrisy than is cynicism.

The notion that a society will render a man more rational or virtuous is precisely the sort of falsehood or half-truth that deceives the members of that society into letting others decide what is true and what is right. A solitary human being is bad enough, but at least his capacity for evil is less than that of a group, particularly one controlled by a demagogue. Society will inevitably resist the allegation that it is the source of our unhappiness. Thus Sherif and Sherif (1970) claim that sociology has refuted Freud, and that society is really a good thing for an individual's mental health. But Freud, who derived his view of society largely from Darwin's theory (Schur and Ritvo 1970; Schur 1972), had every reason to think that pathological behavior has been evoked by reproductive competition. It is only natural that the social battlefield where that struggle occurs may not be the most salutary of environments.

Gladly would the friends of society have us believe that its influence gives rise to right conduct. Yet such objective research as has been done leads to contrary, or at least equivocal, conclusions. We all know about lynchings, and about the worst aspects of the second world war, but naturally pathological individuals are responsible here, persons who somehow were not properly "socialized." Closer to home are the repeated newspaper accounts of crimes being committed in the presence of a crowd, where nobody would aid the victim, even where an anonymous call to the police would have saved a life. Such occurrences led to experiments in which it was shown that a solitary individual was more likely to aid a person in distress than was a member of a group (reviews in Adelson 1969; Sarason and Smith 1971). Indeed, the presence of others led to most irrational behavior in the face of an emergency (Latané and Darley 1970).

Such behavior may not damn the social man so deep as one might think. The presence of others may inhibit taking action of all sorts, or tempt one to pass the buck, or simply confuse. It has been shown that leadership, in the sense of an exemplar or model being presented (see Bryan 1970; Hornstein 1970), will tend to make people more charitable or less charitable, depending upon what the model does (Macaulay 1970). People gamble more when a "shill" is used too. From this we may conclude that although society tends to render man more foolish and depraved, the right kind of leadership will to some degree ameliorate the situation.

The disposition toward conformity cannot be a totally unmixed blessing, for not only does it suppress virtue, but it provides a way of manipulating behavior. Hence we may become the tools of another's wrongdoing, and the only defense is a certain amount of autonomy. Now assuredly Skinner (1971) has denied that individual men can be autonomous, yet his argument perhaps contains more examples per page of the fallacy of false disjunction than anything else so widely read. He tells us that we do not change minds, we change behavior; as well might he argue that we do not clean sparkplugs, we change the rate at which gasoline is consumed. In an argument of the same logical form, he tells us that men are not autonomous because their behavior is influenced by heredity and environment. Yet where is the contradiction? Since when is one's genetic endowment something other than oneself? An organism who obeys his own impulses and lives his own life, rather than allowing other members of his species to determine what he does, is "autonomous" in spite of there being some ultimate determinism to his acts. And in so behaving, he conforms to the natural order of things, as Skinner says he should, but with a difference. He does not surrender that impulse to act in behalf of himself, and in behalf of his family, which is largely responsible for his being here in the first place. He does not become a means to the end of preserving his culture. Skinner may be able to manipulate men so that they will find his *Kulturkampf* "rewarding," but only in the face of some very powerful opposing influences.

Let us grant, then, that we have every reason for wanting to know the real causes of social behavior. How shall we proceed? If we are to understand why men and other animals behave as they do, we need to treat them as the products of reproductive competition. This conclusion becomes inescapable, once we have accepted Darwin's theory. Yet, with such few and only partial exceptions as Freud, it is evident that we do not view brain function in this light. Even where evolution gets dragged in, very few behavioral scientists use selection theory as an important means of solving problems. The reasons for this are rather complicated, and we shall have to consider them at some length as we go.

THE PROBLEM OF REDUCTIONISM

In our remarks on genetics, and on the relationships between the social and the behavioral sciences, we alluded to the futility of treating an integrated whole in terms of the components alone, without due regard for the relationships between them. The term "reductionism" is frequently applied to such approaches, but the issue is rarely made clear. In so far as one may reasonably expect those sciences that deal with phenomena at one level to illuminate what goes on at another, there is every reason for translating the findings of one science into the language of another. This operation, technically called "reduction," greatly aids in unifying and coordinating our thought. Surely it is good that we can relate the genetics of Mendel to the molecular biology of Watson and Crick. But this is not the same as to assert that all we need to know about inheritance is the structure of molecules. This kind of view is what shall here be meant by "reductionism," but note that the term is used by others in different and not always compatible ways. Thus, the epistemological reductionism here rejected is often confounded with an ontological reductionism that asserts that complete reduction of one science to another would be possible, given sufficient knowledge. Similarly, there is nothing wrong with calling a relation between an organism and its environment a relation between a group of molecules-in-relation and its environment. Error arises only when the crucial relations are left out.

Synthesis, rather than analysis, ought to be considered the basic logical operation in scientific discovery. Rather than decompose organisms into their component subunits, place them in their larger social and economic contexts. And this brings out the basic limitation of reductionism as here defined: it ignores context. The meaning of a sentence does not reside in the words. And a fish abstracted from the water in which he dwells is like a word cut out of a book. Reductionists are like that: they concern themselves only with fish out of water.

In the behavioral sciences, the reductionist influence has been particularly strong, partly because physics has so often provided the source of methodological inspiration (see Vale and Vale 1969). It is widely maintained (e.g., Dobzhansky 1968) that Descartes is responsible for the reductionist point of view. But at least Descartes himself was not so naive and his views are better called "determinist" than "reductionist." According to Blake (1960) Descartes realized the limitations of the approach that he developed, or at least some of the limitations. In the seventh chapter of *Le Monde* he writes: "De sorte que ceux qui sauront suffisament examiner les conséquences de ses vérités et de nos règles pourront connaître les effets par leurs causes; et, pour m'expliquer en termes de L'École, pourront avoir des demonstrations a priori de tout ce qui peut être produit en ce nouveau Monde." He does not assert that the laws of nature tell us what will inevitably happen, but only what is possible. The actions of the whole are constrained by the properties of the components. But since the class of possible events is not coextensive with the class of actual ones, and since various configurations are possible, one does have to examine particular things to find out how the world is actually put together. But the ontological principle still stands: given sufficient knowledge of particulars, and given adequate "règles," behavior or anything else could be treated in a strictly deterministic, predictive fashion.

Yet Descartes' attempt to found a psychology according to his epistemological scheme does suffer from excessive emphasis on intrinsic constraint. In L'Homme, he treats animals as if they were simple automata. Subsequent research has shown that they are neither so simple nor so automatic, but the basic notion has been guiding neurophysiological research ever since. However, it would seem that the rationale of Descartes' argument has been widely misconstrued. He appears, at least, to be developing what we would now call a model, justified not on the basis of its "deductive" foundations, nor even on the basis of empirical evidence, but rather exemplifying the kind of psychology he had in mind. The details do not matter all that much. He was founding a "Kuhnian paradigm"—a model, a methodology, and a way of doing research.

THE PROBLEMS OF TELEOLOGY

Perhaps the main paradigm alternative to that of Descartes in psychology has been an approach based upon the teleological world view. It might be better to call it a variety of teleological approaches, or to deny that it deserves the status of a paradigm at all. Nonetheless, the manner of thinking is all too obvious when we look for it, and we can easily appreciate why this is so. Purpose is itself a psychological concept; and indeed we might view the whole teleological cosmology as the result of invalid analogizing from the realm of social psychology to the rest of the universe. The dangers of anthropomorphism as applied to the study of organisms other than man have been so well publicized that we no longer dispute the matter. Yet it may seem paradoxical to maintain that human behavior should not be treated anthropomorphically either. But we can make out an anthropomorphism in the sense of a group of presuppositions about human behavior which have crept into our scientific tradition from "folk psychology." The anthropos here is a mythical construct.

Given our analysis in the first chapter, we should be able to predict the kinds of teleological fallacies that will occur in psychological research. Take the view that the universe is rational; the obvious analogue would be that men are rational. By implication the brain becomes an organ the function of which is to discover the truth and to exercise wise judgment. Thus there exists in anthropology a widespread tradition that the human brain derived its unique features by virtue of its advantage in foresight and planning subservient to hunting, or, less often, to fighting (e.g., Washburn and Lancaster 1968). The not entirely exclusive hypothesis that the brain evolved because of its advantages in obtaining a mate is simply ignored, with certain noteworthy exceptions (for example, Chance and Mead 1953; M. T. Ghiselin 1969a: 73; Fox 1972).

Particularly in the study of motivation, we naturally tend to ascribe rational control where maybe none exists. Thus we may confuse the adaptive significance of a behavioral phenomenon with conscious or unconscious data-processing. So bad has been this error that William James (1890, Vol. I, p. 196) gave it a name—"the psychologist's fallacy"—which he defines as *the confusion of his own standpoint with that of the mental fact* about which he is making his report."

A second form that psychological teleology may assume is the view that organisms will necessarily behave in the "right" way. This leads one to ascribe functions to useless acts or to ascribe the wrong role to what one observes. A non-adaptive feature can be present as a vestigial one, or as the byproduct of something else which does have a function. Some psychologists have been loath to accept one or the other of these phenomena, but we should note that the converse mistake has been rather common too. One should take care to distinguish non-adaptive characters from those the adaptive significance of which is not known. A number of examples will be given later on.

Aristotelian final causes are not lacking in psychology either, particularly in social psychology where the "that for the sake of which" things exist is, of course, society. Or some, particularly advocates of "systems" approaches, advocate a species advantage (e.g., Tiger 1970: 293). We have already documented this point with respect to aggression and other phenomena.

A final version of teleology is goal-seeking. Classical hedonism—the psychological, not the moral doctrine—is a good example. It is not necessarily wrong, but it may be uninformative, to say that an animal does something "for pleasure" (or in what passes for metaphysically untainted terms, "because it is rewarding"). We want to know why pleasure accompanies the act. The goal-seeking approach takes on more of a metaphysical aura in the hands of those who have adopted the crypto-vitalist philosophy (e.g., Tolman 1932; McDougall 1947) or some watered-down version of holism (e.g., Nissen 1958). Again, examples will come later.

THE DARWINIAN REVOLUTION IN COMPARATIVE PSYCHOLOGY

Darwin realized from a very early stage in his research that the theory of evolution would effect profound changes in the study of behavior. He worked out a highly detailed psychological system (see M. T. Ghiselin 1969a, 1973), of which the crowning accomplishment was *The Expression of the Emotions in Man and Animals* (Darwin 1872). Here he showed how muscular movements, especially facial expressions, could be treated from an evolutionary point of view. Selection theory and the comparative data would show how and why the nervous system had evolved, and, equally important, they would reveal the underlying mechanisms. He showed that emotional expressions are a mixture of adaptive and non-adaptive features alike, and that functional constraints had imposed an intelligible order on the course of evolution. He said a great deal more about psychology, of course, but his one great contribution was an epistemological one: to understand behavior find out what caused it to evolve. Likewise, but this is only another way of viewing the same basic topic, he saw that the nervous system is one of the major selective influences determining the course of evolution. Sexual selection by female choice follows from this insight, but that is just one example.

Thus the way lay open for a major paradigm shift. But it never happened, or at least it affected far more the superstructure, than the foundations, of psychology. The reasons are not far to seek. Traditional notions were so far entrenched, and the new way of thinking was so different, that transfer of outlook could be partial at best. The idea of evolution was readily assimilated, but not that of natural selection. Hence such psychologists as John Hughlings Jackson got their views on evolution from Herbert Spencer rather than Darwin (see Magoun 1961; Greenblatt 1965). Freud, again, was somewhat of an exception; yet he was trained as a biologist and his followers did not continue this aspect of their paradigm. George John Romanes, whom Darwin considered his intellectual successor in animal behavior (see E. Romanes 1896), and who published some of Darwin's notes and manuscripts for him (G. J. Romanes 1884), did have a good grasp of evolutionary theory for his day (G. J. Romanes 1896). But he died young, and mostly concerned himself with demonstrating evolutionary continuity rather than with using the theory of natural selection (G. J. Romanes 1882, 1885). Conway Lloyd Morgan began as a very strong advocate of selection theory (Lloyd Morgan 1890–1891), and his well-known canon was important for methodology; but later on he came to advocate holism (Lloyd Morgan 1930). Above, in the second chapter, he was mentioned as having influenced Whitehead; in a late work, he warmly endorsed the views of the Harvard crypto-vitalists and cited with approval the view of Whitehead that a molecule is an organism and its atoms have functions (Lloyd Morgan 1930:64).

One point that did get across, and it got across too well if anything, was the idea that behavior is adaptive. At the time evolution became popular, psychology was still closely linked to philosophy and one important outgrowth of Darwin's discovery was pragmatism. Yet even the pragmatists had a hard time with teleology, especially with its subtler manifestations. John Dewey (1910) argued most convincingly that Darwin had demolished teleology; nonetheless, he had trouble dealing with emotional expression (Dewey 1894, 1895). So did William James (1890), who attempted, in the well-known James-Lange theory of emotion, to show that the peripheral manifestations are where the emotional activity resides. According to this view, long ago refuted, we feel afraid when we start, not the other way round. The teleological nexus is apparent here; there must be some use for our expressions.

Darwin's present-day successors for the most part have a long way to go too. They tend to presuppose that emotional expressions are there for the sake of communication, ignoring Darwin's view that some have a communicative function but others do not (e.g., van Hoof 1962). Some (e.g., Vine 1970) do admit the possibility of "affective" expressions. Many (Tavogla 1970) draw no distinction between communication and other processes in which one organism finds something out about another. It is as if, for example, they were to say that a sunburn is a feature the function of which is to inform others that one has spent too much time out of doors. According to Chevalier-Skolnikoff (1973a), psychologists generally do not make the distinction, because it is so difficult to tell what the subject is thinking. Yet the crucial differentia here is only that the message has to be adaptive for sender and receiver alike; it does not matter whether the animals know what they are doing. And good canons of evidence are available for distinguishing between the two phenomena. For instance, the sender will tend to direct the signal in a way that facilitates its reception. Related to this issue has been the question of whether emotional expressions are innate or learned. Obviously if they are indeed a language, they must be learned, at least to one who endorses the prevailing metaphysics. Very detailed work, with the appropriate crucial experiments, by Ekman and his collaborators (P. Ekman 1972, 1973; Ekman, Sorensen, and Friesen 1969) shows that although experience and cultural context may modify them, Darwin was right: our emotional expressions are in fact innate. We could extend this discussion to show the fallacies that underlie kindred misconceptions—such as the view that language is not inherited or that its function is to help us think (cf. Flavell and Hill 1969 against this)—but the point has been adequately made that teleology is a problem.

AN EXEMPLARY PUZZLE IN EVOLUTIONARY PSYCHOLOGY

An evolutionary psychology that is neither reductionist nor teleological could readily be developed as an extension of the Darwinian paradigm. The approach would be to construct a model in which the way selection operates is related to the context in which it occurs. We have already gone over various passing examples of how such an approach might be used. One should be able to develop a comprehensive system out of it. However, in keeping with what was said in the introductory chapter, let us be content with an exemplary puzzle: what are the selective advantages of our higher moral, spiritual, intellectual, and aesthetic faculties?

The issue to be considered is a momentous one. Individualistic competition seems not to explain all that we do. As we have seen with respect to morals, however, at least a partial answer can be given; and the entire solution hardly seems beyond reach. Yet certain activities upon which we lavish great effort, and from which we derive much of our pleasure in life, have no obvious utility. Art and pure science are good examples. The critics of natural selection have often maintained that since it fails to account for such phenomena, something else must be necessary. This position was maintained not only by the Roman Catholic apologist St. George Mivart (1871) but even by A. R. Wallace (1912). In general, the difficulties are passed over in silence. Or ignorance is confessed by saying that certain activities are carried out "for their own sake" (Morris 1962). Others, such as Herbert Spencer (1857b, 1860b), have erected *ad hoc* hypotheses. Spencer (1904, Vol. I, p. 399) remarked that he was "never puzzled." Too bad. It could be that we are dealing with features of no utility, yet they are so pervasive in our lives that we are led to doubt this. Whatever may be the correct solution, we would gain a great deal by showing either that the features in question do lack a function, or else that they are really advantageous in one way or another.

Let us begin with what Adam Smith (1759) and other traditional philosophers called the "moral sentiments." It is clear that these have adaptive significance, but a teleological view of things would imply that they exist for the sake of society rather than for those who experience them. But that view raised serious problems from the very beginning. What makes us do the right thing? That is to say, what causes virtue to be rewarding? Some invoked a kind of instinct or innate sentiment, but one that ultimately derived from God's will even if His influence were remote and indirect. Alternatively it could come from reason, or from experience—through a system of rewards and punishments, either on earth or in Hell. We can make sense out of the fact that those moralists of the old school who embraced nominalism had to advocate punitive education and retributive justice (e.g., Bain 1884). Evolutionary theory changed all that: the moral sentiments could be both innate and the product of experience—a racial experience. Spencer (e.g., 1871b) thought that our feeling of pain at doing others ill results from the recollected experiences of our ancestors. Darwin stressed inherited habit and natural selection; but their views overlapped considerably. We can now exclude the inheritance of acquired characteristics of all sorts, but the question still remains of just how selection operates here.

The fact that our moral sentiments have an adaptive significance was clearly grasped by Adam Smith, although, being a man of his times, he thought they exist for the good of our species. Yet he clearly went in the right direction by showing that they subserve reproduction. Thus he points out that parents are more attached to their children than children are to their parents, and argues that the former relationship is the more important for the survival of the species. We need only revise the principle, and say that we do that which is expedient in reproduction, to bring our ideas into the proper context. Thus it is easy to see how suicide is adaptively useful, if it helps one's children or grandchildren. Again, one would predict that there should be certain kinds of sexual dimorphism in our ethical attitudes. Females know who are their offspring: hence it is expedient for them to play favorites. Males, in so far as they find it difficult to know who fathered whom, would perhaps benefit more from a general contribution to the welfare of their group. Loyalty should thus be a feminine virtue, justice a masculine one. We need not elaborate upon this topic, but recent research has brought to light quite a number of differences between the sexes in moral attitudes, at least some of which seem to be inherited (reviews in Adelson 1969; Hartup and Yonas 1971; Lindzey, Loehlin, Manosevitz, and Thiessen 1971; Sarason and Smith 1971).

Similar inferences concerning the adaptive meaning of putatively altruistic traits have lately been reviewed by Trivers (1971). He refers to a sort of "reciprocal altruism" in which acts of generosity are done because they are repaid in kind. Hence the transaction may be said to be "altruistic" only in a psychological sense. We really like to please (see Weiss, Buchanan, Altstatt, and Lombardo 1971). But the "ulterior motive" (a term not to be taken in a psychological sense) is clear from certain other features: a preference for the sort of friends that one can use, skill at feigning altruism, means of detecting hypocrisy.

One matter that is mentioned several times by Trivers deserves further emphasis, for it may be extended considerably. This is Darwin's hypothesis, briefly discussed in *The Descent of Man*, that artificial selection, rather than natural selection, might be responsible for our moral traits. That is to say, the virtuous person might be favored through his companions deciding that he, rather than some other individual will survive or reproduce. Darwin somewhat de-emphasized this hypothesis in favor of kin selection and intergroup competition, largely because he thought it required a fair degree of intelligence. It is hard to say just how much brain is necessary to produce such choice; in an earlier chapter we warned against invoking it *a priori* in certain other contexts. The possibility has been suggested for birds, but evidently not for social insects. Yet for man, at least, selection of this kind would be no problem, for he does practice it on other species. As noted above, it has received passing mention in the older literature (see also Gottesman 1965). And it need never have been a deliberate, reflective choice. Darwin (1859:34) refers to "unconscious" selection in the production of domesticated breeds. We need only cherish one kind of individual more than another, and successive generations will come to be more and more like those that we prefer. Or we could cherish his relatives. When Socrates drank the hemlock, his friends could not dissuade him on the grounds that his children would suffer; he knew that they would be cared for. Likewise it becomes intelligible why some kinds of suicide are socially acceptable while others are not. He who kills himself may be taking unfair advantage of his neighbors' charity.

Now let us see how such a mode of artificial selection may apply to the higher intellectual aesthetic and "spiritual" phenomena. If there be any truth in this hypothesis, it should deeply affect our conception of human life in general. Let us begin with what on the face of it appears to be one of the most useless and trivial of activities: play. Some young animals, and to a lesser extent older ones of the same species, devote so much energy to play that an ecologist would almost be compelled to think that there must be a very good reason for their doing so. But its utility is so far from obvious that some persons have taken it for evidence against the theory of natural selection (e.g., Alverdes 1927). The literature on play, whatever its position on evolution, abounds in unfounded hypotheses and teleological posits. Older work has been critically reviewed by Groos (1896), an aesthetician who was strongly selectionist in his outlook. The basic modern review is that of Beach (1945), who perhaps wisely adds nothing constructive. More recent reviews, albeit useful from a descriptive point of view, tend to be inadequate with respect to evolutionary theory (Meyer-Holzapfel 1962; Hutt 1966; Loizos 1966, 1967; Beckoff 1972). However, Gilmore (1966) provides a very useful summary of the psychological literature, and Herron and Sutton-Smith (1971) have compiled an excellent anthology with a representative bibliography. We can sum up the literature in a few words: those who think they know what play is "for" have reasoned from false premises.

One hypothesis for the function of play goes back to Herbert Spencer (1886), perhaps beyond; he thought it to be an outlet for surplus energy. Similar positions have been maintained by Krieg (1937) and Bierens de Haan (1952). From a theoretical point of view, we may note that for every case in which adequate study has been done, surplus energy is used in growth and reproduction or else stored or not used at all. The very idea runs counter to everything we know about ecology. Furthermore, only some animals play, and there is reason to think that those that do and those that do not have more or less the same amount of energy available to them. The hypothesis is obviously invalid, like a number of lesser ones invoking a lack of real adaptive significance (e.g., Morris 1962; Hutt 1966; Sutton-Smith 1967).

Play is sometimes thought to be a form of exploratory behavior (e.g., Brearley 1966). This hypothesis is closely akin to another: that play is a kind of practice for the supposedly serious business of life. This notion would seem to have been, and to remain, the most popular one (Groos 1896; Mitchell 1912; Pycraft 1913; Lorenz 1956; Bolwig 1964; Loizos 1967). The most recent versions emphasize social development (Stone 1965; Piaget 1967, 1969 cf. Sutton-Smith 1966; Eifermann 1971; Beckoff 1972). In turn, it relates to the more general teleological proposition that the earlier developmental stages exist for the sake of adult life. Man's long childhood is widely assumed, but never demonstrated, to exist for the purpose of allowing him to become educated.

Such ideas are flatly contradicted by the empirical evidence. Animals do explore, but their behavior under such cirumstances is much better adapted to obtaining knowledge than it is while they play. Any learning that goes on during play is quite incidental. Similarly, young animals learn, but youth is not set up in a manner appropriate to effective learning. The canons of evidence that have been applied, say, to the hypothesis that the blood circulates, are never used with respect to this one. Our teleological conception of childhood is a metaphysical posit, one so deeply ingrained in our culture that it goes virtually unquestioned. The Aristotelian view which treats the adult stage as having priority over the juvenile (a boy, being only potentially a man, cannot even be happy) has persisted in the form of not really taking children seriously. The Deweyesque notion that education is preparation for life, rather than a part of life itself, is perhaps the worst manifestation of this error.

Some confusion about what is called play has resulted from uncertainties as to how that term should be defined. Thus Lankester (1972) refers to "play mothering" in vervet monkeys, but she makes it abundantly clear that this is work, not play in the usual sense. The adult females aid one another in caring for the infants, and are joined in this activity by juvenile females. Such "aunts" are known in other species (see Crook 1970b), and they may be compared with the "helpers" already mentioned. Selection ought to favor the females who find indiscriminate maternal care rewarding, so long as an effectual "reciprocal baby-sitting" arrangement results.

A second source of difficulty has to do with the problems of isolating play from its natural place in ontogeny. If an animal is prevented from playing, he might develop behavior disorders. And, in fact, social isolation of monkeys and other animals leads to all kinds of pathological states (Harlow, Dodsworth, and Harlow 1965; review in Finger and Mook 1971). Yet by no means does it follow that play is there because it helps the animal to develop. We are presented here with the same fallacy we discussed with respect to the pecking order experiments in chickens. Even were play a necessary condition for normal development, as may well be the case, it does not follow that animals play for that reason. A brainless salamander embryo will not develop eyes, but that does not mean that salamanders have brains in order to provide them with those parts. Of course, the usual activities of the young animal get bound up with its developmental mechanisms, but teratology is a poor guide to functional significance.

That play is less common among "lower" animals is generally accepted (e.g., Thorpe 1966). What is missed is that play is lacking in all but certain social ones, at least social in the sense of having families. However, not all social animals manifest it; for example, modern workers have abandoned the old idea that ants play (E. O. Wilson 1971). It has, furthermore, been inadequately recognized that play has social significance. Although resembling in some ways the serious business of life, it is really antithetical to it, and herein lies its real meaning. It is an intriguing point that behavioral phenomena that are very different on the one hand, or very much alike on the other, tend to become associated with one another. We can find more than a hint of this relationship in Darwin's (1872) principles of associated habits and antithesis. That this phenomenon is a very general one is evident when we consider mechanisms of communication. We habitually stress likeness or contrast, as when we define a term by saying "x is like y" or else "x is the opposite of y." Our idioms show comparable patterns. Thus, when we begin a sentence with "obviously," we really mean that it is not so obvious as to render it unworthy of emphasis. Likewise, one substitute for the subjunctive mood in English is using "doubtless" in a sense more or less opposite to the literal one. Doubtless this maneuver causes the audience to question the assertion.

The same basic principles apply to the communication systems of animals other than man. Their signals tend to be a mixture of contrasts and likenesses, ones which have often seemed rather problematical. Thus the kiss is thought by some to have been derived from the suckling of infant mammals, implying that the association of feeding and social affection accentuates the rewarding effect. Conversely, the smile seems to have originated from an expression of fear, and the signal that is called the "play-face" to be derived from the one for "threat" (Chevalier-Skolnikoff 1973b). If perhaps we should qualify these generalizations by saying that they need further substantiation, it nonetheless makes a great deal of sense that emotional conditions would play a major role in determining how organisms communicate and how their means of communication will evolve.

Indeed, the whole emotional makeup of any animal will reciprocally influence, and be influenced by, the pattern of its social behavior. Thus, Scott and Fuller (1965) point out that the disposition toward imitation and conformity in dogs greatly furthers their cooperation. It is of no small interest that we human beings often seem to vacillate between finding society a pleasure and a bore.

A form of antithesis may be seen in play, and the connection does not seem to be a purely accidental one: animals at play behave very much as if they were at work. They frequently engage in what looks like combat, but all parties know that it is a sham (see Poole 1966). A young animal will induce a much larger, older one to play with him; and signals inform the animals that the "combat" is not in earnest. So long as everyone continues playing, nobody gets hurt. A possible function for play is therefore clear: it protects the members of the society by preventing competitive interactions. Schenkel (1966) notes that a pride of lions becomes more peaceful and harmonious when the cubs are introduced into it. And well it should. This is why juveniles play more than adults: they need more protection. Also it makes sense that animals play more when they are in groups than when alone. It does not follow, however, that solitary play lacks social significance. We do not cease to be members of a group every time we go to the bathroom. Nor does it matter that as we get older our activities in play become increasingly integrated (see, for example, Piaget 1969); the like applies to work as well. Nonetheless it seems likely that play has various functions. Thus, van Lawick-Goodall (1967) saw a chimpanzee preventing her child from hurting himself by playing with him. Parenthetically, we might add that if this hypothesis is correct, many of our educational practices may be seriously misguided as a result of folk-psychology. Little boys playing soldier are not practicing to slaughter their fellow men, but furthering peaceful life within their own society. The way to make a killer out of a child is to put him into a genuinely competitive situation—such as Little League baseball.

Upon the same basis we may understand the function of comedy.

Andrew (1963) and van Hooff (1967) note that smiles and laughter are comparable to the signals that let other primates know that the behavioral context is play rather than combat. And crying is associated with laughter (Darwin 1872; Charlesworth and Kreutzer 1973). Laughter is a social act, even though much of our social life goes on in private. Although humor may be used maliciously, nonetheless we find it very hard to laugh and fight at the same time. (For recent work on humor-reducing aggression, see Adelson 1969.) And we seek out the company of persons with a good sense of humor, not just because they are amusing, but because they are innocuous. Those who are always serious pose a threat. Hence the delight in laughter will be artificially selected because of its great moral utility.

A crucial reason for the social worth of comedy is that it can never be in short supply. Food may run out, there may not be enough mates to satisfy the demand, but our stock of laughs can never be exhausted. So it is with the other higher manifestations of our inner lives, including our aesthetic, intellectual, and purely religious activities. We do not compete for their objects, and nothing is lost by giving them away. By extension of what has already been said, it is evident why the corresponding faculties have evolved. Men who derive pleasure from the less worldly activities are cherished because we find them so much the easier to live with.

The truth of this proposition is obscured, in part, because we achieve the benefits of civilization at some cost. The scholar may neglect his students, or the painter his family, and wealth that might go to the poor is used to build churches. The benefits are not so obvious, and hence we must put up with Philistines and self-righteous moralists. Worse yet, we have little basis for answering those who insist that man should live by bread alone. Likewise we may find that different sorts of higher values tend to commingle and ally themselves in such a manner that their separate realms become confused. If we combine music, ritual, and mysticism into a single experience, their joint effect is greatly enhanced. And the moral impulse is all the more adequately expressed when cast in the form of poetry. But a manifestation of religious insight may readily be confused with an understanding of objective relations on a material plane; or *vice-versa*. Hence an intemperate reaction may lead to intolerance and misunderstanding, especially on the part of those with unbalanced or underdeveloped faculties.

Even those who might have been expected to know better have often denigrated the higher faculties. Plato could see no use for poetry. Aristotle looked down upon comedy, and one might wonder if his book on that subject was lost for no more reason than its not having been worth saving. His notion of catharsis embodies one of the most pervasive forms of anti-intellectualism. Art is viewed as a means of getting rid of pent-up emotions. In subsequent writings we find the sciences and the humanities alike being put down as manifestations of psychopathology, and as the activities of those who cannot succeed in real life. The authors need only drop a few words like "sublimation" and "displacement" to convince themselves that they have argued Jesus down from his cross. For Freud, art was a kind of sex-substitute, and religion a manifestation of neurosis. Darwin, in his later years, tended to equate religion with superstition. In this he has been followed by many biologists. Wallace (1912) could see no value to mathematics, music, or art in the struggle for existence, and treated them as the results of divine intervention. Lorenz (1966b) sees in sport an outlet for man's aggressive impulses, rather than something that can take their place. And so it is with our entire civilization. Although many are happy to praise the life of the mind, nobody seems to know what it is for. We are like the beneficiaries of an anonymous gift, of which the donor has never explained his motives.

CONCLUSIONS

Man's brain, like the rest of him, may be looked upon as a bundle of adaptations. But what it is adapted to has never been self-evident. We are anything but a mechanism set up to perceive the truth for its own sake. Rather, we have evolved a nervous system that acts in the interest of our gonads, and one attuned to the demands of reproductive competition. If fools are more prolific than wise men, then to that degree folly will be favored by selection. And if ignorance aids in obtaining a mate, then men and women will tend to be ignorant. In order for so imperfect an instrument as a human brain to perceive the world as it really is, a great deal of self-discipline must be imposed.

Nonetheless, an ability to come to grips with the world as it really is can hardly be dismissed as maladaptive. Above all else, the truth has ethical significance. We may seriously question whether a parent acts in anyone's true interest when he believes that his child can do no wrong. And it seems equally probable that we suffer more distress from failing to recognize what we have done than from admitting what we judge to be our faults. If, as we have some reason to think, selection within the context of society has somewhat elevated our baser sentiments, and if, as the facts suggest, self-interest and common welfare are not fundamentally beyond reconciliation, we can reasonably hope to develop ethical standards consistent with biological reality. Perhaps the intellect is endowed with sufficient prowess that it yet may make sense out of a world of passion and folly. That the brain is destitute of purpose does not imply that it cannot be used.

REFERENCES

Abbott, D. P. 1955. Larval structure and activity in the ascidian *Metandrocarpa taylori. J. Morphol.* 97:569–594.

Adelson, J. 1969. Personality. *Ann. Rev. Psychol.* 20:217–252.

Agar, W. E. 1938. The concept of purpose in biology. *Quart. Rev. Biol.* 13:255–273.

Åkesson, B. 1958. A study of the nervous system of the Sipunculoideae with some remarks on the development of the two species *Phascolion strombi* Montagu and *Golfingia minuta* Keferstein. *Undersökningar över Öresund* 38:1–249.

———. 1961. Some observations on *Pelagosphaera* larvae. *Galathea Rep.* 5:7–17.

Aldrich, J. M., & L. A. Turley 1899. A balloon-making fly. *Amer. Nat.* 33:809–812.

Alexander, C. I. 1932. Sexual dimorphism in fossil Ostracoda. *Amer. Midl. Nat.* 13:302–311.

Alexander, R. D. 1961. Aggressiveness, territoriality, and sexual behavior in field crickets (Orthoptera: Gryllidae). *Behaviour* 17:130–223.

Allee, W. C. 1951. *Cooperation among animals: with human implications.* New York: Abelard-Schuman. xvi + 233 pp.

———. 1952. Dominance and hierarchy in societies of vertebrates. *Colloques Internat. Centre Nat. Rec. Sci.* 34:157–181.

Allee, W. C., A. E. Emerson, O. Park, T. Park, & K. P. Schmidt 1949. *Principles of Animal Ecology.* Philadelphia: Saunders. xii + 837 pp.

Allen, C. E. 1937. Haploid and diploid generations. *Amer. Nat.* 71:193–205.

Allen, G. E. 1968. Thomas Hunt Morgan and the problem of natural selection. *J. Hist. Biol.* 1:113–139.

Allen, J. A. 1892. The geographical distribution of North American mammals. *Bull. Amer. Mus. Nat. Hist.* 4:199–243.

———. 1958. On the basic form and adaptations to habitat in the Lucinacea (Eulamellibranchia). *Phil. Trans. Roy. Soc. London* (B)241:421–484.

———. 1963. Observations on the biology of *Pandalus montagui* (Crustacea: Decapoda). *J. Mar. Biol. Assoc.* 43:665–682.

———. 1965. Observations on the biology of *Pandalina brevirostris* (Decapoda; Crustacea). *J. Mar. Biol. Assoc.* 45:291–304.

Allen, J. A., & H. L. Sanders 1966. Adaptations to abyssal life as shown by the bivalve *Abra profundorum* (Smith). *Deep-Sea Res.* 13:1175–1184.

Altenberg, E. 1934. A theory of hermaphroditism. *Amer. Nat.* 68:88–91.

Altevogt, R. 1955. Beobachtungen und Untersuchungen an indischen Winkerkrabben. *Z. Morphol. Ökol. Tiere* 43:501–522.

———. 1957a. Untersuchungen zur Biologie, Ökologie und Physiologie indischer Winkerkrabben *Z. Morphol. Ökol. Tiere* 46:1–110.

———. 1957b. Beiträge zur Biologie und Ethologie von *Dotilla blanfordi* Alcock und *Dotilla myctiroides* (Milne-Edwards) (Crustacea Decapoda). *Z. Morphol. Ökol. Tiere* 46:369–388.

———. 1959. Ökologische und ethologische Studien an Europas einziger Winkerkrabbe *Uca tangeri* Eydoux. *Z. Morphol. Ökol. Tiere* 48:123–146.

Alvariño, A. 1965. Chaetognaths. *Oceanogr. Mar. Biol. Ann. Rev.* 3:115–194.

Alverdes, F. 1927. *Social life in the animal world.* London: Kegan Paul, Trench, Trubner. ix + 216 pp.

Amadon, D. 1959. The significance of sexual differences in size among birds. *Proc. Amer. Phil. Soc.* 103:531–536.

———. 1966. Insular adaptive radiation among birds. In R. I. Bowman, *The Galapagos: Proceedings of the symposia of the Galapagos International Scientific Project.* Berkeley and Los Angeles: University of California Press. Pp. 18–30.

Ambrose, J. A. 1965. The study of human social organization: a review of current concepts and approaches. *Symp. Zool. Soc. London* 14:301–314.

Andersen, F. S. 1961. Effect of density on animal sex ratio. *Oikos* 12:1–16.

Anderson, J. M. 1952. Sexual reproduction without cross-copulation in the fresh-water triclad turbellarian, *Curtisia foremanii. Biol. Bull.* 102:1–8.

Andrew, R. J. 1963. The origin and evolution of the calls and facial expressions of the primates. *Behaviour* 20:1–109.

Andrewartha, H. G. 1961. *Introduction to the study of animal populations.* Chicago: University Press. xviii + 281 pp.

Anthony, H. E. 1928–1929. Horns and antlers: their evolution, occurrence and function in the Mammalia. *Bull. New York Zool. Soc.* 31:178–216.

Arcangeli, A. 1932. L'ermafroditismo negli Isopodi terrestri. Ipotesi sopra la natura e l'origine dello stesso e considerazione su quello di altri animali. *Arch. Zool. Ital.* 17:165–256.

Archer, J. 1970. Effects of population density on behaviour in rodents. In J. H. Crook, *Social behaviour in birds and mammals.* New York: Academic Press. Pp. 169–210.

Arkell, W. J. 1957. Introduction to Mesozoic Ammonoidea. *Treatise on Invertebrate Paleontology* (L)4:81–129.

Armstrong, E. A. 1942. *Bird display: an introduction to the study of bird psychology.* 2nd ed. 1947. Cambridge: University Press. xvi + 381 pp.

Armstrong, J. C. 1965. Mating behavior and development of schistosomes in the mouse. *J. Parasitol.* 51:605–616.

Arnold, J. M. 1965. Observations on the mating behavior of the squid *Sepioteuthis sepioidea. Bull. Mar. Sci. Gulf Caribbean* 15:216–222.

Aronson, L. R. 1957. Reproductive and parental behavior. In M. E. Brown, *The physiology of fishes*. New York: Academic Press. 2:271–304.

Arrow, G. J. 1951. *Horned beetles: a study of the fantastic in nature*. The Hague: Junk. vii + 154 pp.

Asher, J. H. Jr., & G. W. Nace 1971. The genetic structure and evolutionary fate of parthenogenetic amphibian populations as determined by Markovian analysis. *Amer. Zool.* 11:381–398.

Ashmole, N. P. 1963. The regulation of numbers of tropical oceanic birds. *Ibis* 103:458–473.

Aspey, W. P. 1971. Inter-species sexual discrimination and approach-avoidance conflict in two species of fiddler crabs, *Uca pugnax* and *Uca pugilator*. *Animal Behav.* 19:669–676.

Assmuth, J. 1923. Ametabolie und Hermaphroditismus bei den Termitoxeniiden (Dipt.). *Biol. Zentralbl.* 43:268–281.

Atkins, D. 1958. A new species and genus of Kraussinidae (Brachiopoda) with a note on feeding. *Proc. Zool. Soc. London* 131:559–581.

Atz, J. W. 1964. Intersexuality in fishes. In C. N. Armstrong & A. J. Marshall, *Intersexuality in vertebrates including man*. New York: Academic Press. Pp. 145–232.

Avel, M. 1959. Classe des Annélides Oligochètes. *Traité de Zoologie* 5(1): 224–470.

Ax, P. 1966. Die Bedeutung der interstitiellen Sandfauna für allgemeine Probleme der Systematik, Ökologie und Biologie. *Veröffentlichungen Inst. Meeresforsch. Bremerhaven* (Sonderband 2): 15–65.

Axelrod, D. I. 1967. Quaternary extinctions of large mammals. *Univ. California Publ. Geol.* 74:1–42.

Axelrod, D. I., & H. P. Bailey 1968. Cretaceous dinosaur extinction. *Evolution* 22:595–611.

Ayala, F. J. 1968. Biology as an autonomous science. *Amer. Sci.* 56:207–221.

———. 1970. Population fitness of geographic strains of *Drosophila serrata* as measured by interspecific competition. *Evolution* 24:483–494.

———. 1972. Competition between species. *Amer. Sci.* 60:348–357.

Bacci, G. 1947a. L'inversione del sesso ed il ciclo stagionale della gonade in *Patella coeruea* L. *Pubbl. Staz. Zool. Napoli* 21:183–217.

———. 1947b. Ricerche preliminari sul sesso di *Patella coerulea* L. *Arch. Zool.* 31:293–310.

———. 1947c. Osservazioni sulla sessualità degli Archaeogastropoda. *Arch. Zool.* 32:329–341.

———. 1951a. Ermafroditismo ed intersessualità nei Gastropodi e Lamellibranchi. *Arch. Zool. Ital.* (*suppl.* 7):57–151.

———. 1951b. L'ermafroditismo di *Calyptraea chinensis* e di altri Calyptraeidae. *Pubbl. Staz. Zool. Napoli* 23:66–90.

———. 1951c. On two sexual races of *Asterina gibbosa* (Penn.). *Experientia* 7:31–32.

———. 1965. *Sex determination*. New York: Pergamon Press. viii + 306 pp.

Baerends, G. P. 1952. Les sociétés et les familles de poissons. *Colloques Internat. Centre Nat. Rec. Sci.* 34:207–219.

Bain, A. 1884. *Mental and moral science: part second: theory of ethics and ethical systems.* London: Longmans, Green. viii + 322 pp.

Baird, R. C. 1965. Ecological implications of the behavior of the sexually dimorphic goby *Microgobius gulosus* (Girard). *Publ. Inst. Mar. Sci. Univ. Texas* 10:1–8.

Baker, H. G. 1959. Reproductive methods as factors in speciation in flowering plants. *Cold Spring Harbor Symp. Quant. Biol.* 24:177–191.

——. 1963. Evolutionary mechanisms in pollination biology. *Science* 139: 877–883.

——. 1965. Characteristics and modes of origin of weeds. In H. G. Baker & G. L. Stebbins, *The genetics of colonizing species.* New York & London: Academic Press. Pp. 147–168.

——. 1967. Support for Baker's law—as a rule. *Evolution* 21:853–856.

——. 1970. Evolution in the tropics. *Biotropica* 2:101–111.

Baker, H. G., & P. D. Hurd, Jr. 1968. Intrafloral ecology. *Ann. Rev. Ent.* 13:385–414.

Balgooyen, T. C. 1972. Behavior and ecology of the American Kestrel (*Falco sparverius*). Ph.D. Dissertation, University of California, Berkeley. ix + 169 pp.

Baltzer, F. 1931. Echiurida. *Handbuch der Zoologie* 2(14):62–168.

——. 1934. Sipunculida. *Handbuch der Zoologie* 2(2:9):15–61.

Banta, A. M. 1937. Population density as related to sex and to evolution in Cladocera. *Amer. Nat.* 71:34–49.

——. 1939. Adaptation and evolution: adaptive character of parthenogenesis. *Carnegie Inst. Washington Publ.* 513:253–264.

Barber, B. 1970. *L. J. Henderson on the social system.* Chicago: University Press. x + 261 pp.

Barbour, M. G., R. B. Craig, F. R. Drysdale, & M. T. Ghiselin 1973. *Coastal ecology: Bodega Head.* Berkeley: University of California Press. xix + 338 pp.

Barghorn, E. S. 1953. Evidence of climatic change in the geologic record of plant life. In H. Shipley, *Climatic change, evidence, causes and effects.* Cambridge: Harvard University Press. Pp. 235–248.

Barnes, H., & D. J. Crisp 1956. Evidence of self-fertilization in certain species of barnacles. *J. Mar. Biol. Assoc.* 35:631–639.

Barnes, N. B., & A. M. Wenner 1968. Seasonal variation in the sand crab *Emerita analoga* (Decapoda, Hippidae) in the Santa Barbara area of California. *Limnol. Oceanogr.* 13:465–475.

Barrington, E. J. W. 1965. *The biology of Hemichordata and Protochordata.* Edinburgh: Oliver & Boyd. vi + 176 pp.

Bartholomew, G. A., Jr. 1952. Reproductive and social behavior of the northern elephant seal. *Univ. California Publ. Zool.* 47:369–471.

Bartholomew, G. A. 1970. A model for the evolution of pinniped polygyny. *Evolution* 24:546–559.

Bartholomew, G. A., & C. L. Hubbs 1960. Population growth and seasonal movements of the northern elephant seal, *Mirounga angustirostris*(1). *Mammalia* 24:313–324.

Bastock, M. 1967. *Courtship: an ethological study*. Chicago: Aldine. viii + 220 pp.

Bateson, W. 1894. *Materials for the study of variation treated with especial regard to discontinuity in the origin of the species*. New York: Macmillan. xvi + 598 pp.

Bateson, W., & H. H. Brindley 1892. On some cases of variation in secondary sexual characters, statistically examined. *Proc. Zool. Soc. London* 1892:585–594.

Battaglia, B., & L. Malesani 1959. Ricerche sulla determinazione del sesso nel Copepode *Tisbe gracilis* (T. Scott). *Bol. Zool.* 26: 423–433.

Battisini, R., & P. Vérin 1967. Ecologic changes in protohistoric Madagascar. In P. S. Martin & H. E. Wright, Jr., *Pleistocene extinctions: the search for a cause*. New Haven: Yale University Press. Pp. 407–424.

Beach, F. A. 1945. Current concepts of play in animals. *Amer. Nat.* 79:523–541.

Beadle, G. W., & V. L. Coonradt 1944. Heterocaryosis in *Neurospora crassa*. *Genetics* 29:291–308.

Beale, G. H. 1954. *The genetics of* Paramecium aurelia. Cambridge: University Press. xi + 179 pp.

Beard, J. 1894. The nature of the hermaphroditism of *Myzostoma*. *Zool. Anz.* 17:399–404.

Beatty, R. A. 1967. Parthenogenesis in vertebrates. In C. B. Metz & A. Monroy, *Fertilization*. New York: Academic Press. 1:413–440.

de Beauchamp, P. 1960. Classe des Brachiopodes: formes actuelles. *Traité de Zoologie* 5(2):1380–1430.

———. 1965. Classe des Rotifères. *Traité de Zoologie* 4(3):1225–1379.

de Beaumont, J. 1945. L'origine et l'évolution des sociétés d'Insectes. *Rev. Suisse Zool.* 52:329–338.

Becher, S. 1919. Flügelfärbung der Kolibris und geschlechtliche Zuchtwahl. *Anat. Hefte* 57:447–482.

Beckoff, M. 1972. The development of social interaction, play, and metacommunication in mammals: an ethological perspective. *Quart. Rev. Biol.* 47:412–434.

Beebe, W. 1925. The variegated Tinamou, *Crypturus variegatus variegatus* (Gmelin). *Zoologica* 6:195–227.

———. 1934. Deep-sea fishes of the Bermuda oceanographic expeditions: family Idiacanthidae. *Zoologica* 16:149–241.

Belding, D. L. 1934. The spawning habits of the Atlantic salmon. *Trans. Amer. Fish. Soc.* 64:211–218.

Bellrose, F. C., T. G. Scott, A. S. Hawkins, & J. B. Low 1961. Sex ratios and age ratios in North American ducks. *Bull. Illinois Nat. Hist. Survey* 27:391–474.

van Beneden, P. J. 1876. *Animal parasites and messmates*. New York: D. Appleton. xii + 274 pp.

Benoit-Smullyan, E. 1948. The sociologism of Émile Durkheim and his school. In H. E. Barnes, *An introduction to the history of sociology*. Chicago: University Press. Pp. 499–537.

Benson, W. W. 1971. Evidence for the evolution of unpalatability through kin selection in the Heliconiinae (Lepidoptera). *Amer. Nat.* 105:213–226.

——. 1972. Natural selection for Müllerian mimicry in *Heliconius erato* in Costa Rica. *Science* 176:936–939.

Berg, K. 1934. Cyclic reproduction, sex determination and depression in the Cladocera. *Biol. Rev.* 9:139–174.

Bergan, P. 1953. On the anatomy and reproduction biology in *Spirorbis* Daudin. *Nytt Mag. Zool.* 1:1–26.

Bergman, G. 1965. Der sexuelle Grössendimorphismus der Anatiden als Anpassung an das Höhlenbrüten. *Comment. Biol. Soc. Sci. Fennica* 28:1–10.

Bergquist, P. R., & M. E. Sinclair 1968. The morphology and behaviour of larvae of some intertidal sponges. *New Zealand J. Mar. Freshw. Res.* 2:426–437.

Bergquist, P. R., M. E. Sinclair, & J. J. Hogg 1970. Adaptation to intertidal existence: reproductive cycles and larval behaviour in Demospongiae. *Symp. Zool. Soc. London* 25:247–271.

Berndt, W. 1907. Studien an bohrenden Cirripedien (Ordnung Acrothoracica Gruvel, Abdominalia Darwin). I Teil: Die Cryptophialidae. *Arch. Biontol.* 1:163–210.

Berrill, N. J. 1931. Studies in tunicate development. II. Abbreviation of development in the Molgulidae. *Phil. Trans. Roy. Soc. London* (B)219:281–346.

——. 1950. *The Tunicata: with an account of the British species*. London: Ray Society. iii + 354 pp.

Bertelsen, E. 1951. The ceratioid fishes: ontogeny, taxonomy, distributions and biology. *Dana-Report* 39:1–276.

Bertram, G. C. L. 1940. The biology of the Weddell and crabeater seals, with a study of the comparative behaviour of the Pinnipedia. *Sci. Rep. British Graham Land Exped. 1934–1937*, 1:1–139. (Seen as MS only.)

Bier, K. 1954. Über den Einfluss der Königin auf die Arbeiterinnen Fertiltät im Ameisenstaat. *In. Soc.* 1:7–19.

Bierens de Haan, J. A. 1926. Die Balz des Argusfasans. *Biol. Zentralbl.* 46:428–435.

——. 1952. Das Spiel eines jungen solitären Schimpansen. *Behaviour* 4:144–156.

Birch, L. C. 1957. The meanings of competition. *Amer. Nat.* 91:5–18.

Birky, C. W., Jr., & J. J. Gilbert 1971. Parthenogenesis in rotifers: the control of sexual and asexual reproduction. *Amer. Zool.* 11:245–266.

Blair, K. G. 1924. Some notes on the luminosity in insects. *Entomol. Monthly Mag.* 60:173–178.

——. 1926. On the luminosity of *Pyrophorus* (Coleoptera). *Entomol. Monthly Mag.* 62:11–15.

Blake, R. M. 1960. The role of experience in Descartes' theory of method. In E. H. Madden, *Theories of scientific method: the Renaissance through the nineteenth century.* Seattle: University of Washington Press. Pp. 75–103.

Blest, A. D. 1957a. The function of eyespot patterns in the Lepidoptera. *Behaviour* 11:209–255.

———. 1957b. The evolution of protective displays in the Saturnioidae and Sphingidae (Lepidoptera). *Behaviour* 11:257–309.

Bliss, L. C. 1971. Arctic and alpine plant life cycles. *Ann. Rev. Ecol. Syst.* 2:405–438.

Bloomer, H. H. 1939. A note on the sex of *Pseudanodonta* Bourguignat and *Anodonta* Lamarck. *Proc. Malacol. Soc. London* 23:285–297.

Blunck, H. 1914. Die Entwicklung des *Dysticus marginalis* L. vom Ei bis zur Imago. *Z. Wiss. Zool.* 111:76–151.

Bock, K. E. 1970. Evolution, function, and change. In S. N. Eisenstadt, *Readings in social evolution and development.* London: Pergamon Press. Pp. 193–209.

Bocquet, C. 1967. Structure génétique du sexe chez les Crustacés. *Année Biol.* 6:225–239.

Bodenheimer, F. S. 1938. *Problems of animal ecology.* London: Oxford University Press. vii + 183 pp.

———. 1958. *Animal ecology to-day.* Den Haag: Junk. 276 pp.

Bodmer, W. F., & A. W. F. Edwards 1960. Natural selection and the sex ratio. *Ann. Human Genet.* 24:239–244.

Bodmer, W. F., & P. A. Parsons 1962. Linkage and recombination in evolution. *Adv. Genet.* 11:1–100.

Bösiger, E. 1960. Sur la rôle de la sélection sexuelle dans l'évolution. *Experientia* 16:270–273.

Boldyrev, B. 1912. Das Liebeswerben und die Spermatophoren bei einigen Locustodeen und Grylloden. *Horæ Soc. Entomol. Rossicae* 40(6): 1–54. (Russian: German summary read.)

Bolwig, N. 1964. Facial expression in primates, with remarks on a parallel development in certain carnivores. *Behaviour* 22:167–190.

Bone, Q. 1961. The origin of the chordates. *J. Linn. Soc. London (Zool.)* 44:252–269.

Bonner, J. 1950. The role of toxic substances in the interactions of higher plants. *Bot. Rev.* 16:51–65.

Bonner, J. T. 1965. *Size and cycle; an essay on the structure of biology.* Princeton: University Press. viii + 219 pp.

Bonnet, P. 1935. La longévité chez les Araignées. *Bull. Soc. Ent. France* 40:272–277.

Boulding, K. E. 1956. General systems theory—the skeleton of science. *Management Sci.* 2:197–208.

Bouligand, Y. 1961. Notes sur la famille des Lamppidae, 2e partie. *Crustaceana* 2:40–52.

Bouvier, E.-L. 1905. Monographie des Onychophores. I. *Ann. Sci. Nat. Zool.* (9)2:1–383.

———. 1907. Monographie des Onychophores. II. *Ann. Sci. Nat. Zool.* (9)5: 61–318.

Bradt, G. W. 1938. A study of beaver colonies in Michigan. *J. Mam.* 19:139–162.

Brady, G. S. 1883. Report on the Copepoda collected by H. M. S. Challenger during the years 1873–76. *Challenger Rep. (Zool.)* 8:1–142.

Braestrup, F. W. 1963. The function of communal displays. *Dansk Ornithol. Foren. Tidsskr.* 57:133–142.

———. 1966. Social and communal display. *Phil. Trans. Roy. Soc. London* (B)251:375–386.

———. 1971. The evolutionary significance of learning. *Vidensk. Med. Dansk Naturh. Foren.* 134:89–102.

Bramlette, M. N. 1965. Massive extinctions in biota at the end of Mesozoic time. *Science* 148:1696–1699.

Brearley, M. 1966. Play in childhood. *Phil. Trans. Roy. Soc. London* (B)251: 321–325.

Breder, C. M. Jr., & D. E. Rosen 1966. *Modes of reproduction in fishes.* Garden City: Natural History Press. xvi + 941 pp.

Bretsky, P. W. 1968. Evolution of Paleozoic marine invertebrate communities. *Science* 159:1231–1233.

Bretsky, P. W., & D. M. Lorenz 1970. An essay on genetic-adaptive strategies and mass extinctions. *Bull. Geol. Soc. America* 81:2449–2456.

Brian, A. D. 1952. Division of labour and foraging in *Bombus agrorum* Fabricius. *J. Animal Ecol.* 21:223–240.

Brian, M. V. 1956. Exploitation and interference in interspecies competition. *J. Animal Ecol.* 25:339–347.

———. 1957. Caste determination in social insects. *Ann. Rev. Ent.* 2:107–120.

———. 1958. The evolution of queen control in the social Hymenoptera. *Proc. X Int. Congr. Ent.* 2:497–502.

———. 1965. *Social insect populations.* New York: Academic Press. vii + 135 pp.

———. 1968. Regulation of sexual production in an ant society. *Colloques Internat. Centre Nat. Rec. Sci.* 173:61–76.

Briggs, J. C. 1953. The behavior and reproduction of salmonid fishes in a small coastal stream. *California Fish Bull.* 94:1–62.

Bristowe, W. S. 1929. The mating habits of spiders, with special reference to the problems surrounding sex dimorphism. *Proc. Zool. Soc. London* 1929:309–358.

Brooks, J. L. 1950. Speciation in ancient lakes. *Quart. Rev. Biol.* 25:30–60, 131–176.

Brower, L. P. 1963. The evolution of sex-limited mimicry in butterflies. *Proc. XVI Int. Congr. Zool.* 4:173–179.

Brower, L. P., J. van Zandt Brower, F. G. Stiles, H. J. Croze, & A. S. Hower 1964. Mimicry: differential advantage of color patterns in the natural environment. *Science* 144:183–185.

Brown, E. S. 1951. The relation between migration-rate and type of habitat in aquatic insects, with special reference to certain species of Corixidae. *Proc. Zool. Soc. London* 121:539–545.

Brown, J. H. 1971. Mammals on mountaintops: nonequilibrium insular biogeography. *Amer. Nat.* 105:467–478.

Brown, J. L. 1970. Cooperative breeding and altruistic behavior in the Mexican jay, *Aphelocoma ultramarina*. *Anim. Behav.* 18:366–378.

Brown, S. W., & G. De Lotto 1959. Cytology and sex ratios of an African species of armored scale insect (Coccoidea—Diaspididae). *Amer. Nat.* 93:369–379.

Browne, F. B. 1922. On the life-history of *Melittobia acasta*, Walker; a chalcid parasite of bees and wasps. *Parasitology* 14:349–370.

Brüll, H. 1937. *Das Leben deutscher Greifvögel; die Umwelt der Raubvögel unter besonderer Berücksichtigung des Habichts, Bussards und Wanderfalken.* Jena: Gustav Fischer Verlag. vii + 144 pp.

Bruhin, H. 1953. Zur Biologie der Stirnaufsätze bei Huftieren. *Physiol. Comp. Oecol.* 3:63–127.

Bruslé, J. 1968. Nouvelles recherches sur l'hermaphrodisme d'*Asterina gibbosa* de Roscoff. *Cah. Biol. Mar.* 9:121–132.

———. 1969a. Les cycles génitaux d'*Asterina gibbosa* P. *Cah. Biol. Mar.* 10:271–287.

———. 1969b. Sexualité d'*Asterina gibbosa*, asteride hermaphrodite, des côtes de Marseille. *Mar. Biol.* 3:276–281.

Bryan, J. H. 1970. Children's reactions to helpers: their money isn't where their mouths are. In J. Macaulay & L. Berkowitz, *Altruism and helping behavior: social psychological studies of some antecedents and consequences.* New York: Academic Press. Pp. 61–73.

Bryk, F. 1919. Bibliotheca sphragidologica. *Arch. Naturges.* 85:102–183.

Buchanan, J. B. 1967. Dispersion and demography of some infaunal echinoderm populations. *Symp. Zool. Soc. London* 20:1–11.

Bütschli, O. 1876. Studien über die ersten Entwicklungsvorgänge der Eizelle, die Zelltheilung und die Conjugation der Infusorien. *Abh. Senckenbergischen Naturf. Gesell.* 10:213–452.

Buffon, G. L. Leclerc, comte de 1812. *Natural history, general and particular.* Translated by W. Smellie. 20 vols. London: Cadell & Davies.

Burdon-Jones, C. 1951. Observations on the spawning behaviour of *Saccoglossus horsti* Brambell & Goodhart, and of other Enteropneusta. *J. Mar. Biol. Assoc.* 29:625–638.

———. 1952. Development and biology of the larva of *Saccoglossus horsti* (Enteropneusta). *Phil. Trans. Roy. Soc. London* (B)236:553–589.

———. 1956. Nachtrag: Enteropneusta. *Handbuch der Zoologie* 3(2:9):57–78.

Burkenroad, M. D. 1947. Production of sound by the fiddler crab, *Uca pugilator* Bosc, with remarks on its nocturnal and mating behavior. *Ecol.* 28:458–462.

Burton, M. 1928. A comparative study of the characteristics of shallow-water and deep-sea sponges, with notes on their external form and reproduction. *J. Quekett Microscop. Club* (2)16:49–70.

———. 1948. Observations on littoral sponges, including the supposed swarming of larvae, movement and coalescence in mature individuals, longevity and death. *Proc. Zool. Soc. London* 118:893–915.

———. 1949. Non-sexual reproduction in sponges, with special reference to a collection of young *Goedia. Proc. Linn. Soc. London* 160:163–178.

Butler, T. H. 1960. Maturity and breeding of the Pacific edible crab, *Cancer magister* Dana. *J. Fish. Res. Board Canada* 17:641–646.

——. 1964. Growth, reproduction, and distribution of pandalid shrimps in British Columbia. *J. Fish. Res. Board Canada* 21:1403–1452.

von Buttel-Reepen, H. 1903. Die phylogenetische Entstehung des Bienenstaates, sowie Mitteilungen zur Biologie der solitären und sozialen Apiden. *Biol. Centralbl.* 23:4–31, 89–108, 129–154, 183–195.

Cable, R. M. 1971. Parthenogenesis in parasitic helminths. *Amer. Zool.* 11:267–272.

Cade, T. J. 1960. Ecology of the peregrine and gyrfalcon populations in Alaska. *Univ. California Publ. Zool.* 63:151–290.

le Calvez, J. 1953. Ordre des Foraminifères. *Traité de Zoologie* 1(2):149–265.

Cameron, A. W. 1967. Breeding behavior in a colony of western Atlantic gray seals. *Canad. J. Zool.* 45:161–173.

——. 1970. Seasonal movements and diurnal activity rhythms of the gray seal (*Halichoerus grypus*). *J. Zool.* 161:15–23.

Campbell, M. H. 1934. The life history and post embryonic development of the copepods, *Calanus tonsus* Brady and *Euchaeta japonica* Manikawa. *J. Biol. Board Canada* 1:1–65.

Campos Rosado, J. M., & A. Robertson 1966. The genetic control of sex ratio. *J. Theoret. Biol.* 13:324–329.

Cannon, W. B. 1939. *The wisdom of the body*. New York: Norton. xviii + 333 pp.

Carlgren, O. 1901. Die Brutpflege der Actinarien. *Biol. Centralbl.* 21:468–484.

Carlisle, D. B. 1959. On the sexual biology of *Pandalus borealis* (Crustacea Decapoda). *J. Mar. Biol. Assoc.* 38:381–394, 481–491, 493–506.

——. 1961. Locomotory powers of adult ascidians. *Proc. Zool. Soc. London* 136:141–146.

Carlquist, S. 1965. *Island life: a natural history of the islands of the world*. Garden City: Natural History Press. viii + 451 pp.

——. 1966. The biota of long-distance dispersal. IV. Genetic systems in the flora of oceanic islands. *Evolution* 20:433–455.

Carnap, R. 1967. *The logical structure of the world and pseudoproblems in philosophy*. Berkeley & Los Angeles: University of California Press. xxvi + 364 pp.

Carneiro, R. L. 1967. *The evolution of society: selections from Herbert Spencer's* Principles of Sociology, edited and with an introduction by Robert L. Carneiro. Chicago: University Press. lvii + 241 pp.

——. 1970. A theory of the origin of the state. *Science* 169:733–738.

Carpenter, C. R. 1958. Territoriality: a review of concepts and problems. In A. Roe & G. G. Simpson, *Behavior and evolution*. New Haven: Yale University Press. Pp. 224–250.

Carrick, R., & S. E. Ingham 1960. Ecological studies of the southern elephant seal, *Mirounga leonina* (L.), at Macquarie Island and Heard Island. *Mammalia* 24:325–342.

Carson, H. L. 1955. Variation in genetic recombination in natural populations. *J. Cell. Comp. Physiol.* 45 (suppl. 2):221–236.

———. 1965. Chromosomal morphism in geographically widespread species of *Drosophila*. In H. G. Baker & G. L. Stebbins, *The genetics of colonizing species*. New York: Academic Press. Pp. 503–531.

Castenholz, R. W. 1961. The effect of grazing on marine littoral diatom populations. *Ecology* 42:783–794.

Castle, W. E. 1896. The early embryology of *Ciona intestinalis*, Flemming (L.). *Bull. Mus. Comp. Zool. Harvard* 27:201–280.

Castro, P. 1971. Nutritional aspects of the symbiosis between *Echinoecus pentagonus* and its host in Hawaii, *Echinothrix calamaris*. In T. C. Cheng, *Aspects of the biology of symbiosis*. Baltimore: University Park Press. Pp. 229–247.

Caullery, M. 1961. Classe des Orthonectides. *Traité de Zoologie* 4(1):695–706.

Caws, P. 1969. The structure of discovery. *Science* 166:1375–1380.

Champy, C. 1929. La croissance dysharmonique des caractères sexuels accessoires. Son importance biologique. Applications pratiques de ses lois. *Ann. Sci. Nat. Zool.* (10)12:193–244.

Chance, M. R. A. 1962. An interpretation of some agonistic postures: the role of "cut-off" acts and postures. *Symp. Zool. Soc. London* 8:71–89.

Chance, M. R. A., & A. P. Mead 1953. Social behaviour and primate evolution. *Symp. Soc. Exper. Biol.* 7:395–439.

Chapman, D. J., & V. J. Chapman 1961. Life histories in the algae. *Ann. Bot.* (N.S.)25:547–561.

Charlesworth, W. R., & M. A. Kreutzer 1973. Facial expressions of infants and children. In P. Ekman, *Darwin and facial expression: a century of research in review*. New York: Academic Press. Pp. 91–168.

Charniaux-Cotton, H. 1963. La gonade de *Pandalus borealis*, crevette à hermaphrodisme protandrique fonctionnel. *Bull. Soc. Zool. France* 88:350–351.

Cheng, T. C. 1971. Enhanced growth as a manifestation of parasitism and shell deposition in parasitized mollusks. In T. C. Cheng, *Aspects of the biology of symbiosis*. Baltimore: University Park Press. Pp. 103–137.

Chevalier-Skolnikoff, S. 1973a. Facial expression of emotion in nonhuman primates. In P. Ekman: *Darwin and facial expression: a century of research in review*. New York: Academic Press. Pp. 11–89.

———. 1973b. Visual and tactile communication in *Macaca speciosa*, and its ontogenetic development. *Amer. J. Phys. Anthropol.* (In press.)

Chewyreuv, I. 1913. Le rôle des femelles dans la détermination du sexe de leur descendance dans le groupe des Ichneumonides. *C. R. Soc. Biol. Paris* 74:695–699.

Christensen, A. M., & B. Kanneworff 1964. *Kronborgia amphipodicola* gen. et sp. nov., a dioecious turbellarian parasitizing ampeliscid amphipods. *Ophelia* 1:147–166.

———. 1965. Life history and biology of *Kronborgia amphipodicola* Christensen & Kanneworff (Turbellaria, Neorhabdocoela). *Ophelia* 2:237–252.

Christensen, A. M., & J. J. McDermott 1958. Life-history and biology of the oyster crab, *Pinnotheres ostreum* Say. *Biol. Bull.* 114: 146–179.

Christie, J. R. 1929. Some observations on sex in the Mermithidae. *J. Exp. Zool.* 53:59–76.

Cifelli, R. 1969. Radiation of Cenozoic planktonic foraminifera. *Syst. Zool.* 18:154–168.

Clark, A. H. 1914. Notes on some specimens of a species of onychophore (*Oroperipatus corradoi*) new to the fauna of Panama. *Smithsonian Misc. Coll.* 63(2):1–2.

Clark, A. H., & J. Zetek 1946. The onychophores of Panama and the Canal Zone. *Proc. United States Nat. Mus.* 96:205–213.

Clark, E. 1959. Functional hermaphroditism and self-fertilization in a serranid fish. *Science* 129:215–216.

Clark, F. N. 1925. The life history of *Leuresthes tenuis*, an atherine fish with tide controlled spawning habits. *California Fish Bull.* 10:1–51.

Clements, F. E. 1916. *Plant succession; an analysis of the development of vegetation.* Washington: Carnegie Institution of Washington. xiii + 512 pp.

Clements, F. E., & V. E. Shelford 1939. *Bio-ecology.* New York: Wiley. viii + 425 pp.

Cleveland, L. R. 1965. Fertilization in *Trichonympha* from termites. *Arch. Protistenk.* 108:1–5.

Cleveland, L. R., S. R. Hall, E. P. Sanders, & J. Collier 1934. The wood-feeding roach *Cryptocercus*, its Protozoa, and the symbiosis between Protozoa and roach. *Mem. Amer. Acad. Arts Sci.* 17:i–x, 185–342.

Cloud, P. E., Jr. 1968. Pre-Metazoan evolution and the origins of the Metazoa. In E. T. Drake, *Evolution and environment.* New Haven: Yale University Press. Pp. 1–72.

Cody, M. L. 1966. A general theory of clutch size. *Evolution* 20: 174–184.

Coe, W. R. 1939. Sexual phases in terrestrial nemerteans. *Biol. Bull.* 76:416–427.

———. 1943. Sexual differentiation in mollusks. I. Pelecypods. *Quart. Rev. Biol.* 18:154–164.

———. 1944. Sexual differentiation in mollusks. II. Gastropods, amphineurans, scaphopods and cephalopods. *Quart. Rev. Biol.* 19:85–97.

———. 1948. Variations in the expression of sexuality in the normally proterandric gastropod *Crepidula plana* Say. *J. Exper. Zool.* 108:155–169.

Cognetti, G. 1954. La proteroginia in una popolazione di *Asterina pancerii* Gassco del Golfo di Napoli. *Boll. Zool.* 21:77–80.

———. 1956. Autofecondazione in *Asterina. Boll. Zool.* 23:275–278.

Cognetti, G., & R. Delavault 1962. La sexualité des Astérides. *Cah. Biol. Mar.* 3:157–182.

Cognetti, G., & A. M. Pagliai 1963. Razze sessuali in *Brevicoryne brassicae* L. (Homoptera Aphididae). *Arch. Zool. Ital.* 48:329–337.

Colbert, E. H. 1955. Giant dinosaurs. *Trans. New York Acad. Sci.* 3:199–209.

Cole, L. C. 1954. The population consequences of life history phenomena. *Quart. Rev. Biol.* 29:103–137.

Cole, L. J. 1901. Notes on the habits of pycnogonids. *Biol. Bull.* 2:195–207.

Collias, N. E., & E. C. Collias 1969. Size of breeding colony related to attraction of mates in a tropical passerine bird. *Ecology* 50:481–488.

———. 1970. The behaviour of the West African village weaverbird. *Ibis* 112:457–480.

Connell, J. H. 1961a. The influence of interspecific competition and other factors on the distribution of the barnacle *Chthamalus stellatus*. *Ecology* 42:710–723.

———. 1961b. Effects of competition, predation by *Thais lapillus*, and other factors on natural populations of the barnacle *Balanus balanoides*. *Ecol. Monogr.* 31:61–104.

———. 1970. A predator-prey system in the marine intertidal region. I. *Balanus glandula* and several predatory species of *Thais*. *Ecol. Monogr.* 40:49–78.

Connell, J. H., & E. Orias 1964. The ecological regulation of species diversity. *Amer. Nat.* 98:399–414.

Conover, R. J. 1965. Notes on the moulting cycle, development of sexual characters and sex ratio in *Calanus hyperboreus*. *Crustaceana* 8:308–320.

Coogan, A. H., C. Dechaseaux, & B. F. Perkins 1969. Superfamily Hippuritacea Gray, 1848. *Treatise on Invertebrate Paleontology* (N)2:749–817.

Cori, C. I. 1939. Phoronidea. *Klassen und Ordnungen des Tier-reichs*, 4(4:1:1):1–183.

Corliss, J. O. 1965. L'autogamie et la sénescence du cilié hyménostome *Tetrahymena rostrata* (Kahl). *Année Biol.* 4:49–69.

Cott, H. B. 1954. Allaesthetic selection and its evolutionary aspects. In J. Huxley, A. C. Hardy, & E. B. Ford, *Evolution as a process*. London: Allen and Unwin. Pp. 47–70.

Coulson, J. C., & G. Hickling 1964. The breeding biology of the grey seal, *Halichoerus grypus* (Fab.), on the Farne Islands, Northumberland. *J. Animal Ecol.* 33:485–512.

Cowles, H. C. 1901. The physiographic ecology of Chicago and vicinity; a study of the origin, development, and classification of plant societies. *Botan. Gaz.* 31:73–108, 145–182.

———. 1911. The causes of vegetative cycles. *Botan. Gaz.* 51:161–183.

Crane, J., 1941a. Crabs of the genus *Uca* from the West Coast of Central America. *Zoologica* 26:145–208.

———. 1941b. On the growth and ecology of Brachyuran crabs of the genus *Ocypode*. *Zoologica* 26:297–310.

———. 1943. Display, breeding and relationships of fiddler crabs (Brachyura, Genus *Uca*) in the northeastern United States. *Zoologica* 28:217–223.

———. 1952. A comparative study of innate defensive behavior in Trinidad mantids (Orthoptera: Mantoidea). *Zoologica* 37:259–293.

———. 1957. Basic patterns of display in fiddler crabs (Ocypodidae, genus *Uca*). *Zoologica* 42:69–82.

————. 1958. Aspects of social behavior in fiddler crabs, with special reference to *Uca maracoani* (Latreille). *Zoologica* 43:113–130.

————. 1966. Combat, display and ritualization in fiddler crabs (Ocypodidae, genus *Uca*). *Phil. Trans. Roy. Soc. London* (B)251:459–472.

Crew, F. A. E. 1937. The sex ratio. *Nature* 140:449–453.

————. 1965. *Sex-determination*. 4th ed. London: Methuen. viii + 188 pp.

Crook, J. H. 1965. The adaptive significance of avian social organizations. *Symp. Zool. Soc. London* 14:181–218.

————. 1970a. Introduction: social behavior and ethology. In J. H. Crook, *Social behaviour in birds and mammals*. New York: Academic Press. Pp. xxi–xl.

————. 1970b. The socio-ecology of primates. In J. H. Crook, *Social behaviour in birds and mammals*. New York: Academic Press. Pp. 103–166.

————. 1970c. Social organization and the environment: aspects of contemporary social ethology. *Anim. Behav.* 18:197–209.

————. 1972. Sexual selection, dimorphism, and social organization in the primates. In B. Campbell, *Sexual selection and The Descent of Man 1871–1971*. Chicago: Aldine. Pp. 231–281.

Crow, J. F., & M. Kimura 1969. Evolution in sexual and asexual populations: a reply. *Amer. Nat.* 103:89–91.

Crowell, K. L. 1961. The effects of reduced competition in birds. *Proc. Nat. Acad. Sci.* 47:240–243.

————. 1962. Reduced interspecific competition among the birds of Bermuda. *Ecology* 43:75–88.

Crozier, R. H. 1970. Coefficients of relationship and identity of genes by descent in the Hymenoptera. *Amer. Nat.* 104:216–217.

Cuénot, L. 1898. Notes sur les Echinodermes. III. L'hermaphroditisme protandrique d'*Asterina gibbosa* Penn. et ses variations suivant les localités. *Zool. Anz.* 21:273–279.

da Cunha, A. B., & Th. Dobzhansky 1954. A further study of chromosomal polymorphism in *Drosophila willistoni* in its relation to the environment. *Evolution* 8:119–134.

Cunningham, J. T. 1900. *Sexual dimorphism in the animal kingdom: a theory of the evolution of secondary sexual characters*. London: Adam and Charles Black. xii + 317 pp.

Cushing, D. H. 1958. The seasonal variation in oceanic production as a problem in population dynamics. *Proc. XV Int. Congr. Zool.* pp. 805–807.

Czaplinski, B. 1965. Separateness of sexes in cestodes. *Wiad. Parazyt.* 11 (Suppl.):237–242.

Darewski, I. S., & W. N. Kulikowa 1961. Natürliche Parthenogenese in der polymorphen Gruppe der kaukasischen Felseidechse (*Lacerta saxicola* Eversmann). *Zool. Jahrb.* (Syst.)89:119–176.

Darling, F. F. 1937. *A herd of red deer; a study in animal behaviour*. Oxford: University Press. x + 215 pp.

Darlington, C. D. 1956. Natural populations and the breakdown of classical genetics. *Proc. Roy. Soc. London* (B)145:350–364.

———. 1958. *The evolution of genetic systems.* 2nd ed. New York: Basic
Books. xii + 265 pp.

———. 1959. *Darwin's place in history.* Oxford: Basil Blackwell. x + 101 pp.

———. 1969. *The evolution of man and society.* London: George Allen &
Unwin. 753 pp.

Darlington, P. J., Jr. 1943. Carabidae of mountains and islands: data on the
evolution of isolated faunas, and on atrophy of wings. *Ecol. Monogr.*
13:37–61.

———. 1957. *Zoogeography: the geographical distribution of animals.* New
York: Wiley. xi + 675 pp.

———. 1959. Area, climate, and evolution. *Evolution* 13:488–510.

———. 1961. Australian carabid beetles V. Transition of wet forest faunas
from New Guinea to Tasmania. *Psyche* 68:1–24.

———. 1970. Carabidae on tropical islands, especially the West Indies. *Biotropica* 2:7–15.

———. 1971a. Interconnected patterns of biogeography and evolution. *Proc.
Nat. Acad. Sci. U.S.A.* 68:1254–1258.

———. 1971b. The carabid beetles of New Guinea. Part IV. General considerations; analysis and history of Fauna; taxonomic supplement.
Bull. Mus. Comp. Zool. Harvard 142:129–337.

———. 1972. Nonmathematical models for evolution of altruism, and for
group selection. *Proc. Nat. Acad. Sci. U.S.A.* 69:293–297.

Darwin, C. 1851. *A monograph on the sub-class Cirripedia, with figures of
all the species: the Lepadidae; or, pedunculated cirripedes.* London:
Ray Society. xii + 400 pp.

———. 1854. *A monograph on the sub-class Cirripedia, with figures of all
the species: the Balanidae, (or sessile cirripedes); the Verrucidae, etc.,
etc., etc.* London: Ray Society. viii + 684 pp.

———. 1858. On the agency of bees in the fertilization of papilionaceous
flowers, and on the crossing of kidney beans. *Ann. Nat. Hist.* (3)2:
459–465.

———. 1859. *On the origin of species by means of natural selection, or the
preservation of favored races in the struggle for life.* London: John
Murray. ix + 490 pp.

———. 1871. *The descent of man, and selection in relation to sex.* New
York: D. Appleton. xiv + 845 pp.

———. 1872. *The expression of the emotions in man and animals.* London:
John Murray. vi + 374 pp.

———. 1873. On the males and complemental males of certain cirripedes,
and on rudimentary structures. *Nature* 8:431–432.

———. 1874. *The descent of man, and selection in relation to sex.* 2nd ed.
New York: Hurst. 705 pp.

———. 1876. *The effects of cross and self fertilisation in the vegetable kingdom.* London: John Murray. viii + 482 pp.

———. 1877a. *The various contrivances by which orchids are fertilised by
insects.* 2nd ed. New York: D. Appleton. xvi + 300 pp.

———. 1877b. *The different forms of flowers on plants of the same species.*
New York: D. Appleton. viii + 352 pp.

————. 1882. *The variation of animals and plants under domestication.* 2nd ed. London: John Murray. xxiv + 968 pp.

Davidson, F. A. 1935. The development of the secondary sexual characters in the pink salmon (*Onchorhynchus gorbuscha*). *J. Morphol.* 57: 169–183.

Dawydoff, C. N. 1930. Quelques observations sur *Pelagosphaera,* larve de sipunculide des côtes d'Annam. *Bull. Soc. Zool. France* 55:88–90.

Day, J. H. 1963. The complexity of the biotic environment. *System. Assoc. Publ.* 5:31–49.

Dayton, P. K. 1971. Competition, disturbance and community organization: the provision and subsequent utilization of space in a rocky intertidal community. *Ecol. Monogr.* 41:351–389.

von Dehn, M. 1950. Experimentelle Untersuchungen über den Generationswechsel der Cladoceren. II. Cytologische Untersuchungen bei *Moina rectirostris. Chromosoma* 3:167–194.

Delavault, R. 1958. L'autofécondation chez les métazoaires. *Année Biol.* 34:5–16.

————. 1960. Les cycles genitaux chez *Asterina gibbosa* de Dinard. *C. R. Acad. Sci. Paris* 251:2240–2241.

Deleurance, E. P. 1952. Le polymorphisme social et son déterminisme chez les guêpes. *Colloques Internat. Centre Nat. Rec. Sci.* 34:141–155.

Dell, R. K. 1965. Marine biology. In T. Hatherton, *Antarctica.* London: Methuen. Pp. 129–152.

De Martini, E. E. 1969. A correlative study of the ecology and comparative feeding mechanism morphology of the Embiotocidae (surf-fishes) as evidence of the family's adaptive radiation into available ecological niches. *Wassman J. Biol.* 27:177–247.

Demoll, R. 1908. Die Bedeutung der Proterandrie bei Insecten. *Zool. Jabrb.* (*Syst.*)26:621–628.

Dendy, A. 1902. On the oviparous species of Onychophora. *Quart. J. Microscop. Sci.* 45:363–416.

De Vore, I. 1962. A comparison of the ecology and behavior of monkeys and apes. In S. L. Washburn, *Classification and human evolution.* Chicago: Aldine. Pp. 301–319.

Dewey, J. 1894. The theory of emotion. I. *Psychol. Rev.* 1:553–569.

————. 1895. The theory of emotion. II. *Psychol. Rev.* 2:13–32.

————. 1910. *The influence of Darwin on philosophy, and other essays in contemporary thought.* New York: Henry Holt. vi + 309 pp.

Diakonov, D. M. 1925. Experimental and biometrical investigations on dimorphic variability of *Forficula. J. Genet.* 15:201–232.

Dilger, W. C., & P. A. Johnsgard 1958. Comments on "species recognition" with special reference to the wood duck and the mandarin duck. *Wilson Bull.* 71:46–53.

Dingle, H. 1972. Migration strategies of insects. *Science* 175:1327–1335.

Dobzhansky, T. 1950. Evolution in the tropics. *Amer. Sci.* 38:209–221.

————. 1968. On some fundamental concepts of Darwinian biology. *Evol. Biol.* 2:1–34.

Dobzhansky, T., H. Burla, & A. B. da Cunha 1950. A comparative study of chromosomal polymorphism in sibling species of the *willistoni* group of *Drosophila. Amer. Nat.* 84:229–246.

Dobzhansky, T., H. Levene, B. Spassky, & N. Spassky 1959. Release of genetic variability through recombination. III. *Drosophila prosaltans. Genetics* 44:75–92.

Dogiel, V. 1925. Die Geschlechtsprozesse bei Infusorien (speziell bei den Ophryoscoleciden), neue Tatsachen und theoretische Erwägungen. *Arch. Protistenk.* 50:283–342.

Dogiel, V. A. 1965. *General protozoology.* Oxford: University Press. xiv + 747 pp.

———. 1966. *General parasitology.* New York: Academic Press. ix + 516 pp.

Dohrn, P. F. R. 1950. Studi sulla *Lysmata seticaudata* Risso (Hyppolitidae). I: Le condizioni normali della sessualità in natura. *Pubbl. Staz. Zool. Napoli* 22:257–272.

Dohrn, P. F. R., & L. B. Holthuis 1950. *Lysmata nilita,* a new species of prawn (Crustacea Decapoda) from the western Mediterranean. *Pubbl. Staz. Zool. Napoli* 22:339–347.

Donner, J. 1966. *Rotifers.* New York: Warne. xii + 80 pp.

Dorier, A. 1965. Classe des Gordiacés. *Traité de Zoologie* 4(3):1201–1222.

Dougherty, E. C. 1955. Comparative evolution and the origin of sexuality. *Syst. Zool.* 4:145–169, 190.

Doutt, R. L. 1959. The biology of parasitic Hymenoptera. *Ann. Rev. Ent.* 4:161–182.

Downes, J. A. 1958. Assembly and mating in the biting Nematocera. *Proc. X Int. Congr. Ent.* 2:425–434.

———. 1969. The swarming and mating flight of Diptera. *Ann. Rev. Ent.* 14:271–298.

Downhower, J. F., & K. B. Armitage 1971. The yellow-bellied marmot and the evolution of polygamy. *Amer. Nat.* 105:355–370.

Drew, G. A. 1911. Sexual activities of the squid, *Loligo pealii* (Les.). I. Copulation, egg-laying and fertilization. *J. Morphol.* 22:327–359.

Driesch, H. 1908. *The science and philosophy of the organism.* London: Adam and Charles Black. xxxii + 710 pp.

Dunbar, M. J. 1968. *Ecological development in polar regions: a study in evolution.* Englewood Cliffs: Prentice-Hall. viii + 119 pp.

Duncan, D. 1908. *Life and letters of Herbert Spencer.* New York: D. Appleton. xx + 858 pp.

Dunn, E. R. 1927. Notes on *Varanus komodoensis. Amer. Mus. Nov.* 286: 1–10.

Dunn, F. L. 1970. Cultural evolution in the late Pleistocene and Holocene of Southeast Asia. *Amer. Anthropol.* 72:1041–1054.

Dunn, G. A. 1917. Development of *Dumontia filiformis.* II. Development of sexual plants and general discussion of results. *Bot. Gaz.* 63:425–467.

Durham, J. W. 1959. Palaeoclimates. In L. H. Ahrens, *Physics and chemistry of the earth.* New York & London: Pergamon Press. 3:1–16.

Durkheim, E. 1933. *The division of labor in society.* New York: Macmillan. xvi + 439 pp.

———. 1938. *The rules of sociological method.* Chicago: University Press. lx + 146 pp.

———. 1968. *Les formes élémentaires de la vie religieuse.* Paris: Presses Universitaires de France. v + 647 pp.

Earhart, C. M., & N. K. Johnson 1970. Size dimorphism and food habits of North American owls. *Condor* 72:251–264.

Eberhard, M. J. W. 1969. The social biology of polistine wasps. *Misc. Publ. Mus. Zool. Univ. Michigan* 140:1–101.

Eberhard, W. G. 1972. Altruistic behavior in a sphecid wasp: support for kin-selection theory. *Science* 175:1390–1391.

Ebling, J. 1969. Introduction. In J. Ebling, *Biology and ethics.* New York: Academic Press. Pp. xiii–xxix.

Edwards, A. W. F. 1960. Natural selection and sex ratio. *Nature* 188:960–961.

Edwards, E. 1966. Mating behaviour in the European edible crab (*Cancer pagurus* L.). *Crustaceana* 10:23–30.

Edwards, W. E. 1967. The late-Pleistocene extinction and diminution in size of many mammalian species. In P. S. Martin and H. E. Wright, Jr., *Pleistocene extinctions: the search for a cause.* New Haven: Yale University Press. Pp. 141–154.

Efford, I. E. 1967. Neoteny in sand crabs of the genus *Emerita* (Anomura, Hippidae). *Crustaceana* 13:81–93.

Ehrendorfer, F. 1965. Dispersal mechanisms, genetic systems, and colonizing abilities in some flowering plant families. In H. G. Baker & G. L. Stebbins, *The genetics of colonizing species.* New York: Academic Press. Pp. 331–351.

Ehrlich, P. R., & L. C. Birch 1967. The "balance of nature" and "population control." *Amer. Nat.* 101:97–107.

Ehrlich, P. R., & R. W. Holm 1962. Patterns and populations. *Science* 137:652–657.

Ehrman, L. 1972. Genetics and sexual selection. In B. Campbell, *Sexual selection and The Descent of Man 1871–1971.* Chicago: Aldine. Pp. 105–135.

Eibl-Eibesfeldt, I. 1955a. Über Symbiosen, Parasitismus und andere besondere zwischenartliche Beziehungen tropischer Meeresfische. *Z. Tierpsychol.* 12:203–219.

———. 1955b. Ethologische Studien am Galápagos-Seelöwen, *Zalophus wollebaeki* Sivertsen. *Z. Tierpsychol.* 12:286–303.

———. 1959. Der Fisch *Aspidontus taeniatus* als Nachahmer des Putzers *Labroides dimidiatus. Z. Tierpsychol.* 16:19–25.

———. 1966. The fighting behaviour of marine iguanas. *Phil. Trans. Roy. Soc. London* (B)251:475–476.

Eifermann, Rivka R. 1971. Social play in childhood. In R. E. Herron & B. Sutton-Smith, *Child's play.* New York: Wiley. Pp. 270–297.

Eigenmann, C. H. 1894. *Cymatogaster aggregatus* Gibbons; a contribution to the ontogeny of viviparous fishes. *Bull. United States Fish Comm.* 1892:381–478.

———. 1896. Sex-differentiation in the viviparous teleost *Cymatogaster. Arch. Entwickelungsmech. Organismen* 4:125–179.

Eiseley, L. 1972. The intellectual antecedents of *The Descent of Man*. In B. Campbell, *Sexual selection and* The Descent of Man *1871–1971*. Chicago: Aldine. Pp. 1–16.

Eisenberg, J. F., N. A. Muckenhirn, & R. Rudran 1972. The relation between ecology and social structure in primates. *Science* 176:863–874.

Ekman, P. 1972. Universals and culture differences in facial expressions of emotion. In J. Cole, *Nebraska symposium on motivation*, 1971. Lincoln: University of Nebraska Press. Pp. 207–283.

———. 1973. Darwin and cross cultural studies of facial expression. In P. Ekman, *Darwin and facial expression: a century of research in review*. New York: Academic Press. Pp. 169–222.

Ekman, P., E. R. Sorensen, & W. V. Friesen 1969. Pan-cultural elements in facial displays of emotions. *Science* 164:86–88.

Ekman, S. 1905. Die Phyllopoden, Cladoceren und freilebenden Copepoden der nordschwedischen Hochgebirge. *Zool. Jahrb. (Syst.)* 21:1–168.

El Mofty, M., & J. D. Smyth 1960. Endocrine control of sexual reproduction in *Opalina ranarum* parasitic in *Rana temporaria*. *Nature* 186:559.

Elton, C. 1930. *Animal ecology and evolution*. Oxford: University Press. 96 pp.

———. 1958. *The ecology of invasions by animals and plants*. London: Methuen. 181 pp.

Eltringham, H. 1925. On the source of the sphragidal fluid in *Parnassius apollo* (Lepidoptera). *Trans. Ent. Soc. London* 73:11–14.

Emerson, A. E. 1939a. Social coordination and the superorganism. *Amer. Midl. Nat.* 21:182–209.

———. 1939b. Populations of social insects. *Ecol. Monogr.* 9:287–300.

———. 1949. The organization of insect societies. See Allee, Emerson, Park, Park, & Schmidt (1949:419–435).

———. 1952. The supraorganismic aspects of the society. *Colloques Internat. Centre Nat. Rec. Sci.* 34:333–353.

———. 1958. The evolution of behavior among social insects. In A. Roe & G. G. Simpson, *Behavior and evolution*. New Haven: Yale University Press. Pp. 311–335.

———. 1960. The evolution of adaptation in population systems. In S. Tax, *Evolution after Darwin*. Chicago: University Press. 1:307–348.

Emiliani, C. 1958. Ancient temperatures. *Sci. Amer.* 198:54–63.

Emmel, T. C. 1972. Mate selection and balanced polymorphism in the tropical nymphalid butterfly, *Anartia fatima*. *Evolution* 26:96–107.

Ernst, A. 1953. Primärer und sekundärer Blütenmonomorphismus bei Primeln. *Österreichische Bot. Z.* 100:235–255.

Ernst-Schwarzenbach, M. 1939. Zur Kenntnis des sexuellen Dimorphismus der Laubmoose. *Arch. Julius Klaus-Stiftung für Verb.-Forsch. Socialanthropol. Rassenhygene Zürich*, 14:361–474.

Espinas, A. 1924. *Des sociétés animales; études de psychologie comparée*. 3rd ed. Paris: Alcan. 454 pp.

Evans, H. E. 1958. The evolution of social life in wasps. *Proc. X Int. Congr. Ent.* 2:449–457.

———. 1966. *The comparative ethology and evolution of the sand wasps*. Cambridge: Harvard University Press. xviii + 526 pp.

Evans, M. A., & H. E. Evans 1970. *William Morton Wheeler, biologist.* Cambridge: Harvard University Press. xii + 363 pp.

Evans-Pritchard, E. E. 1962. *Social anthropology and other essays.* New York: Free Press. viii + 354 pp.

Ewer, R. F. 1971. The biology and behaviour of a free-living population of black rats (*Rattus rattus*). *Animal Behav. Monogr.* 4:127–174.

Faegri, K., & L. van der Pijl 1966. *The principles of pollination ecology.* New York: Pergamon Press. ix + 248 pp.

Farran, G. P. 1926. Biscayan plankton collected during a cruise of H. M. S. "Research," 1900. Part XIV. The Copepoda. *J. Linn. Soc. London* (*Zool.*)36:219–310.

Fauré-Fremiet, E. 1953. L'hypothèse de la sénescence et les cycles de réorganisation nucléaire chez les Ciliés. *Rev. Suisse Zool.* 60:426–438.

Feder, H. M. 1966. Cleaning symbiosis in the marine environment. In S. M. Henry, *Symbiosis.* New York: Academic Press. 1:327–380.

Fedotov, D. M. 1938. Spezialisation und Degradation im Körperbau der Myzostomiden in Abhängigkeit von der Lebensweise. *Acta Zool.* 19:353–385.

Feinberg, E. H., & D. Pimentel 1966. Evolution of increased "female sex ratio" in the blowfly (*Phaenica sericata*) under laboratory competition with the housefly (*Musca domestica*). *Amer. Nat.* 100: 235–244.

Feldmann, J. 1957. La reproduction des algues marines dans ses rapports avec leur situation géographique. *Année Biol.* 33:49–56.

Ferguson, E., Jr. 1967. Morphology and taxonomy of freshwater Ostracoda. *Proc. Symp. on Crustacea, Marine Biol. Assoc. India* 2:497–505.

Fields, W. G. 1965. The structure, development, food relations, reproduction, and life history of the squid *Loligo opalescens* Berry. *California Fish Bull.* 131:1–108.

Finger, F. W., & D. G. Mook 1971. Basic drives. *Ann. Rev. Psychol.* 22:1–38.

Finley, H. E. 1943. The conjugation of *Vorticella microstoma.* *Trans. Amer. Microscop. Soc.* 62:97–121.

———. 1952. Sexual differentiation in peritrichous ciliates. *J. Morphol.* 91: 569–605.

Fischer, A. G. 1960. Latitudinal variations in organic diversity. *Evolution* 14:64–81.

Fishelson, L. 1970. Protogynous sex reversal in the fish *Anthias squamipinnis* (Teleostei, Anthiidae) regulated by the presence or absence of a male fish. *Nature* 227:90–91.

Fisher, K. W. 1957. The nature of the endergonic processes in conjugation in *Escherichia coli* K–12. *J. Gen. Microbiol.* 16:136–145.

———. 1961. Environmental influences on genetic recombination in bacteria and their viruses. *Symp. Soc. Gen. Microbiol.* 11:272–295.

Fisher, R. A. 1930. *The genetical theory of natural selection.* Oxford: University Press. xiv + 272 pp.

Fisher, W. K., & G. E. MacGinitie 1928. The natural history of an echiuroid worm. *Ann. Mag. Nat. Hist.* (10)1:204–213.

Flanders, S. E. 1939. Environmental control of sex in hymenopterous insects. *Ann. Ent. Soc. America* 32:11–26.

————. 1958. The regulation of caste ratios in the social Hymenoptera. *Proc. X Int. Congr. Ent.* 2:495.

Flavell, J. H., & J. P. Hill 1969. Developmental psychology. *Ann. Rev. Psychol.* 20:1–56.

Fleischer, M. 1920. Über die Entwicklung der Zwergmännechen aus sexuell differenzierten Sporen bei den Laubmoosen. *Ber. Deutsch. Bot. Ges.* 38:84–92.

Föyn, B. 1927. Studien über Geschlecht und Geschlechtzellen bei Hydroiden. 1. Ist *Clava squamata* (Müller) eine gonochoristische oder hermaphrodite Art. *Ark. Entwicklungsmech. Organ.* 109:513–534.

Forbes, S. A. 1887. The lake as a microcosm. *Bull. Sci. Acad. Peoria* 1887: 77–87. Seen as reprint in *Bull. Illinois State Nat. Hist. Survey* 15: 537–550.

Ford, E. B. 1953. The genetics of polymorphism in the Lepidoptera. *Adv. Genet.* 5:43–87.

Forel, A. 1928. *The social world of the ants compared with that of man.* New York: Putnam's Sons lxv + 996 pp.

Forsman, B. 1956. Notes on the invertebrate fauna of the Baltic. *Ark. Zool.* (2)9:389–419.

Fowler, G. H. 1909. Biscayan plankton. Part XII. The Ostracoda. *Trans. Linn. Soc. London* (*Zool.*) 10:219–336.

Fox, R. 1972. Alliance and constraint: sexual selection in the evolution of human kinship systems. In B. Campbell, *Sexual selection and* The Descent of Man *1871–1971.* Chicago: Aldine. Pp. 282–331.

Fraser, A. F. 1968. *Reproductive behaviour in ungulates.* London: Academic Press. x + 202 pp.

Frechkopf, S., 1946. Notes sur les mammifères XXIX. De l'okapi et des affinités des Giraffidés avec les antilopes. *Bull. Mus. Roy. Hist. Nat. Belgique* 22:1–28.

Free, J. B. 1965. The allocation of duties among worker honeybees. *Symp. Zool. Soc. London* 14:39–59.

————. 1966. Seasonal regulation in the honeybee colony. In H. Kalmus, *Regulation and control in living systems.* New York: Wiley. Pp. 351–379.

Free, J. B., & C. G. Butler 1959. *Bumblebees.* New York: Macmillan. xiv + 208 pp.

Frey, D. G. 1965. Gynandromorphism in the chydorid Cladocera. *Limnol. Oceanogr.* 10(Suppl.):R103–R114.

————. 1966. Cladocera in space and time. *Proc. Symp. on Crustacea, Marine Biol. Assoc. India* 1:1–9.

Friederichs, K. 1927. Grundsätzliches über die Lebenseinheiten höher Ordnung und den ökologischen Einheitfaktor. *Naturwissenschaften* 15: 153–157, 182–186.

————. 1937. Ökologie als Wissenschaft von der Natur, oder biologische Raumforschung. *Bios* 7:i–viii, 1–108.

von Frisch, K. 1952. Die wechselseitigen Beziehungen und die Harmonie im Bienenstaat. *Colloques Internat. Centre Nat. Rec. Sci.* 34: 271–292.

———. 1967. *The dance language and orientation of bees.* Cambridge: Harvard University Press. xiv + 566 pp.

Fritsch, F. E. 1965. *The structure and reproduction of the algae.* Vol. II. Cambridge: University Press. xiv + 939 pp.

Frochot, B. 1967. Reflexions sur les rapports entre prédateurs et proies chez les rapaces II. L'influence des proies sur les rapaces. *Terre Vie* 21: 33–62.

Fryxell, P. A. 1957. Mode of reproduction of higher plants. *Botan. Rev.* 23:135–233.

Fuhrmann, O. 1904. Ein getrenntgeschlechtiger Cestode. *Zool. Jahrb.* (*Syst.*) 20:131–150.

Funke, W. 1957. Zur Biologie und Ethologie einheimischer Lamiinen (Cerambycidae, Coleoptera). *Zool. Jahrb.* (*Syst.*)85:73–176.

Furgeson, E. 1967. Morphology and taxonomy of freshwater Ostracoda. *Proc Symp. on Crustacea,* Marine Biol. Assoc. India (2):497–505.

Gabbutt, P. D. 1954. Notes on the mating behaviour of *Nemobius sylvestris* (Bosc) (Orth., Gryllidae). *British J. Animal Behav.* 2:84–88.

Gadgil, M. 1972. Male dimorphism as a consequence of sexual selection. *Amer. Nat.* 106:576–580.

Gadgil, M., & W. H. Bossert 1970. Life historical consequences of natural selection. *Amer. Nat.* 104:1–24.

Gadgil, M., & O. T. Solbrig 1972. The concept of *r* and *K* selection: evidence from wildflowers and some theoretical considerations. *Amer. Nat.* 106:14–31.

Gaffron, E. 1885. Beiträge zur Anatomie und Histologie von *Peripatus.* i, ii. *Zool. Beitr.* 1:33–60, 145–163.

Gallien, L., & M. de Larambergue 1938. Biologie et sexualité de *Lacuna pallidula* Da Costa (Littorinidae). *Trav. Stat. Zool. Wimereux* 13: 293–306.

Gandolfi, G. 1971. Sexual selection in relation to the social status of males in *Poecilia reticulata* (Teleostei: Poeciliidae). *Bol. Zool.* 38:35–48.

Garstang, W. 1928. The morphology of the Tunicata, and its bearings on the phylogeny of the Chordata. *Quart. J. Microscop. Sci.* 72: 51–187.

Gasman, D. 1971. *The scientific origins of national socialism: social Darwinism in Ernst Haeckel and the German Monist League.* London: Macdonald, and New York: American Elsevier. xxxii + 208 pp.

Gatenby, J. B. 1920. The germ-cells, fertilisation, and early development of *Grantia* (*Sycon*) *compressa. J. Linn. Soc. London* (*Zool.*) 34:261–297.

Gause, G. F. 1934. *The struggle for existence.* Reprint (1969): New York: Hafner. ix + 163 pp.

Geddes, P., & J. A. Thomson 1901. *The evolution of sex.* 2nd ed. London: Scott. xx + 342 pp.

———. 1911. *Evolution.* New York: Holt. 256 pp.

Gee, J. M., & G. B. Williams 1965. Self and cross-fertilization in *Spirorbis borealis* and *S. pagenstecheri. J. Mar. Biol. Assoc.* 45:275–285.

Geiser, S. W. 1923. Evidences of a differential death rate of the sexes among animals. *Amer. Midl. Nat.* 8:153–163.

Geist, V. 1966. The evolution of horn-like organs. *Behaviour* 27:175–214.

———. 1971. The relation of social evolution and dispersal in ungulates during the Pleistocene, with emphasis on the Old World deer and the genus *Bison. Quarternary Res.* 1:285–315.

Gerhardt, U. 1913–1914. Copulation und Spermatophoren von Grylliden und Locustiden. I, II. *Zool. Jahrb. (Syst.)* 35:415–532, 37:1–64.

———. 1921. Neue Studien ueber Copulation und Spermatophoren von Grylliden und Locustiden. *Acta Zool.* 2:293–327.

———. 1924. Neue Studien zur Sexualbiologie und zur Bedeutung des Grossendimorphismus der Spinnen. *Z. Morphol. Ökol. Tiere* 1:507–538.

———. 1928. Biologische Studien an greichischen, corsischen und deutschen Spinnen. *Z. Morphol. Ökol. Tiere* 10:576–675.

Gerhardt, U., & A. Kästner 1937. Araneae=echte Spinnen=Webespinnen. *Handbuch der Zoologie* 3(2):394–656.

Ghiselin, B. 1952. *The creative process: a symposium.* Berkeley: University of California Press. 259 pp.

———. 1963. Ultimate criteria for two levels of creativity. In C. W. Taylor & F. Barron, *Scientific creativity: its recognition and development.* New York: Wiley. Pp. 30–43.

Ghiselin, J. 1970. Prey population: a parsimonious model for evolution of response to predator species diversity. *Science* 170:649–650.

Ghiselin, M. T. 1963. On the functional and comparative anatomy of *Runcina setoensis* Baba, an opisthobranch gastropod. *Publ. Seto Mar. Biol. Lab.* 11:390–398.

———. 1966a. On semantic pitfalls of biological adaptation. *Phil. Sci.* 33:147–153.

———. 1966b. On psychologism in the logic of taxonomic controversies. *Syst. Zool.* 15:207–215.

———. 1966c. Reproductive function and the phylogeny of opisthobranch gastropods. *Malacologia* 3:327–378.

———. 1969a. *The triumph of the Darwinian method.* Berkeley & Los Angeles: University of California Press. xi + 287 pp.

———. 1969b. The evolution of hermaphroditism among animals. *Quart. Rev. Biol.* 44:189–208.

———. 1971. The individual in the Darwinian revolution. *New Lit. Hist.* 3:113–134.

———. 1973. Darwin and evolutionary psychology. *Science* 179:964–968.

Ghiselin, M. T., & B. R. Wilson 1966. On the anatomy, natural history, and reproduction of *Cyphoma*, a marine prosobranch gastropod. *Bull. Mar. Sci.* 16:132–141.

Giese, A. C., L. Greenfield, H. Huang, A. Farmanfarmaian, R. Boolootian, & R. Lasker 1959. Organic productivity in the reproductive cycle of the purple sea urchin. *Biol. Bull.* 116:49–58.

Gilbert, J. J. 1968. Dietary control of sexuality in the rotifer *Asplanchna brightwelli* Gosse. *Physiol. Zool.* 41:14–43.

Gill, E. D. 1955. The problem of extinction, with special references to Australian marsupials. *Evolution* 9:87–92.

Gilliard, E. T. 1962. On the breeding behavior of the cock-of-the-rock (Aves, *Rupicola rupicola*). *Bull. Amer. Mus. Nat. Hist.* 124: 31–68.
———. 1969. *Birds of paradise and bower birds.* London: Weidenfeld and Nicholson. xxii + 485 pp.

Gilmore, J. Barnard 1966. Play: a special behavior. In R. N. Haber, *Current research in motivation.* New York: Holt, Rinehart & Winston. Pp. 343–355.

Ginsberg, M. 1970. Social change. In S. N. Eisenstadt, *Readings in social evolution and development.* London: Pergamon Press. Pp. 37–69.

Gleason, H. A. 1926. The individualistic concept of the plant association. *Bull. Torrey Bot. Club* 53:7–26.

———. 1939. The individualistic concept of the plant association. *Amer. Mid. Nat.* 21:92–110. (Includes discussion comments.)

Goebel, K. 1910. Über sexuellen Dimorphismus bei Pflanzen. *Biol. Zentralbl.* 30:657–679, 692–718, 721–737.

Goetsch, W. 1953. *Vergleichende Biologie der Insekten-Staaten.* 2nd ed. Leipzig: Akademische Verlagsgesellschaft Geest & Portig. vi + 482 pp.

Goetsch, W., & Br. Käthner 1937. Die Koloniegrundung der Formicinen und ihre experimentelle Beeinflussung. *Z. Morphol. Ökol. Tiere* 33:201–260.

Goldschmidt, R. 1906. *Amphioxides* und *Amphioxus. Zool. Anz.* 30:443–448.

———. 1933. A note on *Amphioxides* from Bermuda based on Dr. W. Beebe's collections. *Biol. Bull.* 64:321–325.

Gontcharoff, M. 1961. Embranchement des Némertiens. *Traité de Zoologie* 4(1):783–886.

Gooch, J. L., & T. J. M. Schopf 1972. Genetic variability in the deep sea: relation to environmental variability. *Evolution* 26:545–552.

Goodbody, I. 1961. Continuous breeding in three species of tropical ascidian. *Proc. Zool. Soc. London* 136:403–409.

Gottesman, I. I. 1965. Personality and natural selection. In S. G. Vandenberg, *Methods and goals in human behavior genetics.* New York: Academic Press. Pp. 63–80.

Gould, S. J. 1971. Muscular mechanics and the ontogeny of swimming in scallops. *Paleontology* 14:61–94.

Grant, P. R. 1965. Plumage and the evolution of birds on islands. *Syst. Zool.* 14:47–52.

———. 1968. Bill size, body size, and the ecological adaptations of bird species to competitive situations on islands. *Syst. Zool.* 17:319–333.

Grant, V. 1958. The regulation of recombination in plants. *Cold Spring Harbor Symp. Quant. Biol.* 23:337–363.

Grassé, P. P. 1952a. Le fait social: ses critères biologiques, ses limites. *Colloques Internat. Centre Nat. Rec. Sci.* 34:7–17.

———. 1952b. La régulation sociale chez les Isoptères et les Hyménoptères. *Colloques Internat. Centre Nat. Rec. Sci.* 34:323–331.

Grassé, P. P., & C. Noirot 1951. La sociotomie: migration et fragmentation de la termitière chez les *Anoplotermes* et les *Trinervitermes. Behaviour* 3:146–166.

Gravier, C. J. 1916. Sur l'incubation chez l'*Actinia equina* L. a l'île San Thomé (golfe de Guinée). *C. R. Acad. Sci. Paris* 162:986–988.

Gray, D. 1882. Notes on the characters and habits of the bottlenose whale (*Hyperodoon rostratus*). *Proc. Zool. Soc. London* (1882):726–731.

Greenblatt, S. H. 1965. The major influences on the early life and work of John Hughlings Jackson. *Bull. Hist. Med.* 39:346–376.

Gregg, R. E. 1942. The origin of castes in ants with special reference to *Pheidole morrisi* Forel. *Ecology* 23:295–308.

Grell, K. G. 1967. Sexual reproduction in Protozoa. *Res. Protozool.* 2:147–213.

Grimpe, G. 1928. Über zwei jugendliche Männchen von *Argonauta argo* L. *Zool. Jahrb.* (*Allg.*)45:79–98.

Grisebach, A. 1880. *Gesammelte Abhandlungen und kleinere Schriften zur Pflanzengeographie.* Leipzig: Engelmann. vi + 628 pp.

Groos, K. 1896. *Die Spiele der Thiere.* Jena: Fischer. xvi + 359 pp.

Gruhl, K. 1924. Paarungsgewohnheiten der Dipteren. *Z. Wiss. Zool.* 122:205–280.

Gruvel, A. 1900. Etude du mâle complémentaire du "*Scalpellum vulgare.*" *Arch. Biol.* 16:27–47.

Guber, A. L. 1970. Problems of sexual dimorphism, population structure and taxonomy of the Ordovician genus *Tetradella* (Ostracoda). *J. Paleontol.* 45:6–22.

Guenther, K. 1905. Zur geschlechtlichen Zuchtwahl. *Arch. Rassen-u. Gesellschaftsbiol.* 2:321–335.

Guhl, A. M., & W. C. Allee 1944. Some measurable effects of social organization in flocks of hens. *Physiol. Zool.* 17:320–347.

Guilday, J. E. 1967. Differential extinction during late-Pleistocene and Recent times. In P. S. Martin & H. E. Wright, Jr., *Pleistocene extinctions: the search for a cause.* New Haven: Yale University Press. Pp. 121–140.

Guinot-Dumortier, D., & B. Dumortier 1960. La stridulation chez les crabes. *Crustaceana* 1:117–155.

Gulick, A. 1932. Biological peculiarities of oceanic islands. *Quart. Rev. Biol.* 7:405–427.

Gurney, R. 1929. Dimorphism and rate of growth in Copepoda. *Int. Rev. Ges. Hydrobiol.* 21:189–207.

Guthrie, R. D. 1970. Evolution of human threat display organs. *Evol. Biol.* 4:257–302.

Guthrie, R. D., & R. G. Petocz 1970. Weapon automimicry among mammals. *Amer. Nat.* 104:585–588.

Haeckel, E. 1869. Ueber Arbeitstheilung in Natur- und Menschenleben. *Samml. gemeinverständ. wiss. Vortr.* (4):78:195–232.

von Hagen, H. O. 1962. Freilandstudien zur Sexual- und Fortpflanzungsbiologie von *Uca tangeri* in Andalusien. *Z. Morphol. Ökol. Tiere* 51:611–725.

Hagen, Y. 1942. Totalgewichts-Studien bei norwegischen Vogelarten. *Arch. Naturgesch.* 11:1–173.

———. 1952. Birds of Tristan da Cunha. *Results Norwegian Sci. Exped. Tristan da Cunha 1937–1938* (20):1–248.

Hairston, N. G. 1959. Species abundance and community organization. *Ecology* 40:404–416.

Hairston, N. G., F. E. Smith, & L. B. Slobodkin 1960. Community structure, population control, and competition. *Amer. Nat.* 94:421–425.

Haldane, J. B. S. 1964. A defense of beanbag genetics. *Persp. Biol. Med.* 7:343–359.

Hall, K. R. L. 1964. Aggression in monkey and ape societies. In J. D. Carthy & F. J. Ebling, *The natural history of aggression*. New York: Academic Press. Pp. 51–64.

Hallam, A. 1963. Major epeirogenic and eustatic changes since the Cretaceous, and their possible relationship to crustal structure. *Amer. J. Sci.* 261:397–423.

Hamilton, J. E. 1934. The southern sea lion, *Otaria byronia* (De Blainville). *Discovery Rep.* 8:269–318.

Hamilton, T. H. 1961. The adaptive significances of intraspecific trends of variation in wing length and body size among bird species. *Evolution* 15:180–195.

Hamilton, T. H., & R. H. Barth, Jr. 1962. The biological significance of season change in male plumage appearance in some new world migratory bird species. *Amer. Nat.* 96:129–144.

Hamilton, W. D. 1964. The genetical evolution of social behaviour, I., II. *J. Theoret. Biol.* 7:1–16, 17–52.

———. 1966. The moulding of senescence by natural selection. *J. Theoret. Biol.* 12:12–45.

———. 1967. Extraordinary sex ratios. *Science* 156:477–488.

———. 1971. Geometry for the selfish herd. *J. Theoret. Biol.* 31:295–311.

———. 1972. Altruism and related phenomena, mainly in social insects. *Ann. Rev. Ecol. Syst.* 3:193–232.

Hamilton, W. J. III, & K. E. F. Watt 1970. Refuging. *Ann. Rev. Ecol. Syst.* 1:263–286.

Hansen, B. 1953. Brood protection and sex ratio of *Transennella tantilla* (Gould) a Pacific bivalve. *Videnskab. Medd. Dansk Naturhist. Foren.* 115:313–324.

Hanson, N. R. 1958. *Patterns of discovery: an inquiry into the conceptual foundations of science.* Cambridge: University Press. x + 240 pp.

Haq, S. M. 1965. Development of the copepod *Enterpina acutifrons* with special reference to dimorphism in the male. *Proc. Zool. Soc. London* 144:175–201.

Hardy, A. C. 1956. *The open sea, its natural history: the world of plankton.* Boston: Houghton Mifflin. xv + 335 pp.

Harlow, H. F. 1958. The evolution of learning. In A. Roe & G. G. Simpson, *Behavior and evolution.* New Haven: Yale University Press. Pp. 269–290.

Harlow, H. F., R. O. Dodsworth, & M. K. Harlow 1965. Total social isolation in monkeys. *Proc. Nat. Acad. Sci. U.S.A.* 54:90–97.

Harms, J. W. 1927. Koloniegründung bei *Macrotermes gilvus* Hag. *Zool. Anz.* 74:221–236.

Harper, J. L., P. H. Lovell, & K. G. Moore 1970. The shapes and sizes of seeds. *Ann. Rev. Ecol. Syst.* 1:327–356.

Harrington, R. W., Jr. 1961. Oviparous hermaphroditic fish with internal self-fertilization. *Science* 134:1749–1750.

——. 1967. Environmentally controlled induction of primary male gonochorists from eggs of the self-fertilizing hermaphroditic fish, *Rivulus marmoratus* Poey. *Biol. Bull.* 132:174–199.

Harrington, R. W., Jr., & K. D. Kallman 1968. The homozygosity of clones of the self-fertilizing hermaphroditic fish *Rivulus marmoratus* Poey (Cyprinodontidae, Atheriniformes). *Amer. Nat.* 102:337–343.

Harris, W. V. 1958. Colony formation in the Isoptera. *Proc. X Int. Congr. Ent.* 2:435–439.

Harrison Matthews, L. 1929. The natural history of the elephant seal: with notes on other seals found at South Georgia. *Discovery Rep.* 1:233–256.

——. 1964. Overt fighting in mammals. In J. D. Carthy & F. J. Ebling, *The natural history of aggression*. New York: Academic Press. Pp. 23–32.

Hart, T. J. 1942. Phytoplankton periodicity in Antarctic surface waters. *Discovery Rep.* 21:261–356.

Hartl, D. L. 1971. Some aspects of natural selection in arrhenotokous populations. *Amer. Zool.* 11:309–325.

Hartl, D. L., & S. W. Brown 1970. The origin of male haploid genetic systems and their expected sex ratio. *Theoret. Pop. Biol.* 1:165–190.

Hartmann, M. 1956a. *Die Sexualität: das Wesen und die Grundgesetzlichkeiten des Geschlechts und der Geschlechtsbestimmung im Tier- und Pflanzenreich.* 2nd ed. Stuttgart: Fischer. xv + 463 pp.

——. 1956b. *Gesammelte Vorträge und Aufsätze.* Stuttgart: Fischer. 649 pp.

Hartnoll, R. G. 1969. Mating in the Brachyura. *Crustaceana* 16:161–181.

Hartup, W. W., & A. Yonas 1971. Developmental psychology. *Ann. Rev. Psychol.* 22:337–392.

Harvey, E. B. 1956. Sex in sea urchins. *Pubbl. Staz. Zool. Napoli* 28:127–135.

Haskins, C. P., E. F. Haskins, & R. E. Hewitt 1960. Pseudogamy as an evolutionary factor in the poeciliid fish *Mollienisia formosa*. *Evolution* 14:473–483.

Hauenschild, D. C. 1954. Zur Frage der Geschlechtsbestimmung bei *Asterina gibbosa* (Echinoderm. Asteroïd). *Zool. Jahrb. (Allg.)* 65:43–53.

Haven, N. D. 1971. Temporal patterns of sexual and asexual reproduction in the colonial ascidian *Metandrocarpa taylori* Huntsman. *Biol. Bull.* 140:400–415.

Hawes, R. S. J. 1963. The emergence of asexuality in Protozoa. *Quart. Rev. Biol.* 38:234–242.

Hazlett, B. A. 1968. Sexual behavior of some European hermit crabs (Anomuri: Paguridae). *Pubbl. Staz. Zool. Napoli* 36:238–252.

Healey, M. C. 1967. Aggression and self-regulation of population size in deermice. *Ecology* 48:377–392.

Heath, H. 1908. The gonad in certain species of chiton. *Zool. Anz.* 32:10–12.

Hecht, M. K. 1952. Natural selection in the lizard genus *Aristelliger*. *Evolution* 6:112–124.

Hedgpeth, J. W. 1955. Pycnogonida. *Treatise on Invertebrate Paleontology* (P):163–170.

———. 1971a. Philosophy on Cannery Row. In R. Astro & T. Hayashi, *Steinbeck: the man and his work*. Corvallis: Oregon State University Press. Pp. 89–129.

———. 1971b. Perspectives of benthic ecology in Antarctica. In *Research in the Antarctic*. Washington: American Association for the Advancement of Science. Pp. 93–136.

Hediger, H. 1952. Beiträge zur Säugetier-Soziologie. *Colloques Internat. Centre Nat. Rec. Sci.* 34:297–321.

———. 1970. Zum Fortpflanzungsverhalten des kanadischen Bibers (*Castor fiber canadensis*). *Forma et Functio* 2:336–351.

Heinrich, B., & P. H. Raven 1972. Energetics and pollination ecology. *Science* 176:597–602.

Helfer, H., & E. Schlottke 1935. Pantopoda. *Klassen und Ordnungen des Tier-reichs* 5(4:2):1–314.

Henderson, L. J. 1913. *The fitness of the environment: an inquiry into the biological significance of the properties of matter*. New York: Macmillan. xv + 317 pp.

Hendler, G., & D. R. Franz 1971. Population dynamics and life history of *Crepidula convexa* Say (Gastropoda: Prosobranchia) in Delaware Bay. *Biol. Bull.* 141:514–526.

Henry, D. P., & P. A. McLaughlin 1965. Unique occurrence of complemental males in a sessile barnacle. *Nature* 207:1107–1108.

Herbst, C. G., & G. R. Johnstone 1937. Life history of *Pelagophycus porra*. *Bot. Gaz.* 99:339–354.

Herron, R. E., & B. Sutton-Smith 1971. *Child's play*. New York: Wiley. xiii + 386 pp.

Hertwig, O. 1909. *The cell: outlines of general anatomy and physiology*. London: Swan Sonnenschein. xvi + 368 pp.

Hertwig, R. 1912. Über derzeitigen Stand des Sexualitätsproblems nebst eigenen Untersuchungen. *Biol. Centralb.* 32:1–45, 65–111, 129–146.

Hessler, A. Y., R. R. Hessler, & H. L. Sanders 1970. Reproductive system of *Hutchinsoniella macracantha*. *Science* 168:1464.

Hessler, R. R. 1969. Cephalocarida. *Treatise on Invertebrate Paleontology* (R):120–128.

Hessler, R. R., & H. L. Sanders 1967. Faunal diversity in the deep sea. *Deep-Sea Res.* 14:65–78.

Hester, J. J. 1960. Late Pleistocene extinction and radiocarbon dating. *Amer. Antiquity* 26:58–77.

Hewer, H. R. 1960. Behaviour of the grey seal (*Halichoerus grypus* Fab.) in the breeding season. *Mammalia* 24:400–421.

———. 1964. The determination of age, sexual maturity, longevity and a life-table in the grey seal (*Halichoerus grypus*). *Proc. Zool. Soc. London* 142:593–624.

Heymons, R. 1935. Pentastomida. *Klassen und Ordnungen des Tier-reichs* 5(4):1–268.

Heymons, R., & H. von Lengerken 1929. Biologische Untersuchungen an coprophagen Lamellicorniern. I. Nahrungserwerb und Fortpflanzungsbiologie der Gattung *Scarabaeus* L. *Z. Morphol. Ökol. Tiere* 14:531–613.

Hildebrand, F. 1867. *Die Geschlechter-Vertheilung bei den Pflanzen und das Gesetz der vermiedenen und unvortheilhaften stetigen Selbstbefruchtung.* Leipzig: Verlag von Wilhelm Engelmann. iv + 92 pp.

Hill, N. P. 1944. Sexual dimorphism in the Falconiformes. *Auk* 61:228–234.

Hille Ris Lambers, D. 1966. Polymorphism in Aphididae. *Ann. Rev. Ent.* 11:47–78.

Hingston, R. W. G. 1933. *The meaning of animal colour and adornment: being a new explanation of the colours, adornments and courtships of animals, their songs, moults, extravagant weapons, the differences between their sexes, the manner of formation of their geographical varieties, and other allied problems.* London: Edward Arnold. 411 pp.

Hoek, P. P. C. 1883. Report on the Cirripedia collected by H. M. S. Challenger during the years 1873–1876.—Systematic part. *Challenger Rep. (Zool.)*8(3):1–168.

———. 1884. Report on the Cirripedia collected by H. M. S. Challenger, during the years 1873–1876.—Anatomical part. *Challenger Rep. (Zool.)* 10(3):1–47.

Hoffmann, D. L. 1968. Seasonal eyestalk inhibition on the androgenic glands of a protandric shrimp. *Nature* 218:170–172.

Hoffman, H. 1927. Pulmonata. *Klassen und Ordnungen des Tier-reichs* 3(2:2):965–1219.

Hofmeister, K. T. 1939. Untersuchungen über Zwitterbildungen beim Stint (*Osmerus eperlanus* L.). *Z. Morphol. Ökol. Tiere* 35:221–245.

Hofstadter, R. 1955. *Social Darwinism in American thought.* 2nd ed. Boston: Beacon Press. x + 248 pp. (1st ed. 1944. Philadelphia: University of Pennsylvania Press.)

Hogan-Warburg, A. J. 1966. Social behaviour of the ruff, *Philomachus pugnax* (L.). *Ardea* 54:109–229.

Holgate, P. 1967. Population survival and life history phenomena. *J. Theoret. Biol.* 14:1–10.

Honda, H. 1925. Experimental and cytological studies on bisexual and hermaphrodite free-living nematodes, with special reference to problems of sex. *J. Morphol.* 40:191–233.

Honigberg, B. M. 1970. Protozoa associated with termites and their role in digestion. In K. Krishna & F. M. Weesner, *Biology of termites.* New York: Academic Press. 2:1–36.

van Hooff, J. A. R. A. M. 1962. Facial expressions in higher primates. *Symp. Zool. Soc. London* 8:97–125.

———. 1967. The facial displays of the catarrhine monkeys and apes. In D. Morris, *Primate ethology.* Chicago: Aldine. Pp. 7–68.

Hopkins, T. S. 1964. The host relations of a pinnotherid crab, *Opisthopus transversus*. *Bull. Southern California Acad. Sci.* 63:175–180.

Horn, H. S. 1968. The adaptive significance of colonial nesting in the Brewer's blackbird (*Euphagus cyanocephalus*). *Ecology* 49:682–694.

Hornstein, H. A. 1970. The influence of social models on helping. In J. Macaulay & L. Berkowitz, *Altruism and helping behavior: social psychological studies of some antecedents and consequences*. New York: Academic Press. Pp. 29–41.

van der Horst, C. J. 1927–1939. Hemichordata. *Klassen u. Ordnungen des Tier-reichs* 4(4:2:2):xii + 737 pp.

———. 1932–1956. Enteropneusta. *Handbuch der Zoologie* 3(2:9):1–56.

———. 1936. *Planktosphaera* and Tornaria. *Quart. J. Microscop. Sci.* 78: 605–613.

van der Horst, C. J., & J. G. Helmcke 1956a. Cephalodiscidae. *Handbuch der Zoologie* 3(2:8):33–66.

———. 1956b. *Planktosphaera. Handbuch der Zoologie* 3(2:9):79–82.

Hotchkiss, R. D. 1954. Cyclic behavior in pneumococcal growth and transformability occasioned by environmental changes. *Proc. Nat. Acad. Sci.* 40:49–55.

Howard, H. E. 1920. *Territory in bird life*. London: John Murray. xiii + 308 pp.

Howard, H. W. 1940. The genetics of *Armadillidium vulgare* Latr. I. A general survey of the problems. *J. Genet.* 40:83–108.

Howat, G. R. 1945. Variations in the composition of the sea in West-African waters. *Nature* 155:415–417.

Hubbard, H. G. 1897. Ambrosia beetles. *United States Dept. Agric. Yearb.* (1896):421–430.

Hubbs, C. L. 1921. The ecology and life-history of *Amphigonopterus aurora* and other viviparous perches of California. *Biol. Bull.* 40:181–209.

Hubbs, C. L., & A. N. Wick 1951. Toxicity of the roe of the cabezon, *Scorpaenichthys marmoratus. California Fish & Game* 37:195–196.

Hubendick, B. 1952. On the evolution of the so-called thalassoid molluscs of Lake Tanganyika. *Ark. Zool.* (2)3:319–323.

———. 1962. Aspects on the diversity of the fresh-water fauna. *Oikos* 13: 249–261.

Hughes, D. A. 1966. Behavioural and ecological investigations of the crab *Ocypode ceratophthalmus* (Crustacea: Ocypodidae). *J. Zool. London* 150:129–143.

Hughes-Schrader, S. 1930. Contributions to the life history of the iceryine coccids, with special reference to parthenogenesis and hermaphroditism. *Ann. Ent. Soc. America* 23:359–380.

Hull, D. L. 1965. The effect of essentialism on taxonomy—two thousand years of stasis. *British J. Phil. Sci.* 15:314–326, 16:1–18.

———. 1968. The operational imperative: sense and nonsense in operationism. *Syst. Zool.* 17:438–457.

———. 1970. Systemic dynamic social theory. *Sociol. Quart.* 11:351–363.

von Humboldt, Alexander 1806. Ideen zu einer Physiognomik der Gewächse.

Tübingen: Cotta. In Ostwald's *Klassiker der Exakten Wissenschaften*, No. 247 (1959). 46 pp.

von Humboldt, A. 1807. Ideen zu einer Geographie der Pflanzen nebst einem Naturgemälde der Tropenländer etc. Tübingen: Cotta. In Ostwald's *Klassiker der Exakten Naturwissenschaften*, No. 248 (1960). 180 pp. Leipzig: Akademische Verlagsgesellschaft Geest & Portig.

Hutchinson, G. E. 1948. Circular causal systems in ecology. *Ann. New York Acad. Sci.* 50:221–246.

———. 1959. Homage to Santa Rosalia *or* Why are there so many kinds of animals? *Amer. Nat.* 93:145–159.

———. 1965. *The ecological theater and the evolutionary play.* New Haven: Yale University Press. xiii + 139 pp.

Hutt, C. 1966. Exploration and play in children. *Symp. Zool. Soc. London* 18:61–81.

Huxley, J. S. 1914. The court-ship habits of the great crested grebe (*Podiceps cristatus*); with an addition to the theory of sexual selection. *Proc. Zool. Soc. London* (1914):491–562.

———. 1923. Courtship activities in the red-throated diver (*Colymbus stellatus* Pontopp); together with a discussion of the evolution of courtship in birds. *J. Linn. Soc. London* (*Zool.*) 35:253–292.

———. 1932. *Problems of relative growth.* New York: Dial Press. xix + 276 pp.

———. 1938a. The present standing of the theory of sexual selection. In G. R. de Beer, *Evolution: essays on aspects of evolutionary biology presented to Professor E. S. Goodrich on his seventieth birthday.* Oxford: University Press. Pp. 11–42.

———. 1938b. Darwin's theory of sexual selection and the data subsumed by it, in the light of recent research. *Amer. Nat.* 72:416–433.

———. 1958. Cultural process and evolution. In A. Roe & G. G. Simpson, *Behavior and evolution.* New Haven: Yale University Press. Pp. 437–454.

———. 1966. Introduction: a discussion on ritualization of behaviour in animals and man. *Phil. Trans. Roy. Soc. London* (B)251:249–271.

Huxley, T. H., & J. Huxley 1947. *Evolution and ethics: 1893–1943.* London: Pilot Press. viii + 235 pp.

Hyman, L. H. 1940. *The invertebrates: Protozoa through Ctenophora.* New York: McGraw-Hill. x + 726 pp.

———. 1951a. *The invertebrates: Platyhelminthes and Rhynchocoela.* New York: McGraw-Hill. vii + 550 pp.

———. 1951b. *The invertebrates: Acanthocephala, Aschelminthes, and Entoprocta.* New York: McGraw-Hill. vii + 572 pp.

———. 1955. *The invertebrates: Echinodermata.* New York: McGraw-Hill. vii + 763 pp.

———. 1959. *The invertebrates: smaller coelomate groups.* New York: McGraw-Hill. viii + 783 pp.

Ichikawa, A., & R. Yanagimachi 1960. Studies on the sexual organization of the Rhizocephala. II. The reproductive function of the larval (Cypris) males of *Peltogaster* and *Sacculina*. *Annot. Zool. Japon.* 33:42–56.

von Ihering, H. 1896. Zur Biologie der socialen Wespen Brasiliens. *Zool. Anz.* 19:449–453.

Istock, C. A. 1967. The evolution of complex life cycle phenomena: an ecological perspective. *Evolution* 21:592–605.

Itô, Y. 1970. Groups and family bonds in animals in relation to their habitat. In L. R. Aronson, E. Tobach, D. S. Lehrman, & J. S. Rosenblatt, *Development and evolution of behavior: essays in memory of T. C. Schneirla.* San Francisco: Freeman. Pp. 389–415.

Ivanov, A. V. 1963. *Pogonophora.* New York: Consultants Bureau. xvi + 479 pp.

Jackson, H. G. 1928. Hermaphroditism in *Rhyscotus*, a terrestrial isopod. *Quart. J. Microscop. Sci.* 71:527–539.

Jacob, F., & E. L. Wollman 1961. *Sexuality and the genetics of bacteria.* New York: Academic Press. xv + 374 pp.

Jacobs, M. E. 1955. Studies on territorialism and sexual selection in dragonflies. *Ecology* 36:566–586.

Jägersten, G. 1963. On the morphology and behaviour of *Pelagosphaera* larvae (Sipunculoidea). *Zool. Bidr. Uppsala* 36:27–35.

James, F. C. 1970. Geographic size variation in birds and its relationship to climate. *Ecology* 51:365–390.

James, W. 1890. *The principles of psychology.* New York: Henry Holt. xii + 1,389 pp.

Janzen, D. H. 1971. Euglossine bees as long-distance pollinators of tropical plants. *Science* 171:203–205.

Jarecka, L. 1960. Separation of sexes and quantitative regulation in cestodes genus *Diploposthe* (Cestoda). *Bull. Acad. Polon. Sci. (Biol.)* 8:155–157.

———. 1961. Morphological adaptations of tapeworm eggs and their importance in the life cycles. *Acta Parasitol. Polon.* 9:409–426.

Jeannel, R. 1925. L'aptérisme chez les Insectes insulaires. *C. R. Acad. Sci. Paris* 180:1222–1224.

Jenni, D. A. & G. Collier 1972. Polyandry in the American Jaçaná. *Auk* 84:743–765.

Jennings, H. S. 1910. What conditions induce conjugation in *Paramecium? J. Exper. Zool.* 9:279–299.

———. 1913. The effect of conjugation in *Paramecium. J. Exper. Zool.* 14:279–391.

———. 1929. Genetics of the Protozoa. *Bibl. Genet.* 5:105–330.

Johnson, C. 1961. Breeding behaviour and oviposition in *Hetaerina americana* (Fabricius) and *H. titia* (Drury) (Odonata: Agriidae). *Canad. Ent.* 93:260–266.

———. 1962a. A study of territoriality and breeding behavior in *Pachydiplax longipennis* Burmeister (Odonata: Libellulidae). *Southwest. Nat.* 7:191–197.

———. 1962b. Breeding behavior and oviposition in *Calopteryx maculatum* (Beauvais) (Odonata: Calopterygidae). *Amer. Midl. Nat.* 68:242–247.

———. 1962c. A description of territorial behavior and a quantitative study of its function in males of *Hetaerina americana* (Fabricius) (Odonata: Agriidae). *Canad. Ent.* 94:178–190.

———. 1963. Interspecific territoriality in *Hetaerina americana* (Fabricius) and *H. titia* (Drury) (Odonata: Calopterygidae) with a preliminary analysis of the wing color pattern variation. *Canad. Ent.* 95:575–582.

Johnson, C. G. 1966. A functional system of adaptive dispersal by flight. *Ann. Rev. Ent.* 11:233–260.

Johnson, G. 1961. Contribution à l'étude de la détermination du sexe chez les Oniscoïdes: phénomènes d'hermaphroditisme et de monogénie. *Bull. Biol. France Belgique* 95:177–267.

Johnson, R. G. 1970. Variations in diversity within benthic marine communities. *Amer. Nat.* 104:285–300.

———. 1971. Animal-sediment relations in shallow water benthic communities. *Marine Geol.* 11:93–104.

———. 1972. Conceptual models of benthic marine communities. In T. J. M. Schopf, *Models in paleobiology*. San Francisco: Freeman, Cooper. Pp. 148–159.

Johri, G. N. 1963. On a new protogynous cestode with remarks on certain species of the genus *Progynotaenia* Fuhrmann, 1909. *J. Helminthol.* 37:39–46.

Jordan, P. A., P. C. Shelton, & D. L. Allen 1967. Numbers, turnover, and social structure of the Isle Royale wolf population. *Amer. Zool.* 7:233–252.

Juchault, P. 1967. Contribution a l'étude de la différenciation sexuelle male chez les Crustacés Isopodes. *Année Biol.* 6:191–212.

Kabata, Z. 1964a. *Clavellisa emarginata* (Krøyer, 1873): morphological study of a parasitic copepod. *Crustaceana* 7:1–10.

———. 1964b. The morphology and the taxonomy of *Clavellodes pagelli* (Krøyer, 1863) (Copepoda, Lernaeopodidae). *Crustaceana* 7:103–112.

Kallman, K. D. 1962a. Gynogenesis in the teleost, *Mollienesia formosa* (Girard), with a discussion of the detection of parthenogenesis in vertebrates by tissue transplantation. *J. Genet.* 58:7–24.

———. 1962b. Population genetics of the gynogenetic teleost, *Mollienesia formosa* (Girard). *Evolution* 16:497–504.

Kalmus, H. 1932. Über den Erhaltungswert der phänotypischen (morphologischen) Anisogamie und die Entstehung der ersten Geschlechtsunterschiede. *Biol. Zentralbl.* 52:716–726.

———. 1966. Introduction—control and regulation as interactions within systems. In H. Kalmus, ed., *Regulation and control in living systems*. New York: Wiley. Pp. 3–11.

Kalmus, H., & S. Maynard Smith 1966. Some evolutionary consequences of pegmatypic mating systems (imprinting). *Amer. Nat.* 100:619–635.

Kalmus, H., & C. A. B. Smith 1960. Evolutionary origin of sexual differentiation and the sex ratio. *Nature* 186:1004–1006.

Kammerer, P. 1912. Ursprung der Geschlechtsunterschiede. *Fortsch. Naturwiss. Forschung* 5:1–240.

Kao, C. Y. 1966. Tetrodotoxin, saxitoxin and their significance in the study of excitation phenomena. *Pharmacol. Rev.* 18:997–1049.

Karlin, S., & F. M. Scudo 1969. Assortative mating based on phenotype: II. Two autosomal alleles without dominance. *Genetics* 63:499–510.

Kaston, B. J. 1965. Some little known aspects of spider behavior. *Amer. Midl. Nat.* 73:336–356.

Kauffman, E. G. 1969. Form, function and evolution. *Treatise on Invertebrate Paleontology* (N): 129–205.

Kear, J. 1970. The adaptive radiation of parental care in waterfowl. In J. H. Crook, *Social behaviour in birds and mammals*. New York: Academic Press. Pp. 357–392.

Keferstein, W. 1868. Ueber eine Zwitternemertine (*Borlasia hermaphroditica*) von St. Malo. *Arch. Naturges.* 34:102–105.

Kemp, S. 1914. Onychophora. *Rec. Indian Mus.* 8:471–492.

Kendeigh, S. C. 1972. Energy control of size limits in birds. *Amer. Nat.* 106:79–88.

Kenk, R. 1937. Sexual and asexual reproduction in *Euplanaria tigrina* (Girard). *Biol. Bull.* 73:280–294.

Kennedy, J. S., & H. L. G. Stroyan 1959. Biology of aphids. *Ann. Rev. Ent.* 4:139–160.

Kenyon, K. W. 1960. Territorial behavior and homing in the Alaska fur seal. *Mammalia* 24:431–444.

———. 1962. History of the Steller sea lion at the Pribilof Islands, Alaska. *J. Mammal.* 43:68–75.

Kenyon, K. W., & D. W. Rice 1959. Life history of the Hawaiian monk seal. *Pacific Sci.* 13:215–252.

Kenyon, K. W., & F. Wilke 1953. Migration of the northern fur seal, *Callorhinus ursinus*. *J. Mammal.* 34:86–98.

Kerner [von Marilaun,] A. 1863. *Das Pflanzenleben der Donauländer*. Translated by Henry S. Conard, 1951 as *The background of plant ecology*. Ames: Iowa State College Press. x + 238 pp.

———. 1878. *Flowers and their unbidden guests: with a prefatory letter by Charles Darwin, M.A., F.R.S.* London: Kegan Paul. xvi + 164 pp.

Kessel, E. L. 1955. The mating activities of balloon flies. *Syst. Zool.* 4:97–104.

Kilham, L. 1964. Differences in feeding behavior of male and female hairy woodpeckers. *Wilson Bull.* 77:134–145.

Kimura, M., & J. F. Crow 1963. On the maximum avoidance of inbreeding. *Genet. Res.* 4:399–415.

King, P. E., & J. H. Jarvis 1970. Egg development in a littoral pycnogonid *Nymphon gracile*. *Mar. Biol.* 7:294–304.

Kistner, D. H. 1969. The biology of termitophiles. In K. Krishna & F. M. Weesner, *Biology of termites*. New York: Academic Press. 1:525–557.

Klomp, H. 1970. The determination of clutch-size in birds: a review. *Ardea* 58:1–124.

Kniep, H. 1928. *Die Sexualität der niederen Pflanzen; Differenzierung, Verteilung, Bestimmung und Vererbung des Geschlechts bei den Thallophyten*. Jena: Fischer. vi + 544 pp.

Knight-Jones, E. W., & J. Moyse 1961. Intraspecific competition in sedentary marine animals. *Symp. Soc. Exp. Biol.* 15:72–95.

Knudsen, J. 1950. Egg capsules and development of some marine proso-branchs from tropical West Africa. *Atlantide Report*, 1:85–130.

———. 1970. The systematics and biology of abyssal and hadal Bivalvia. *Galathea Rep.* 11:1–241.

Knuth, P. 1906. *Handbook of flower pollination based upon Hermann Müller's work 'The fertilisation of flowers by insects.'* Transl. J. R. Ainsworth Davis. *Vol. I: Introduction and literature.* Oxford: Clarendon Press. xix + 382 pp.

———. 1908. *Handbook of flower pollination based upon Hermann Müller's work 'The fertilisation of flowers by insects.' Vol. II. Observations on flower pollination made in Europe and the Arctic regions on species belonging to the natural orders* Ranunculaceae *to* Stylidieae. Transl. J. R. Ainsworth Davis. Oxford: Clarendon Press. viii + 703 pp.

———. 1909. *Handbook of flower pollination based upon Hermann Müller's work 'The fertilisation of flowers by insects.' Vol. III. Observations on flower pollination made in Europe and the Arctic regions on species belonging to the natural orders* Goodenovieae *to* Cycadae. Transl. J. R. Ainsworth Davis. Oxford: Clarendon Press. iv + 664 pp.

Kofoid, C. A. 1941. The life cycle of the Protozoa. In G. N. Calkins and F. M. Summers, *Protozoa in biological research.* New York: Columbia University Press. Pp. 565–582.

Kohn, A. J. 1959. The ecology of *Conus* in Hawaii. *Ecol. Monogr.* 29:47–90.

Kolman, W. A. 1960. The mechanism of natural selection for the sex ratio. *Amer. Nat.* 94:373–377.

Komai, T. 1922. *Studies on two aberrant ctenophores,* Coeloplana *and* Gastrodes. Kyoto: Published by the author. 102 pp.

Kormos, J., & K. Kormos 1957. Neue Untersuchungen über den Geschlechts-dimorphismus der Prodiscophryen. *Acta Biol. Hung.* 7:109–125.

———. 1958. Äussere und innere Konjugation. *Acta Biol. Hung.* 8:103–126.

Kowalski, K. 1967. The Pleistocene extinction of mammals in Europe. In P. S. Martin & H. E. Wright, Jr., *Pleistocene extinctions: the search for a cause.* New Haven: Yale University Press. Pp. 349–364.

Kramer, G. 1951. Body proportions of mainland and island lizards. *Evolution* 5:193–206.

Krieg, H. 1937. Luxusbildungen bei Tieren unter besonderer Berücksichtigung der luftlebenden Wirbeltiere. *Zool. Jahrb. (Syst.)* 69:303–318.

Kropotkin, P. 1904. *Mutual aid, a factor of evolution.* London: W. Heinemann. xix + 384 pp.

Krüger, P. 1940. Ascothoracica. *Klassen und Ordnungen des Tier-reichs* 5(1:3:4):1–46.

Kruijt, J. P., & J. A. Hogan 1967. Social behavior on the lek in black grouse, *Lyrurus tetrix tetrix* (L.). *Ardea* 55:203–240.

Krumbach, T. 1925. Ctenophora. *Handbuch der Zoologie* 1:905–995.

Kudinova-Pasternak, R. K. 1965. Deep-sea Tanaidacea from the Bougainville trench of the Pacific. *Crustaceana* 8:75–91.

Kuhn, T. S. 1962. *The structure of scientific revolutions.* Chicago: University Press. xii + 172 pp.

————. 1970. *The structure of scientific revolutions.* 2nd ed. Chicago: University Press. xii + 210 pp.

Kurtén, B. 1953. On the variation and population dynamics of fossil and recent mammal populations. *Acta Zool. Fenn.* 76:1–122.

————. 1965. The Carnivora of the Palestine caves. *Acta Zool. Fenn.* 107: 1–74.

————. 1968. *Pleistocene mammals of Europe.* Chicago: Aldine. vii + 317 pp.

Kusnezov, N. 1957. Numbers of species of ants in faunae of different latitudes. *Evolution* 11:298–299.

Lack, D. 1947. *Darwin's finches: an essay on the general biological theory of evolution.* Cambridge: University Press. x + 208 pp.

————. 1954. *The natural regulation of animal numbers.* Oxford: University Press. viii + 343 pp.

————. 1966. *Population studies of birds.* Oxford: University Press. v + 341 pp.

————. 1968. *Ecological adaptations for breeding in birds.* London: Methuen. xii + 409 pp.

Lameere, A. 1904. L'évolution des ornements sexuels. *Bull. Classe Sci. Acad. Roy. Belgique* (1904):1327–1364.

————. 1920. The origin of insect societies. *Smithsonian Inst. Ann. Rep.* (1920):511–521.

Lang, A. 1888. *Ueber den Einfluss der festsitzenden Lebensweise auf die Thiere und über den Ursprung der ungeschlechtlichen Fortpflanzung durch Theilung und Knospung.* Jena: Fischer. 166 pp.

Lang, K. 1952. *Apseudes hermaphroditicus* n. sp. a hermaphroditic tanaide from the Antarctic. *Ark. Zool.* (2)4:341–350.

————. 1958. Protogynie bei zwei Tanaidaceen-Arten. *Ark. Zool.* (2)11: 535–540.

Lankester, J. B. 1972. Play-mothering: the relations between juvenile females and young infants among free-ranging vervet monkeys. In F. E. Poirier, *Primate socialization.* New York: Random House. Pp. 83–104.

Lantz, L. A., & O. Cyrén 1936. Contribution à la connaissance de *Lacerta saxicola* Eversmann. *Bull. Soc. Zool. France* 61:159–181.

Latané, B., & J. M. Darley 1970. Social determinants of bystander intervention in emergencies. In J. Macaulay & L. Berkowitz, *Altruism and helping behavior: social psychological studies of some antecedents and consequences.* New York: Academic Press. Pp. 13–27.

de Lattin, G. 1952. Über die Bestimmung und Vererbung des Geschlechts einiger Oniscoiden (Crust., Isop.). II. Zur Vererbung der Monogenie von *Cylisticus convexus* (Deg.). *Z. Ind. Abst. Vererbungslehre* 84: 536–567.

van Lawick-Goodall, J. 1967. Mother-offspring relationships in free-ranging chimpanzees. In D. Morris, *Primate ethology.* Chicago: Aldine. Pp. 287–346.

Laws, R. M. 1958. Growth rates and ages of crabeater seals, *Lobodon carcinophagus* Jacquinot & Pucheran. *Proc. Zool. Soc. London* 130:275–288.

Leakey, L. S. B. 1966. Africa and Pleistocene overkill? *Nature* 212:1615–1616.

Lebedinsky, N. G. 1932. Darwin's Theorie der geschlechtlichen Zuchtwahl im Lichte der heutigen Forschung. Zugleich eine Untersuchung über das "Manometerprinzip" der Sexualselektion. *Bibliogr. Genet.* 9:183–419.

Lederberg, J. 1955. Recombination mechanisms in bacteria. *J. Cell. Comp. Physiol.* 45 (Suppl. 2):75–107.

Le Fevre, G., & W. C. Curtis 1912. Studies on the reproduction and artificial propagation of fresh-water mussels. *Bull. Bur. Fisheries* 30:105–204.

Legewie, H. 1925. Zur Theorie der Staatenbildung. *Z. Morphol. Ökol. Tiere* 3:619–684; 4:246–300.

Legrand, J. J. 1967. Contribution à l'étude du contrôle génétique et humoral de l'inversion du sexe chez les Crustacés Isopodes. Notion de balance génique sexuelle. *Année Biol.* 6:241–258.

Lehman, H. 1965. Functional explanation in biology. *Phil. Sci.* 32:1–20.

Lehrman, D. S. 1953. A critique of Konrad Lorenz's theory of instinctive behavior. *Quart. Rev. Biol.* 28:337–363.

———. 1970. Semantic and conceptual issues in the nature-nurture problem. In L. R. Aronson, E. Tobach, D. S. Lehrman, & J. S. Rosenblatt, *Development and evolution of behavior: essays in memory of T. C. Schneirla.* San Francisco: Freeman. Pp. 17–52.

Leigh, E. G., Jr. 1970a. Natural selection and mutability. *Amer. Nat.* 104:301–305.

———. 1970b. Sex ratio and differential mortality between the sexes. *Amer. Nat.* 104:205–210.

Lemche, H., & K. G. Wingstrand 1959. The anatomy of *Neopilina galatheae* Lemche, 1957 (Mollusca Tryblidiacea). *Galathea Rep.* 3:9–71.

von Lengerken, H. 1939. *Die Brutfürsorge- und Brutpflegeinstinkte der Käfer.* Leipzig: Akademische Verlagsgesellschaft M. B. H. viii + 285 pp.

Lenz, F. 1917. Einschüchterungsauslese und weibliche Wahl bei Tier und Mensch. *Arch. Rassen- und Gesellschaftsbiol.* 12:129–150.

Leopold, E. B. 1967. Late-Cenozoic patterns of plant extinction. In P. S. Martin & H. E. Wright, Jr., *Pleistocene extinctions: the search for a cause.* New Haven: Yale University Press. Pp. 203–246.

Lévi, C. 1956. Étude des *Halisarca* de Roscoff, embryologie et systématique des démosponges. *Arch. Zool. Expér. Gén.* 93:1–181.

Levin, D., 1972. Competition for pollinator service: a stimulus for the evolution of autogamy. *Evolution* 26:668–669.

Levins, R. 1968. *Evolution in changing environments: some theoretical explorations.* Princeton: University Press. ix + 120 pp.

Lewin, R. A. 1954. Sex in unicellular algae. In D. H. Wenrich, *Sex in microorganisms.* Washington: American Association for the Advancement of Science. Pp. 100–133.

———. 1960. Genetics and marine algae. In A. A. Buzzati-Traverso, *Per-*

spectives in marine biology. Berkeley & Los Angeles: University of California Press. Pp. 547–557.

Lewis, D. 1942. The evolution of sex in flowering plants. *Biol. Rev.* 17:46–67.

Lewis, D., & L. K. Crowe 1956. The genetics and evolution of gynodioecy. *Evolution* 10:115–125.

Lewontin, R. C. 1957. The adaptations of populations to varying environments. *Cold Spring Harbor Symp. Quant. Biol.* 22:395–408.

———. 1961. Evolution and the theory of games. *J. Theoret. Biol.* 1:382–403.

———. 1962. Interdeme selection controlling a polymorphism in the house mouse. *Amer. Nat.* 96:65–78.

———. 1965. Selection for colonizing ability. In H. G. Baker & G. L. Stebbins, *The genetics of colonizing species.* New York & London: Academic Press. Pp. 77–91.

———. 1970. The units of selection. *Ann. Rev. Ecol. Syst.* 1:1–18.

L'Héritier, P., Y. Neefs, & G. Teissier 1937. Aptérisme des Insectes et sélection naturelle. *C. R. Acad. Sci. Paris* 204:907–909.

Lidicker, W. Z., Jr. 1962. Emigration as a possible mechanism permitting the regulation of population density below carrying capacity. *Amer. Nat.* 96:29–33.

Lieder, U. 1955. Männchenmangel und natuerliche Parthenogenese bei der Silberkarausche *Carassius auratus gibelio* (Vertebrata, Pisces). *Naturwissenschaften* 42:590.

———. 1959. Über die Eientwicklung bei mannchenlosen Stämmen der Silberkarausche *Carassius auratus gibelio* (Bloch) (Vertebrata, Pisces). *Biol. Zentralbl.* 78:284–291.

Ligon, J. D. 1968. Sexual differences in foraging behavior in two species of *Dendrocopos* woodpeckers. *Auk* 85:203–215.

Lillelund, K. 1965. Weitere Untersuchungen über den Hermaphroditismus bei *Osmerus eperlanus* (L.) aus der Elbe. *Z. Morphol. Ökol. Tiere* 55:410–424.

Lillie, F. R. 1911. Charles Otis Whitman. *J. Morphol.* 22:xv–lxxvii.

Limbaugh, C. 1961. Cleaning symbiosis. *Sci. Amer.* 205:42–49.

Limoges, C. 1970. *La sélection naturelle: étude sur la première constitution d'un concept (1839–1859).* Paris: Presses Universitaires de France. 184 pp.

Lin, N., & C. D. Michener 1972. Evolution of sociality in insects. *Quart. Rev. Biol.* 47:131–159.

Lincicome, D. R. 1971. The goodness of parasitism: a new hypothesis. In T. C. Cheng, *Aspects of the biology of symbiosis.* Baltimore: University Park Press. Pp. 139–227.

Lindauer, M. 1952. Ein Beitrag zur Frage der Arbeitsteilung im Bienenstaat. *Z. Vergl. Physiol.* 34:299–345.

———. 1953. Division of labour in the honeybee colony. *Bee World* 34:63–73, 85–90.

———. 1965. Social behavior and mutual communication. M. Rockstein, *The Physiology of Insecta.* New York: Academic Press. 2:123–186.

Lindeman, R. L. 1942. The trophic-dynamic aspect of ecology. *Ecology* 23: 399–418.

Lindroth, C. H. 1946. Inheritance of wing dimorphism in *Pterostichus anthracinus* Ill. *Hereditas* 32:37–40.

Lindzey, G. 1967. Some remarks concerning incest, the incest taboo, and psychoanalytic theory. *Amer. Psychol.* 22:1051–1059.

Lindzey, G., J. Loehlin, M. Manosevitz, & D. Thiessen 1971. Behavioral genetics. *Ann. Rev. Psychol.* 22:39–94.

Linsley, E. G. 1959. Ecology of Cerambycidae. *Ann. Rev. Entomol.* 4:99–138.

———. 1961. The Cerambycidae of North America, part I: introduction. *Univ. California Publ. Entomol.* 18:1–135.

Lipps, J. H. 1970a. Plankton evolution. *Evolution* 24:1–22.

———. 1970b. Ecology and evolution of silicoflagellates. *Proc. North American Paleontol. Conv.* 1969G:965–993.

Lipps, J. H., & E. Mitchell 1969. Climatic regulation of factors controlling otariid pinniped origins and diversification. *Geol. Soc. Amer. Sp. Paper* 101:176.

Lloyd, J. E. 1971. Bioluminescent communication in insects. *Ann. Rev. Ent.* 16:97–122.

Lloyd Morgan, C. 1890–1891. *Animal life and intelligence.* London: Edward Arnold. xvi + 503 pp.

———. 1926. *Emergent evolution.* New York: Holt. xii + 313 pp.

———. 1930. *The animal mind.* London: Edward Arnold. xii + 275 pp.

Löve, A., & D. Löve 1943. The significance of differences in the distribution of diploids and polyploids. *Hereditas* 29:145–163.

Loizos, C. 1966. Play in mammals. *Symp. Zool. Soc. London* 18:1–9.

———. 1967. Play behaviour in higher primates: a review. In D. Morris, *Primate ethology.* Chicago: Aldine. Pp. 176–218.

Longhurst, A. R. 1955a. The reproduction and cytology of the Notostraca (Crustacea, Phyllopoda). *Proc. Zool. Soc. London* 125:671–680.

———. 1955b. A review of the Notostraca. *Bull. British Mus. (Nat. Hist.)* 3:1–57.

Lorbeer, G. 1930. Geschlechtsunterschiede im Chromosomensatz und in der Zellgrösse bei *Sphaerocarpus donnellii* Aust. *Z. Bot.* 23: 932–956.

Lorenz, K. Z. 1939. Vergleichende Verhaltensforschung. *Zool. Anz. (Suppl.)* 12:69–102.

———. 1950. The comparative method in studying innate behaviour patterns. *Symp. Soc. Exper. Biol.* 4:221–268.

———. 1952. *King Solomon's ring: new light on animal ways.* New York: Thomas Y. Crowell. xxii + 202 pp.

———. 1956. Plays and vacuum activities. Anonymous, *L'instinct dans le comportement des animaux et de l'Homme.* Paris: Fondation Singer-Polignac, Masson et Cie. Pp. 633–645.

———. 1964. Ritualized fighting. In J. D. Carthy & F. J. Ebling, *The natural history of aggression.* New York: Academic Press. Pp. 39–50.

———. 1966a. The psychobiological approach: methods and results. *Phil. Trans. Roy. Soc. London* (B)251:273–284.

————. 1966b. *On aggression*. New York: Harcourt, Brace & World. xiv + 306 pp.

Losey, G. S., Jr. 1971. Communication between fishes in cleaning symbiosis. In T. C. Cheng, *Aspects of the biology of symbiosis*. Baltimore: University Park Press. Pp. 45–76.

zur Loye, J. G. 1908. Die Anatomie von *Spirorbis borealis* mit besonderer Berücksichtigung der Unregelmässigkeiten des Körperbaues und deren Ursachen. *Zool. Jahrb. (Anat.)* 26:305–354.

Lubbock, J. 1873. *Monograph of the Collembola and Thysanura.* London: Ray Society. x + 276 pp.

————. 1874–1882. Observations on ants, bees and wasps. *J. Linn. Soc. London (Zool.)* 12:110–139, 227–251, 445–514; 13:217–258; 14:265–290, 607–626; 15:167–187, 362–387; 16:110–121.

Lüscher, M. 1951. Beobachtungen über Koloniegründung bei verschiedenen afrikanischen Termitenarten. *Acta Tropica* 8:36–43.

————. 1961. Social control of polymorphism in termites. *Symp. Roy. Ent. Soc. London* 1:57–67.

McAlester, A. L. 1966. Evolutionary and systematic implications of a transitional Ordovician lucinoid bivalve. *Malacologia* 3:433–439.

————. 1970. Animal extinctions, oxygen consumption, and atmospheric history. *J. Paleontol.* 44:405–409.

MacArthur, R. H. 1965. Patterns of species diversity. *Biol. Rev.* 40:510–533.

MacArthur, R. H., & E. O. Wilson 1963. An equilibrium theory of insular zoogeography. *Evolution* 17:373–387.

————. 1967. *The theory of island biogeography.* Princeton: University Press. xi + 203 pp.

Macaulay, J. R. 1970. A shill for charity. In J. Macaulay & L. Berkowitz, *Altruism and helping behavior: social psychological studies of some antecedents and consequences.* New York: Academic Press. Pp. 43–59.

McCullough, D. R. 1969. The Tule elk: its history, behavior and ecology. *Univ. California Publ. Zool.* 88:1–209.

McDougall, W. 1947. *An outline of psychology.* 11th ed. London: Methuen. xxii + 456 pp.

MacGinitie, G. E. 1938. Movements and mating habits of the sand crab, *Emerita analoga. Amer. Midl. Nat.* 19:471–481.

McGowan, J. A. 1954. Observations on the sexual behavior and spawning of the squid, *Loligo opalescens*, at La Jolla, California. *California Fish & Game* 40:47–54.

McIlhenny, E. A. 1940. Sex ratio in wild birds. *Auk* 57:85–93.

McIntosh, R. P. 1970. Community, competition, and adaptation. *Quart. Rev. Biol.* 45:259–280.

McLaren, I. A. 1958. Some aspects of growth and reproduction of the bearded seal, *Erignathus barbatus* (Erxleben). *J. Fish. Res. Board Canada* 15:219–227.

————. 1967. Seals and group selection. *Ecology* 48:104–110.

McLaughlin, P. A., & D. P. Henry 1972. Comparative morphology of complemental males in four species of *Balanus* (Cirripedia Thoracica). *Crustaceana* 22:13–30.

McNab, B. K. 1971. On the ecological significance of Bergmann's rule. *Ecology* 52:845–854.

Magnus, D. B. E. 1958. Experimental analysis of some "overoptimal" sign-stimuli in the mating-behaviour of the Fritillary butterfly *Argynnis paphia* L. (Lepidoptera: Nymphalidae). *Proc. X Int. Congr. Entomol.* 2:405–418.

———. 1963. Sex limited mimicry II—visual selection in the mate choice of butterflies. *Proc. XVI Int. Congr. Zool.* 4:179–183.

Magoun, H. W. 1961. Darwin and concepts of brain function. In J. F. Delafresnaye, *Brain mechanisms and learning*. Springfield: Charles C. Thomas. Pp. 1–20.

Mainardi, D., F. M. Scudo, & D. Barbieri 1965. Assortative mating based on early learning: population genetics. *Acta Biomedica L'Ateneo Parmense* 36:583–605.

Malaquin, A. 1901. Le parasitisme évolutif des monstrillides (Crustacés, Copépodes). *Arch. Zool. Expér. Gén.* (3)9:81–232.

Malthus, T. R. 1798. *An essay on the principle of population, as it affects the future improvement of society, with remarks on the speculations of Mr. Godwin, M. Condorcet, and other writers*. London: Johnson. v + ix + 396 pp. (Facsimile reprint, 1966, New York: St. Martin's Press.)

Maly, E. J. 1970. The influence of predation on the adult sex ratios of two copepod species. *Limnol. Oceanogr.* 15:566–573.

Mandelli, E. F., & P. R. Burkholder 1966. Primary productivity in the Gerlache and Bransfield Straits of Antarctica. *J. Marine Res.* 24:15–27.

Mangold-Wirz, K., & P. Fioroni 1970. Die Sonderstellung der Cephalopoden. *Zool. Jahrb.* (*Syst.*)97:522–631.

Mann, K. H. 1962. *Leeches (Hirudinea): their structure, physiology, ecology and embryology*. New York: Pergamon Press. x + 201 pp.

Manton, S. M. 1949. Studies on the Onychophora. VII. The early embryonic stages of *Peripatopsis*, and some general considerations concerning the morphology and phylogeny of the Arthropoda. *Phil. Trans. Roy. Soc. London* (B) 233:483–580.

Marcus, E. 1929. Tardigrada. *Klassen und Ordnungen des Tier-reichs*, 5(4:3):i–viii, 1–608.

Marcus, E. du Bois-Reymond 1949. *Phoronis ovalis* from Brazil. *Bol. Fac. Fil. Ci. Letr. São Paulo* (*Zool.*)14:157–171.

———. 1952. On south American Malacopoda. *Bol. Fac. Fil. Ci. Letr. São Paulo* (*Zool.*)17:189–209.

Margalef, R. 1958. Mode of evolution of species in relation to their places in ecological succession. *Proc. XV Internat. Congr. Zool.* pp. 787–789.

———. 1968. *Perspectives in ecological theory*. Chicago: University Press. viii + 111 pp.

Marion, A. F. 1873. Recherches sur les animaux inférieurs du Golfe de Marseille. *Ann. Sci. Nat. Zool.* (5)17(6):1–23.

Marshall, A. J. 1954. *Bower-birds: their displays and breeding cycles: a preliminary statement*. Oxford: University Press. x + 208 pp.

Marshall, N. B. 1953. Egg size in Arctic, Antarctic and deep-sea fishes. *Evolution* 7:328–341.

———. 1954. *Aspects of deep sea biology.* New York: Philosophical Library. 380 pp.

Marshall, S. M., & A. P. Orr 1955. *The biology of a marine copepod,* Calanus finmarchicus (*Gunnerus*). Edinburgh: Oliver and Boyd. viii + 188 pp.

Martin, F. W. 1967. Distyly, self-incompatibility, and evolution in *Melochia*. *Evolution* 21:493–499.

Martin, P. S. 1966. Africa and Pleistocene overkill. *Nature* 212:339–342.

———. 1967. Prehistoric overkill. In P. S. Martin & H. E. Wright, Jr., *Pleistocene extinctions: the search for a cause.* New Haven: Yale University Press. Pp. 75–120.

———. 1973. The discovery of America. *Science* 179:969–974.

Martin, P. S., & J. E. Guilday 1967. A bestiary for Pleistocene biologists. In P. S. Martin & H. E. Wright, Jr., *Pleistocene extinctions: the search for a cause.* New Haven: Yale University Press. Pp. 1–62.

Maslin, T. P. 1962. All-female species of the lizard genus *Cnemidophorus*, Teiidae. *Science* 135:212–213.

———. 1971. Parthenogenesis in reptiles. *Amer. Zool.* 11:361–380.

Mason, H. L., & J. H. Langenheim 1957. Language analysis and the concept environment. *Ecology* 38:325–340.

———. 1961. Natural selection as an ecological concept. *Ecology* 42:158–165.

Mather, K. 1940. Outbreeding and separation of the sexes. *Nature* 145:484–486.

———. 1943. Polygenic inheritance and natural selection. *Biol. Rev.* 18: 32–64.

Maupas, E. 1889. La rajeunissement chez les ciliés. *Arch. Zool. Expér. Gén.* (2)7:149–517.

———. 1900. Modes et formes de reproduction des Nématodes. *Arch. Zool. Expér. Gén.* (3)8:463–624.

Mayer, P. 1882. Zur Naturgeschichte der Feigeninsecten. *Mitt. Zool. Stat. Neapel.* 3:551–590.

Maynard Smith, J. 1964. Group selection and kin selection. *Nature* 201: 1145–1147.

———. 1965. The evolution of alarm calls. *Amer. Nat.* 99:59–63.

———. 1968. Evolution in sexual and asexual populations. *Amer. Nat.* 102: 469–473.

———. 1971. What use is sex? *J. Theoret. Biol.* 30:319–335.

Maynard Smith, J., & M. G. Ridpath 1972. Wife sharing in the Tasmanian native hen, *Tribonyx mortierii:* a case of kin selection? *Amer. Nat.* 106:447–452.

Mayr, E. 1935. Bernard Altum and the territory theory. *Proc. Linn. Soc. New York* (45–46):24–38.

———. 1939. The sex ratio in wild birds. *Amer. Nat.* 73:156–179.

———. 1954. Change of genetic environment and evolution. In J. Huxley, A. C. Hardy & E. B. Ford, *Evolution as a process.* London: Allen & Unwin. Pp. 157–280.

———. 1956. Geographical character gradients and climatic adaptation. *Evolution* 10:105–108.

———. 1959. Darwin and the evolutionary theory in biology. In B. J. Meggers, *Evolution and anthropology: a centennial appraisal.* Washington: The Anthropological Society of Washington. Pp. 1–10.

———. 1961. Cause and effect in biology. *Science* 134:1501–1506.

———. 1963. *Animal species and evolution.* Cambridge: Harvard University Press. xiv + 797 pp.

———. 1965. Avifauna: turnover on islands. *Science* 150:1587–1588.

———. 1972a. The nature of the Darwinian revolution. *Science* 176:981–989.

———. 1972b. Sexual selection and natural selection. In B. Campbell, *Sexual selection and* The Descent of Man *1871–1971*. Chicago: Aldine. Pp. 87–104.

Mead, G. W. 1960. Hermaphroditism in archibenthic and pelagic fishes of the order Iniomi. *Deep-Sea Res.* 6:234–235.

Mead, M. 1958. Cultural determinants of behavior. In A. Roe & G. G. Simpson, *Behavior and Evolution.* New Haven: Yale University Press. Pp. 480–503.

Mednikov, B. M. 1961. On the sex ratio in deep water Calanoida. *Crustaceana* 3:105–109.

Megelsberg, O. 1935. Über die postimaginale Entwicklung (Physogastrie) und den Hermaphroditismus bei afrikanischen Termitoxenien (Dipt.). (Zugleich ein Beitrag zur Entwicklung der Eizelle.) *Zool. Jahrb.* (*Anat.*)60:345–398.

Meltzer, B. 1970. Generation of hypotheses and theories. *Nature* 225:972.

Mendel, G. 1865. Versuche über Pflanzenhybriden. *Verhandl. Naturf. Vereines Brünn* 4:3–47.

Mertens, R. 1934. Die Insel-Reptilien, ihre Ausbreitung, Variation, und Artbildung. *Zoologica* 32:1–209.

Mesnil, F., & M. Caullery 1898a. Sur la viviparité d'une Annélide polychète (*Dodecaceria concharum* Oerst., forme A). *C. R. Soc. Biol. Paris* 50:905–908.

———. 1898b. Sur la viviparité d'une Annélide polychète (*Dodecaceria concharum* Oerst., forme A). *C. R. Acad. Sci. Paris* 127:486–489.

Mettrick, D. F. 1962. A re-description of the female and a description of the male form of a dioecious cestode, *Gyrocoelia kiewietti* Ortlepp, 1937. *J. Helminthol.* 36:149–156.

Meyer, A. 1932–1933. Acanthocephala. *Klassen und Ordnungen des Tierreichs* 4(2:2:1):1–332.

Meyer, E. 1888. Studien über Körperbau der Anneliden. *Mitt. Zool. Stat. Neapel.* 8:462–662.

Meyer-Holzapfel, M. 1962. Das Spiel der Säugetieren. *Handbuch der Zoologie* 8(10:17):1–36.

Michener, C. D. 1958. The evolution of social behavior in bees. *Proc. X Int. Congr. Ent.* 2:441–447.

———. 1961. Social polymorphism in Hymenoptera. *Symp. Roy. Ent. Soc. London* 1:43–56.

————. 1964. Reproductive efficiency in relation to colony size in hymenopterous societies. *In. Soc.* 11:317–342.

————. 1969. Comparative social behavior of bees. *Ann. Rev. Ent.* 14:299–342.

Mileikovsky, S. A. 1971. Types of larval development in marine bottom invertebrates, their distribution and ecological significance: a re-evaluation. *Mar. Biol.* 10:193–213.

Millar, R. H. 1966. Evolution in ascidians. In H. Barnes, *Some contemporary studies in marine science*. London: George Allen & Unwin. Pp. 519–534.

Milne, A. 1961. Definition of competition among animals. *Symp. Soc. Exper. Biol.* 15:40–61.

Mitchell, P. C. 1912. *The childhood of animals*. New York: Stokes. xiv + 269 pp.

Mivart, St. George 1871. *On the genesis of species*. London: Macmillan. xv + 296 pp.

Mockford, E. L. 1971. Parthenogenesis in psocids (Insecta: Psocoptera). *Amer. Zool.* 11:327–339.

Montgomery, T. H., Jr. 1903. Studies on the habits of spiders, particularly those of the mating period. *Proc. Acad. Nat. Sci. Philadelphia* 55:59–149.

————. 1908. The sex ratio and cocooning habits of an aranead and the genesis of sex ratios. *J. Exper. Zool.* 5:429–452.

————. 1909. Further studies on the activities of araneads, II. *Proc. Acad. Nat. Sci. Philadelphia* 61:548–569.

————. 1910. The significance of the courtship and secondary sexual characters of araneads. *Amer. Nat.* 34:151–177.

Moore, H. B. 1937. The biology of *Littorina littorea*. Part I. Growth of the shell and tissues, spawning, length of life and mortality. *J. Mar. Biol. Assoc.* 21:721–742.

Moore, N. W. 1952. On the so-called "territories" of dragonflies (Odonata—Anisoptera). *Behaviour* 4:85–100.

————. 1957. Territory in dragonflies and birds. *Bird Study* 4:125–130.

————. 1964. Intra- and interspecific competition among dragonflies (Odonata). *J. Animal Ecol.* 33:49–71.

Moore, O. K., & D. J. Lewis 1953. Purpose and learning theory. *Psychol. Rev.* 60:149–156.

Moore, R. C. 1954. Evolution of Late Paleozoic invertebrates in response to major oscillations of shallow seas. *Bull. Mus. Comp. Zool. Harvard* 112:259–286.

————. 1955. Expansion and contraction of shallow seas as a causal factor in evolution. *Evolution* 9:482–483.

Moraitou-Apostolópoulou, M. 1972. Sex ratio in the pelagic copepods *Temora stylifera* Dana and *Centropages typicus* Krøyer. *J. Exper. Mar. Biol. Ecol.* 8:83–87.

Mordukhai-Boltovskoi, Ph.D. 1967. On the males and gamogenetic females of the Caspian Polyphemidae (Cladocera). *Crustaceana* 12:113–123.

Mordvilko, A. 1928. The evolution of cycles and the origin of heteroecy (migrations) in plant-lice. *Ann. Mag. Nat. Hist.* (10)2:570–582.

Morgan, T. H. 1903. *Evolution and adaptation.* New York: Macmillan. xiii + 470 pp.

———. 1914. *Heredity and sex.* New York: Columbia University Press. x + 284 pp.

———. 1942a. Cross- and self-fertilization in the ascidian *Styela. Biol. Bull.* 82:161–171.

———. 1942b. Cross- and self-fertilization in the ascidian *Molgula manhattensis. Biol. Bull.* 82:172–177.

Morris, D. 1952. Homosexuality in the ten-spined stickleback (*Pygosteus pugnitius* L.). *Behaviour* 4:233–261.

———. 1956. The function and causation of courtship ceremonies. *Fondation Singer-Polignac, Colloq. Internat. Instinct* pp. 261–286.

———. 1957. The reproductive behaviour of the bronze mannikin, (*Longchura cucullata*). *Behaviour* 11:156–201.

———. 1958. The reproductive behaviour of the ten-spined stickleback (*Pygosteus pugnitius* L.). *Behaviour* (Suppl. 6):1–154.

———. 1962. *The biology of art: a study of the picture-making behaviour of the great apes and its relationship to human art.* London: Methuen. 176 pp.

Morse, D. H. 1971. The insectivorous bird as an adaptive strategy. *Ann. Rev. Ecol. Syst.* 2:177–200.

Mortensen, T. 1933. Papers from Dr. Th. Mortensen's Pacific Expedition 1914–1916. lxvi. The echinoderms of St. Helena (other than crinoids). *Vidensk. Medd. Naturh. Foren. Kjøbenhavn* 93:401–472.

———. 1936. Echinoidea and Ophiuroidea. *Discovery Rep.* 12:199–348.

Moser, H. G. 1967. Reproduction and development of *Sebastodes paucispinis* and comparison with other rockfishes off Southern California. *Copeia* 1967:773–797.

Mosimann, J. E. 1958. The evolutionary significance of rare matings in animal populations. *Evolution* 12:246–261.

Mosley, H. N. 1874. On the structure and development of *Peripatus capensis. Phil. Trans. Roy. Soc. London* 164:757–782.

Mosquin, T. 1971. Competition for pollinators as a stimulus for the evolution of flowering time. *Oikos* 22:398–402.

Mottram, J. C. 1915. The distribution of secondary sexual characters amongst birds, with relation to their liability to the attack of enemies. *Proc. Zool. Soc. London* (1915):663–678.

Mountford, M. D. 1968. The significance of litter-size. *J. Animal Ecol.* 37: 363–367.

Moynihan, M. 1955. Types of hostile display. *Auk* 72:247–259.

Müller, F. 1885. Die Zwitterbildung im Tierreich. *Kosmos* (N.F.)2:321–334.

Müller, H. 1853. Ueber das Männchen von *Argonauta argo* und die Hectocotylen. *Z. Wiss. Zool.* 4:1–35.

———. 1873. *Die Befruchtung der Blumen durch Insekten und die gegen-*

seitigen Anpassungen beider: ein Beitrag zur Erkenntnis des ursäch-lichen Zusammenhanges in der organischen Natur. Leipzig: Verlag Wilhelm Engelmann. viii + 478 pp.

Mulcahy, D. L., & D. Caporello 1970. Pollen flow within a tristylous species: *Lythrum salicaria. Amer. J. Bot.* 57:1027–1030.

Muller, H. J. 1932. Some genetic aspects of sex. *Amer. Nat.* 66:118–138.

Murdock, G. P. 1971. Anthropology's mythology. *Proc. Roy. Anthropol. Inst. Great Britain, Ireland* 1971:17–24.

Murdock, W. W. 1966. Population stability and life history phenomena. *Amer. Nat.* 100:5–11.

Murphy, G. I. 1968. Pattern in life history and the environment. *Amer. Nat.* 102:391–403.

Murray, J. 1964. Multiple mating and effective population size in *Cepaea nemoralis. Evolution* 18:283–291.

Murtha, E. F. 1960. Pharmacological study of poisons from shellfish and puffer fish. *Ann. New York Acad. Sci.* 90:820–836.

Myers, M. E. 1925. Contributions toward a knowledge of the life histories of the Melanophyceae. *Univ. California Publ. Bot.* 13:109–124.

Naef, A. 1921–1923. Die Cephalopoden. *Fauna Flora G. Neapel* 35(1:1&2): 1–863.

Naisse, J. 1966. Contrôle endocrinien de la différenciation sexuelle chez l'Insecte *Lampyris noctiluca* (Coléoptère Malacoderme Lampyride). I.—Rôle androgène des testicules. *Arch. Biol.* 77:139–201.

Nero, R. W., & J. T. Emlen, Jr. 1951. An experimental study of territorial behavior in breeding red-wing blackbirds. *Condor* 53:105–116.

Newell, N. D. 1949. Phyletic size increase, an important trend illustrated by fossil invertebrates. *Evolution* 3:103–124.

———. 1956. Catastrophism and the fossil record. *Evolution* 10:97–101.

———. 1965. Mass extinctions at the end of the Cretaceous period. *Science* 149:922–924.

———. 1971. An outline history of tropical organic reefs. *Amer. Mus. Nov.* 2465:1–37.

———. 1972. The evolution of reefs. *Sci. Amer.* 226:54–65.

Nice, M. M. 1957. Nesting success in altricial birds. *Auk* 74:305–321.

Nicholls, A. G. 1933. On the biology of *Calanus finmarchicus.* III. Vertical distribution and diurnal migrations in the Clyde Sea area. *J. Marine Biol. Assoc.* 19:139–164.

Nichols, G. E. 1923. A working basis for the ecological classification of plant communities. *Ecology* 4:11–23, 154–179.

———. 1929. Plant associations and their classification. *Proc. Int. Congr. Plant Sci., Ithaca* 1926 1:629–641.

Nicol, D. 1961. Biotic associations and extinction. *Syst. Zool.* 10:35–41.

———. 1964. An essay on size of marine pelecypods. *J. Paleontol.* 38:968–974.

Nicol, D., & A. P. Gavenda 1964. Inferences derived from general analysis of Recent and fossil marine pelecypod faunas. *J. Paleontol.* 38:975–983.

Nielsen, E. S. 1963. Productivity, definition and measurement. *The Sea* 2:129–164.

Nielsen, J. G. 1966. Synopsis of the Ipnopidae (Pisces, Iniomi) with description of two new abyssal species. *Galathea Rep.* 8:49–75.

Nigon, V. 1949a. Modalités de la reproduction et déterminisme du sexe chez quelques Nématodes libres. *Ann. Sci. Nat. Zool.* (11)11:1–132.

———. 1949b. Modifications expérimentales de la proportion des sexes chez un Nématode pseudogame. *C. R. Acad. Sci. Paris* 234:2568–2570.

———. 1965. Développement et reproduction des Nématodes. *Traité de Zoologie* 4(2):219–386.

Nigon, V., & E. C. Dougherty 1949. Reproductive patterns and attempts at reciprocal crossing of *Rhabditis elegans* Maupas, 1900, and *Rhabditis briggsae* Dougherty and Nigon, 1949 (Nematoda: Rhabditidae). *J. Exp. Zool.* 112:485–503.

Nisbet, R. A. 1966. *The sociological tradition.* New York: Basic Books, xii + 349 pp.

———. 1969. *Social change and history: aspects of the western theory of development.* New York: Oxford University Press. x + 335 pp.

Nissen, H. W. 1958. Axes of behavioral comparison. In A. Roe & G. G. Simpson, *Behavior and evolution.* New Haven: Yale University Press. Pp. 183–205.

Noble, G. K. 1936. Courtship and sexual selection of the flicker (*Colaptes auratus luteus*). *Auk* 53:269–282.

———. 1938. Sexual selection among fishes. *Biol. Rev.* 13:133–158.

Noble, G. K., & M. T. Bradley 1933. The mating behavior of lizards: its bearing on the theory of sexual selection. *Ann. New York Acad. Sci.* 35:25–100.

Noirot, C. 1952: Le polymorphisme social chez les termites et son déterminisme. *Colloques Internat. Centre Nat. Rec. Sci.* 34:103–116.

———. 1956. Les sexués de remplacement chez les termites supérieurs (Termitidae). *In. Soc.* 3:145–158.

Nouvel, H., & L. Nouvel 1937. Recherches sur l'accouplement et la ponte chez les Crustacés Décapodes Natantia. *Bull. Soc. Zool. France* 62:208–221.

Nur, U. 1971. Parthenogenesis in coccids (Homoptera). *Amer. Zool.* 11:301–308.

Nutting, C. C. 1891. Some of the causes and results of polygamy among the Pinnipedia. *Amer. Nat.* 25:103–112.

Nutting, W. L. 1969. Flight and colony foundation. In K. Krishna & F. M. Weesner, *Biology of termites.* New York: Academic Press. 1:233–282.

Ockelmann, K. W. 1965. Developmental types in marine bivalves and their distribution along the Atlantic coast of Europe. *Proc. First European Malacol. Congr., London* 1962:25–35.

O'Donald, P. 1962. The theory of sexual selection. *Heredity* 17:541–552.

———. 1972. Natural selection of reproductive rates and breeding times and its effect on sexual selection. *Amer. Nat.* 106:368–379.

Odum, E. P. 1963. *Ecology.* New York: Holt, Rinehart & Winston. vii + 139 pp.

————. 1969. The strategy of ecosystem development. *Science* 164:262–270.

Odum, H. T., J. E. Cantlon, & L. S. Kornicker 1960. An organizational hierarchy postulate for the interpretation of species-individual distributions, species entropy, ecosystem evolution, and the meaning of a species-variety index. *Ecology* 41:395–399.

Odum, H. T., & R. C. Pinkerton 1955. Time's speed regulator, the optimum efficiency for maximum power output in physical and biological systems. *Amer. Sci.* 43:331–343.

Oehlert & Deniker (sic) 1883. Analyse d'observations sur le développement des Brachiopodes par M. Kowalevski. *Arch. Zool. Expér. Gén.* (2)1: 58–76.

Öklund, F. 1930. Studien über die Arbeitsteilung und die Teilung des Arbeitsgebietes bei der roten Waldameise (*Formica rufa* L.). *Z. Morphol. Ökol. Tiere* 20:63–131.

Oken, "Dr.," (i.e., L.) 1805. *Die Zeugung.* Bamberg: Goebhardt. viii + 216 pp.

Oliver, J. H., Jr. 1971. Parthenogenesis in mites and ticks (Arachnida: Acari). *Amer. Zool.* 11:283–299.

Olofsson, O. 1918. Studien über die Süsswasserfauna Spitzbergens: Beitrag zur Systematik, Biologie und Tiergeographie der Crustaceen und Rotatorien. *Zool. Bidr. Uppsala* 6:183–648.

Orians, G. 1969. On the evolution of mating systems in birds and mammals. *Amer. Nat.* 103:589–603.

Ornduff, R. 1964. The breeding system of *Oxalis suksdorfii. Amer. J. Bot.* 51:307–314.

————. 1966a. The breeding system of *Pontederia cordata* L. *Bull. Torrey Bot. Club* 93:407–416.

————. 1966b. The origin of dioecism from heterostyly in *Nymphoides* (Menyanthaceae). *Evolution* 20:309–314.

————. 1970. The systematics and breeding system of *Gelsemium* (Loganiaceae). *J. Arnold Arboretum* 51:1–17.

————. 1972. The breakdown of trimorphic incompatibility in *Oxalis* section Corniculatae. *Evolution* 26: 52–65.

Ornduff, R., & J. D. Perry 1964. Reproductive biology of *Piriqueta caroliniana* (Turneraceae). *Rhodora* 66:100–109.

Orton, J. H. 1920. Mode of feeding and sex-phenomena in the pea crab *Pinnotheres pisum. Nature* 106:533–534.

Osborn, H. F. 1906. The causes of extinction of Mammalia. *Amer. Nat.* 40: 769–795.

Osche, G. 1962. Ökologie des Parasitismus und der Symbiose. *Fortschr. Zool.* 15:125–164.

Oshima, H. 1929. Hermafrodita marstelo, *Asterina batheri* Goto. *Annot. Zool. Japon.* 12:333–349.

————. 1935. A further note on *Nymphonella tapetis:* the egg-carrying mature male (Eurycycdidae, Pantopoda). *Annot. Zool. Japon.* 15:95–102.

Paine, R. T. 1962. Ecological diversification in sympatric gastropods of the genus *Busycon. Evolution* 16:515–523.

―――. 1969. The *Pisaster-Tegula* interaction: prey patches, predator food preference, and intertidal community structure. *Ecology* 50:950–961.

Pajunen, V. I. 1963a. Reproductive behaviour in *Leucorrhinia dubia* v. d. Lind. and *L. rubicunda* L. (Odon. Libellulidae). *Ann. Ent. Fenn.* 29:106–118.

―――. 1963b. On the threat display of resting dragonflies. *Ann. Ent. Fenn.* 29:236–239.

―――. 1964. Aggressive behaviour in *Leucorrhinia caudalis* Charp. (Odon., Libellulidae). *Ann. Zool. Fenn.* 1:357–369.

―――. 1966a. The influence of population density on the territorial behaviour of *Leucorrhinia rubicunda* L. (Odon., Libellulidae). *Ann. Zool. Fenn.* 3:40–52.

―――. 1966b. Aggressive behaviour and territoriality in a population of *Calopteryx virgo* L. (Odon., Calopterygidae). *Ann. Zool. Fenn.* 3: 201–214.

Parascandola, J. 1971. Organismic and holistic concepts in the thought of L. J. Henderson. *J. Hist. Biol.* 4:63–113.

Pardi, L. 1948. Dominance order in *Polistes* wasps. *Physiol. Zool.* 21:1–13.

―――. 1952. Dominiazione e gerarchia in alcuni invertebrati. *Colloques Internat. Centre Nat. Rec. Sci.* 34:183–197.

Park, T. 1942. Integration in infra-social insect populations. *Biol. Symp.* 8:121–138.

Parker, G. A. 1970. Sperm competition and its evolutionary consequences in the insects. *Biol. Rev.* 45:525–567.

Parr, A. E. 1931. Sex dimorphism and schooling behavior among fishes. *Amer. Nat.* 65:173–180.

Parsons, T. 1970. On building social system theory: a personal history. *Daedalus* 99:826–881.

Patel, B., & D. J. Crisp 1961. Relation between the breeding and moulting cycles in cirripedes. *Crustaceana* 2:89–107.

Patrick, R. 1954. Sexual reproduction in diatoms. In D. H. Wenrich, *Sex in microorganisms.* Washington: American Association for the Advancement of Science. Pp. 82–91.

Patten, B. C. 1959. Introduction to the cybernetics of the ecosystem: the trophic-dynamic aspect. *Ecology* 40:221–231.

Pearce, J. B. 1962. Adaptation in symbiotic crabs of the family *Pinnotheridae*. *Biologist* 45:11–15.

―――. 1966a. On *Pinnixa faba* and *Pinnixa littoralis* (Decapoda: Pinnotheridae) symbiotic with the clam *Tresus capax* (Pelecypoda: Mactridae). In H. Barnes, *Some contemporary studies in marine science.* New York: Hafner. Pp. 565–589.

―――. 1966b. The biology of the mussel crab, *Fabia subquadrata*, from the waters of the San Juan Archipelago, Washington. *Pacific Sci.* 20: 3–35.

Pearl, R., & J. R. Miner 1935. Experimental studies on the duration of life. XIV. The comparative mortality of certain lower organisms. *Quart. Rev. Biol.* 10:60–79.

Pearse, A. S. 1914. Habits of fiddler crabs. *Smithsonian Rep.* 1913:415–428.

Pearse, J. S. 1969. Slow developing demersal embryos and larvae of the Antarctic sea star *Odontaster validus*. *Mar. Biol.* 3:110–116.

Pearse, J. S., & A. C. Giese 1966. Food, reproduction and organic constitution of the common Antarctic echinoid *Sterechinus neumayeri* (Meissner). *Biol. Bull.* 130:387–401.

Pearson, A. K., & O. P. Pearson 1955. Natural history and breeding behavior of the tinamou, *Nothoprocta ornata*. *Auk* 72:113–127.

Pelseneer, P. 1912. L'hermaphroditisme chez les Lamellibranches. *Verh. VIII Internat. Zool.-Kongr., Graz* 1910. Pp. 444–446.

————. 1926. La proportion relative des sexes chez les animaux et particulièrement chez les Mollusques. *Mém. Acad. Roy. Belgique* 8:1–258.

————. 1933. La durée de la vie et l'âge de la mâturité sexuelle chez certains Mollusques. *Ann. Soc. Roy. Zool. Belgique* 64:93–104.

Peters, H. M. 1955. Die Winkgebärde von *Uca* und *Minuca* (Brachyura) in vergleichend-ethologischer, -ökologischer und -morphologischanatomischer Betrachtung. *Z. Morphol. Ökol. Tiere* 43:425–500.

Petersen, W. 1892. Ueber die Ungleichzeitigkeit in der Erscheinung der Geschlechter bei Schmetterlingen. *Zool. Jahrb.* (*Syst.*)6:671–679.

Peterson, R. S. 1968. Social behavior in pinnipeds with particular reference to the northern fur seal. In C. E. Rice, *The behavior and physiology of pinnipeds.* New York: Appleton-Century Crofts. Pp. 3–53.

Peterson, R. S., & G. A. Bartholomew 1967. The natural history and behavior of the California sea lion. *Amer. Soc. Mammal. Special Publ.* No. 1. xi + 79 pp.

Petit, C. 1958. Le déterminisme génétique et psycho-physiologique de la compétition sexuelle chez *Drosophila melanogaster*. *Bull. Biol. France Belgique* 92:248–329.

————. 1967. L'isolement sexuel: déterminisme et role dans la spéciation. *Année Biol.* 6:271–285.

Petrunkevitch, A. 1910. Courtship in *Dysdera crocata*. *Biol. Bull.* 19:127–129.

————. 1911. Sense of sight, courtship and mating in *Dugesiella hentzi* (Gerard), a Theraphosid spider from Texas. *Zool. Jahrb.* (*Syst.*) 31:355–376.

Phillips, J. 1931. The biotic community. *J. Ecol.* 19:1–24.

————. 1934. Succession, development, the climax, and the complex organism: an analysis of concepts. Part I. *J. Ecol.* 22:554–571.

————. 1935a. Succession, development, the climax, and the complex organism: an analysis of concepts [·] Part II. Development and the climax. *J. Ecol.* 23:210–246.

————. 1935b. Succession, development, the climax, and the complex organism: an analysis of concepts [·] Part III. The complex organism: conclusions. *J. Ecol.* 23:488–508.

Piaget, J. 1967. *Biologie et connaissance: essai sur les relations entre les régulations organiques et les processus cognitifs.* Paris: Gallimard 430 pp.

————. 1969. *Le jugement moral chez l'enfant.* Paris: Presses Universitaires de France. viii + 334 pp.

Pianka, E. R. 1966. Latitudinal gradients in species diversity: a review of concepts. *Amer. Nat.* 100:33–46.

———. 1972. *r* and *K* selection or *b* and *d* selection? *Amer. Nat.* 106:581–588.

Piel, G. 1970. The comparative psychology of T. C. Schneirla. In L. R. Aronson, E. Tobach, D. S. Lehrman, & J. S. Rosenblatt, *Development and evolution of behavior: essays in memory of T. C .Schneirla.* San Francisco: W. H. Freeman. Pp. 1–13.

van der Pijl, L. 1969. *Principles of dispersaly in higher plants.* New York: Springer Verlag. vii + 153 pp.

Pillsbury, R. W. 1957. Avoidance of poisonous eggs of the marine fish *Scorpaenichthys marmoratus* by predators. *Copeia* (3):251–252.

Pitelka, F. A. 1942. Territoriality and related problems in North American hummingbirds. *Condor* 44:189–204.

Pittendrigh, C. S. 1958. Adaptation, natural selection and behavior. In A. Roe & G. G. Simpson, *Behavior and evolution.* New Haven: Yale University Press. Pp. 390–416.

Plateaux-Quénu, C. 1961. Les sexués de remplacement chez les insectes sociaux. *Année Biol.* 37:177–216.

———. 1967. Tendances évolutives et degré de socialisation chez les *Halictinae* [*Hym., Apoidea*]. *Ann. Soc. Ent. Fr.* (N.S.) 3:859–866.

Platt, J. R. 1964. Strong Inference. *Science* 146:347–352.

Poole, T. B. 1966. Aggressive play in polecats. *Symp. Zool. Soc. London* 18:23–44.

Popper, K. R. 1962. *Conjectures and refutations: the growth of scientific knowledge.* New York: Basic Books. xi + 412 pp.

———. 1963. *The open society and its enemies.* New York: Harper & Row. 2 Vols., xii + 351 + 420 pp. (Also Princeton: University Press, 1962 & earlier editions.)

———. 1964. *The poverty of historicism.* 3rd ed. New York: Harper. 166 pp.

———. 1965. *The logic of scientific discovery.* 3rd ed. New York: Harper. London: Hutchinson. 480 pp.

Portal, M. 1924. Breeding habits of the red-legged partridge. *British Birds* 17:315–316.

Portmann, A. 1961. *Animals as social beings.* New York: Viking Press. 249 pp.

Potswald, H. E. 1967. Observations on the genital segments of *Spirorbis* (Polychaeta). *Biol. Bull.* 132:91–107.

Prenant, M. 1959. Classe des Myzostomides. *Traité de Zoologie* 5(1):714–784.

Preston, F. W. 1969. Diversity and stability in the biological world. *Bookhaven Symp. Biol.* (22):1–12.

Pringsheim, E. G. 1964. Entwicklung und biologische Bedeutung der Sexualität. *Biol. Zentralbl.* 83:739–756.

Provine, W. B. 1971. *The origins of theoretical population genetics.* Chicago: University Press. xi + 201 pp.

Pukowski, E. 1933. Ökologische Untersuchungen an *Necrophorus* F. *Z. Morphol. Ökol. Tiere* 27:518–586.

Puri, H. S. 1966. Ecological distribution of Recent Ostracoda. *Proc. Symp. on Crustacea, Marine Biol. Assoc. India* (1):457–495.

Pycraft, W. P. 1913. *The infancy of animals*. New York: Holt. xiv + 272 pp.
——. 1914. *The courtship of animals*. London: Hutchinson. xvi + 318 pp.
Rádl, E. 1930. *The history of biological theories*. Oxford: University Press. xii + 408 pp.
Rand, A. L. 1952. Secondary sexual characters and ecological competition. *Fieldiana* (*Zool.*)34:65–70.
——. 1954. Social feeding behavior of birds. *Fieldiana* (*Zool.*) 36(1):1–71.
Rand, R. W. 1955. Reproduction in the female cape fur seal, *Arctocephalus pusillus* (Schreber). *Proc. Zool. Soc. London* 124:717–740.
Randall, J. E. 1958. A review of ciguatera, tropical fish poisoning, with a tentative explanation of its cause. *Bull. Mar. Sci. Gulf Caribbean* 8:236–267.
Raper, J. R. 1954. Life cycles, sexuality, and sexual mechanisms in the Fungi. In D. H. Wenrich, *Sex in Microorganisms*. Washington: American Association for the Advancement of Science. Pp. 42–81.
——. 1966. *Genetics of sexuality in higher fungi*. New York: Ronald Press. viii + 283 pp.
Rappaport, R. A. 1971. The sacred in human evolution. *Ann. Rev. Ecol. Syst.* 2:23–44.
Rapport, D. J. 1971. An optimization model of food selection. *Amer. Nat.* 105:575–587.
Rasmussen, B. 1967. Note on growth and protandric hermaphroditism in the deep sea prawn, *Pandalis borealis. Proc. Symp. on Crustacea, Mar. Biol. Assoc. India* 2:701–714.
Rattenbury, J. C. 1953. Reproduction in *Phoronopsis viridis*. The annual cycle in the gonads, maturation and fertilization of the ovum. *Biol. Bull.* 104:182–196.
Raven, P. H., & H. J. Thompson 1964. Haploidy and angiosperm evolution. *Amer. Nat.* 98:251–252.
Ray, C. 1960. The application of Bergmann's and Allen's rules to poikilotherms. *J. Morphol.* 106:85–108.
Redfield, R. 1942. Introduction. *Biol. Symp.* 8:1–26.
Reid, R. G. B., & A. Reid 1969. Feeding processes of members of the genus *Macoma* (Mollusca: Bivalvia). *Canad. J. Zool.* 47:649–657.
Reinhard, E. G. 1949. Experiments on the determination and differentiation of sex in the bopyrid *Stegophryxus hyptius* Thompson. *Biol. Bull.* 96:17–31.
Reish, D. J. 1957. The life history of the polychaetous annelid *Neanthes caudata* (delle Chiaje), including a summary of development in the family Nereidae. *Pacific Sci.* 11:216–228.
Remane, A. 1929–1933. Rotatoria. *Klassen und Ordnungen des Tier-reichs* 4(2:1:1–4):1–576.
——. 1935. Gastrotricha. *Klassen und Ordnungen des Tier-reichs* 4(2:1): 1–242.
Rensch, B. 1924. Das Dépéretsche Gesetz und die Regel von der Kleinheit der Inselformen als Spezialfall des Bergmannschen Gesetzes und ein Erklärungsversuch desselben: eine Hypothese. *Z. Ind. Abst.-Vererb. Lehre* 35:139–155.

———. 1950. Die abhängigkeit der relativen Sexualdifferenz von der Körper-grösse. *Bonner Zool. Beitr.* 1:58–69.

———. 1960. *Evolution above the species level.* New York: Columbia University Press. xviii + 419 pp.

Rettenmeyer, C. W. 1970. Insect mimicry. *Ann. Rev. Ent.* 15:43–74.

Reverberi, G., & G. Ortolani 1965. Nuove ricerche sullo sviluppo dell'uovo dei Ctenofori. *Riv. Biol.* 58:113–129.

Rice, M. E. 1967. A comparative study of the development of *Phascolosoma agassizii, Golfingia pugettensis,* and *Themiste pyroides* with a discussion of developmental patterns in the Sipuncula. *Ophelia* 4:143–171.

Richards, O. W. 1927. Sexual selection and allied problems in the insects. *Biol. Rev.* 2:298–364.

———. 1961. An introduction to the study of polymorphism in insects. *Symp. Roy. Ent. Soc. London* 1:1–10.

Richards, O. W., & M. J. Richards 1951. Observations on the social wasps of South America (Hymenoptera, Vespidae). *Trans. Roy. Ent. Soc. London* 102:1–170.

Ricklefs, R. E. 1968. On the limitation of brood size in passerine birds by the ability of adults to nourish their young. *Proc. Nat. Acad. Sci.* 61:847–851.

———. 1969. Natural selection and the development of mortality rates in young birds. *Nature* 223:922–925.

———. 1970. Clutch size in birds: outcome of opposing predator and prey adaptations. *Science* 168:599–600.

du Rietz, G. E. 1921. *Zur methodologischen Grundlage der modernen Pflanzensociologie.* Wien: Holzhausen. 272 pp.

———. 1929. The fundamental units of vegetation. *Proc. Internat. Congr. Plant Sci., Ithaca 1926* 1:623–627.

Rink, W. 1935. Zur Entwicklungsgeschichte, Physiologie und Genetik der Lebermoosgattungen *Anthoceros* und *Aspiromitus. Flora* 130:87–130.

Riverberi, G. & G. Ortolani 1965. Nuove ricerche sullo sviluppo dell'uovo dei Ctenofori. *Riv. Biol.* 58:113–129.

Robel, R. J. 1966. Booming territory size and mating success of the greater prairie chicken (*Tympanachus cupido pinnatus*). *Animal Behaviour* 14:328–331.

Robertson, D. R. 1972. Social control of sex reversal in a coral-reef fish. *Science* 177:1007–1009.

Robson, G. C. 1932. *A monograph of the Recent Cephalopoda based on the collections in the British Museum (Natural History). Part II. The Octopoda (excluding the Octopodinae).* London: British Museum. xii + 359 pp.

Rösch, G. A. 1925. Untersuchungen über die Arbeitsteilung im Bienenstaat. *Z. Vergl. Physiol.* 2:571–631.

———. 1927. Über die Bautätigkeit im Bienenvolk und das Alter der Bau-bienen. *Z. Vergl. Physiol.* 6:264–298.

le Roi, O. 1907. *Dendrogaster arborescens* und *Dendrogaster ludwigi,* zwei entoparasitische Ascothoraciden. *Z. Wiss. Zool.* 86:100–133.

Romanes, E. 1896. *The life and letters of George John Romanes M.A.,*

LL.D., F.R.S. late Honorary Fellow of Gonville and Caius College, Cambridge: written and edited by his wife. London: Longmans, Green. xii + 391 pp.

Romanes, G. J. 1882. *Animal intelligence.* London: Kegan Paul, Trench. xiv + 520 pp.

———. 1884. *Mental evolution in animals: with a posthumous essay on instinct by Charles Darwin, M.A., LL.D., F.R.S.* New York: D. Appleton. iv + 411 pp.

———. 1885. *Jelly-fish, star-fish and sea-urchins: being a research on primitive nervous systems.* London: Kegan Paul, Trench, vii + 323 pp.

———. 1896. *Darwin, and after Darwin: an exposition of the Darwinian theory and a discussion of post-Darwinian questions: I: the Darwinian theory.* Chicago: Open court. xiv + 460 pp.

Romer, A. S. 1958. Phylogeny and behavior with special reference to vertebrate evolution. In A. Roe & G. G. Simpson, *Behavior and evolution.* New Haven: Yale University Press. Pp. 48–75.

Rosenblueth, A., N. Wiener, & J. Bigelow 1943. Behavior, purpose, and teleology. *Phil. Sci.* 10:18–24.

Rosenwald, K. 1926. Beeinflussung des Geschlechtswechsels von *Limax laevis. Z. Ind. Abstam. Vererbungslehre* 43:238–251.

Rosenzweig, M. L. 1968. The strategy of body size in mammalian carnivores. *Amer. Midl. Nat.* 80:299–315.

Rowan, M. K. 1965. Regulation of sea-bird numbers. *Ibis* 107:54–59.

Royama, T. 1966. A re-interpretation of courtship feeding. *Bird Study* 13:116–129.

Rudwick, M. J. S. 1964. The inference of function from structure in fossils. *British J. Phil. Sci.* 15:27–40.

———. 1970. *Living and fossil brachiopods.* London: Hutchinson University Library. 199 pp.

Rübel, E. 1921. Über Die Entwicklung der Gesellschaftsmorphologie. *J. Ecol.* 8:18–40.

Rützler, K. 1965. Substratstabilität im marinen Benthos als ökologischer Faktor, dargestellt am Beispiel adriatischer Porifera. *Int. Rev. Ges. Hydrobiol.* 50:281–292.

Russell, B. 1960. *Authority and the individual.* Boston: Beacon Press. vi + 101 pp.

Russell, F. E. 1965. Marine toxins and venomous and poisonous marine animals. *Adv. Mar. Biol.* 3:255–384.

Rutten, M. G. 1955. Evolution and oscillations of shallow seas. *Evolution* 9:481–482.

Ryther, J. H. 1963. Geographic variations in productivity. *The Sea* 2:347–380.

Sahlins, M. D., & E. R. Service 1960. *Evolution and culture.* Ann Arbor: University of Michigan Press. xiv + 131 pp.

Sakagami, S. F. 1954. Occurrence of an aggressive behaviour in queenless hives, with considerations on the social organization of honeybee. *In. Soc.* 1:331–343.

———. 1960. Ethological peculiarities of the primitive social bees, *Allodape* Lepeltier and allied genera. *In. Soc.* 7:231–249.

Salmon, M. 1965. Waving display and sound production in the courtship behavior of *Uca pugilator*, with comparisons to *U. minax* and *U. pugnax*. *Zoologica* 50:123–150.

Salmon, M., & S. P. Atsaides 1968. Visual and acoustical signalling during courtship by fiddler crabs (Genus *Uca*). *Amer. Zool.* 8:623–639.

Sanderson, A. R. 1940. Maturation in the parthenogenetic snail *Potamopyrgus jenkinsi* (Smith), and in the snail *Peringia ulvae* (Pennant). *Proc. Zool. Soc. London* (A)110:11–15.

Sarà, M. 1961. Ricerche sul gonocorismo ed ermafroditismo nei Poriferi. *Bol. Zool.* 28:47–59.

Sarason, I. G., & R. E. Smith 1971. Personality. *Ann. Rev. Psychol.* 22:393–446.

Sato, T. 1936. Vorläufige Mitteilung über *Atubaria heterolopha* gen. nov. spec. nov., einen in freiem Zustand aufgefundenen Pterobranchier aus dem Stillen Ozean. *Zool. Anz.* 115:97–106.

Schaeffer, P. 1956. Transformation interspécifique chez des Bactéries du genre *Hemophilus*. *Ann. Inst. Pasteur* 91:192–211.

Schaffer, W. H. 1968. Intraspecific combat and the evolution of the Caprini. *Evolution* 22:817–825.

van der Schalie, H. 1966. Hermaphroditism among North American freshwater mussels. *Malacologia* 5:77–78.

Scheffer, V. B. 1958. *Seals, sea lions, and walruses: a review of the Pinnipedia*. Stanford: University Press. x + 179 pp.

Scheltema, R. S. 1966. Evidence for trans-Atlantic transport of gastropod larvae belonging to the genus *Cymatium*. *Deep-Sea Res.* 13:83–95.

Schenkel, R. 1966. Play, exploration and territoriality in the wild lion. *Symp. Zool. Soc. London* 18:11–22.

Schmalhausen, I. I. 1949. *Factors of evolution: the theory of stabilizing selection*. Philadelphia: Blakiston. xiv + 327 pp.

Schmidt, G. D. 1969. *Dioecotaenia cancellata* (Linton 1890) gen. et comb. n., a dioecious cestode (Tetraphyllidea) from the cow-nosed ray, *Rhinoptera bosanus* (Mitchell), in Chesapeake Bay, with the proposal of a new family, *Dioecotaeniidae*. *J. Parasitol.* 55:271–275.

Schmidt-Bey, W. 1913. Neckereien der Raubvögel nebst Gedanken über die Entstehung iherer sekundären Geschlechtsunterschiede. *Ornithol. Monatschr.* 38:400–416.

Schneider, F. 1962. Dispersal and migration. *Ann. Rev. Ent.* 7:223–242.

Schneirla, T. C. 1952. Basic correlations and coordinations in insect societies with special reference to ants. *Colloques Internat. Centre Nat. Rec. Sci.* 34:247–269.

———. 1957. A comparison of species and genera in the ant subfamily Dorylinae with respect to functional pattern. *In. Soc.* 4:259–298.

Schneirla, T. C., & R. Z. Brown 1952. Sexual broods and the production of young queens in two species of army ants. *Zoologica* 37:5–32.

Schöne, H. 1968. Agonistic and sexual display in aquatic and semiterrestrial brachyuran crabs. *Amer. Zool.* 8:641–654.

Schoener, T. W. 1967. The ecological significance of sexual dimorphism in size in the lizard *Anolis conspersus*. *Science* 155:474–477.

———. 1968. Sizes of feeding territories among birds. *Ecology* 49:123–141.

Scholander, P. F. 1955. Evolution of climatic adaptation in homeotherms. *Evolution* 9:15-26.

———. 1956. Climatic rules. *Evolution* 10:339-340.

Schopf, T. J. M. 1972. Varieties of paleobiologic experience. In T. J. M. Schopf, *Models in paleobiology*. San Francisco: Freeman, Cooper. Pp. 8-25.

Schopf, T. J. M., A. Farmanfarmaian, & J. L. Gooch 1971. Oxygen consumption rates and their paleontological significance. *J. Paleontol.* 45:247-252.

Schopf, T. J. M., & J. L. Gooch 1972. A natural experiment to test the hypothesis that loss of genetic variability was responsible for mass extinctions of the fossil record. *J. Geol.* 80:481-483.

Schultz, R. J. 1961. Reproductive mechanism of unisexual and bisexual strains of the viviparous fish *Poeciliopsis*. *Evolution* 15:302-325.

———. 1969. Hybridization, unisexuality, and polyploidy in the teleost *Poeciliopsis* (Poeciliidae) and other vertebrates. *Amer. Nat.* 103:605-619.

———. 1971. Special adaptive problems associated with unisexual fishes. *Amer. Zool.* 11:351-360.

Schur, M. 1972. *Freud: living and dying*. New York: International Universities Press. xiii + 587 pp.

Schur, M., & L. B. Ritvo 1970. The concept of development and evolution in psychoanalysis. In L. R. Aronson, E. Tobach, D. S. Lehrman, & J. S. Rosenblatt, *Development and evolution of behavior: essays in memory of T. C. Schneirla*. San Francisco: Freeman. Pp. 600-619.

Schuster, R. 1962. Nachweis eines Paarungszeremoniells bei der Hornmilben (Orobatei, Acari). *Naturwissenschaften* 21:502.

Schwalb, H. H. 1961. Beiträge zur Biologie der einheimischen Lampyriden *Lampyris noctiluca* Geoffr. und *Phausis splendidula* Lec. und experimentelle Analyse ihres Beutefang und sexualverhaltens. *Zool. Jahrb.* (*Syst.*)88:399-550.

Scott, J. P. 1958. *Aggression*. Chicago: University Press. xi + 149 pp.

———. 1967. Comparative psychology and ethology. *Ann. Rev. Psychol.* 18:65-86.

Scott, J. P., & J. L. Fuller 1965. *Genetics and the social behavior of the dog*. Chicago: University Press. xviii + 468 pp.

Scott, J. W. 1942. Mating behavior of the sage grouse. *Auk* 59:477-498.

Scudo, F. M. 1967a. The adaptive value of sexual dimorphism. I. Anisogamy. *Evolution* 21:285-291.

———. 1967b. L'accoppiamento assortativo basato sul fenotipo di parenti. Alcune conseguenze in popolazioni. *Inst. Lombardo* (*Rend. Sc.*), (B)101:435-455.

———. 1967c. Criteria for the analysis of multifactorial sex determination. *Monit. Zool. Ital.* (N.S.)1:1-21.

———. 1969. On the adaptive value of sexual dimorphism: II. Unisexuality. *Evolution* 23:36-49.

Scudo, F. M., & S. Karlin 1969. Assortative mating based on phenotype: I. Two alleles with dominance. *Genetics* 63:479-498.

Sedgwick, A. 1888. A monograph on the species and distribution of the genus *Peripatus* (Guilding). *Quart. J. Microscop. Sci.* 28:431–494.

Seed, R. 1969. The incidence of the pea-crab, *Pinnotheres pisum* in two types of *Mytilus* (Mollusca: Bivalvia) from Padstow, South-west England. *J. Zool. Lond.* 158:413–420.

Selander, R. K. 1965. On mating systems and sexual selection. *Amer. Nat.* 99:129–141.

———. 1966. Sexual dimorphism and differential niche utilization in birds. *Condor* 68:113–151.

———. 1969. The ecological aspects of the systematics of animals. *Systematic Biology: Proceedings of an International Conference. National Academy of Sciences Publ.* 1692:213–244.

———. 1972. Sexual selection and dimorphism in birds. In B. Campbell, *Sexual selection and* The Descent of Man *1871–1971*. Chicago: Aldine. Pp. 180–230.

Sellmer, G. P. 1967. Functional morphology and ecological life history of the gem clam, *Gemma gemma* (Eulamellibranchia: Veneridae). *Malacologia* 5:137–223.

Selous, E. 1906–1907. Observations tending to throw light on the question of sexual selection in birds, including a day-to-day diary on the breeding habits of the ruff (*Machetes pugnax*). *Zoologist* (4)10:201–219, 285–294, 419–428; 11:60–65, 161–182, 367–381.

Semler, D. E. 1971. Some aspects of adaptation in a polymorphism for breeding colours in the threespine stickleback (*Gasterosteus aculeatus*). *Proc. Zool. Soc. London* 165:291–302.

Senn, E. 1934. Die Geschlechtsverhältnisse der Brachiopoden, im besonderen die Spermato- und Oogenese der Gattung *Lingula*. *Acta Zool.* 15:1–154.

Serban, M. 1960. La néotenie et le problème de la taille chez les Copépodes. *Crustaceana* 1:77–83.

Setchell, W. A. 1914. Parasitic Florideae I. *Univ. California Publ. Bot.* 6:1–34.

Seth-Smith, D. 1925. On the display of the Argus pheasant (*Argusianus argus*). *Proc. Zool. Soc. London* 1925:323–325.

Sewell, R. B. S. 1912. Notes on the surface-living Copepoda of the Bay of Bengal, I & II. *Rec. Indian Mus.* (2)7:313–382.

———. 1929–1932. The Copepoda of Indian seas. Calanoida. *Mem. Indian Mus.* 10:1–407.

Seyrig, A. 1935. Relations entre le sexe de certains Ichneumonides (Hym.) et l'hôte aux dépens duquel els ont vécu. *Bull. Soc. Ent. France* 40:67–70.

Shapovalov, L., & A. C. Taft 1954. The life histories of the steelhead rainbow trout (*Salmo gairdneri gairdneri*) and silver salmon (*Onchorhynchus kisutch*) with special reference to Waddell Creek, California, and recommendations regarding their management. *California Fish Bull.* 98:1–375.

Shaver, R. H. 1953. Ontogeny and sexual dimorphism in *Cytherella bullata*. *J. Paleontol.* 27:471–480.

Shaw, E. 1971. Evidence of sexual maturation in young adult shiner perch, *Cymatogaster aggregata* Gibbons (Perciformes, Embiotocidae). *Amer. Mus. Nov.* 2479:1–10.

Shelford, V. E. 1911. Physiological animal geography. *J. Morphol.* 22:551–618.

————. 1915. Principles and problems of ecology as illustrated by animals. *J. Ecol.* 3:1–23.

————. 1931. Some concepts of bioecology. *Ecology* 12:455–467.

————. 1932. Basic principles of the classification of communities and habitats and the use of terms. *Ecology* 13:105–120.

Sherif, M., & C. W. Sherif 1970. Motivation and intergroup aggression: a persistent problem in levels of analysis. In L. R. Aronson, E. Tobach, D. S. Lehrman, & J. S. Rosenblatt, *Development and evolution of behavior: essays in memory of T. C. Schneirla.* San Francisco: Freeman. Pp. 563–579.

Shuel, R. W., & S. E. Dixon 1960. The early establishment of dimorphism in the female honeybee, *Apis mellifera* L. *In. Soc.* 7:265–282.

Shull, A. F. 1925. Sex and the parthenogenetic-bisexual cycle. *Amer. Nat.* 59:138–154.

Sibley, C. G. 1957. The evolutionary and taxonomic significance of sexual dimorphism and hybridization in birds. *Condor* 59:166–191.

Siegel, R. W. 1958. Hybrid vigor, heterosis and evolution in *Paramecium aurelia. Evolution* 12:402–416.

Silén, L. 1954. Developmental biology of Phoronidoidea of the Gullmar Fiord area (West Coast of Sweden). *Acta Zool.* 35:215–257.

Simberloff, D. 1972. Models in biogeography. In T. J. M. Schopf, *Models in paleobiology.* San Francisco: Freeman, Cooper Pp. 160–191.

Simon, J. L. 1968. Occurrence of pelagic larvae in *Spio setosa* Verrill, 1873 (Polychaeta: Spionidae). *Biol. Bull.* 134:503–515.

Simon, J. R. 1940. Mating performance of the sage grouse. *Auk* 57:467–471.

Simpson, G. G. 1931. Origin of mammalian faunas as illustrated by that of Florida. *Amer. Nat.* 65:258–276.

————. 1953. *The major features of evolution.* New York: Columbia University Press. xx + 434 pp.

————. 1958. The study of evolution: methods and present status of the theory. In A. Roe & G. G. Simpson, *Behavior and evolution.* New Haven: Yale University Press. Pp. 7–26.

————. 1963. Historical science. In C. C. Albritton, *The fabric of geology.* Reading, Mass.: Addison-Wesley. Pp. 24–48.

————. 1964. Species density of North American Recent Mammals. *Syst. Zool.* 13:57–73.

————. 1966. The biological nature of man. *Science* 152:472–478.

————. 1968. Evolutionary effects of cosmic radiation. *Science* 162:140–141.

————. 1972. The evolutionary concept of man. In B. Campbell, *Sexual selection and* The Descent of Man *1871–1971.* Chicago: Aldine. Pp. 17–39.

Simpson, T. L. 1968. The biology of the marine sponge *Microciona prolifera* (Ellis and Solander) II. Temperature-related, annual changes in functional and reproductive elements with a description of larval metamorphosis. *J. Exper. Mar. Biol. Ecol.* 2:252–277.

Skaife, S. H. 1953. Subsocial bees of the genus *Allodape* Lep. & Serv., *J. Ent. Soc. Southern Africa* 16:3–16.

Skinner, B. F. 1971. *Beyond freedom and dignity.* New York: Knopf. 225 pp.

Skinner, K. L. 1922. More light on the habits of the Cuckoo. *Oologist's Record* 2:64–65.

Skutch, A. F. 1935. Helpers at the nest. *Auk* 52:257–273.

———. 1949. Do tropical birds rear as many young as they can nourish? *Ibis* 91:430–455.

———. 1967. Adaptive limitation of the reproductive rate of birds. *Ibis* 109:579–599.

Slaughter, B. H. 1967. Animal ranges as a clue to Late-Pleistocene extinction. In P. S. Martin & H. E. Wright, Jr., *Pleistocene extinctions: the search for a cause.* New Haven: Yale University Press. Pp. 155–167.

Slijper, E. J. 1958. Das Verhalten der Wale (Cetacea). *Handbuch der Zoologie* 8(10:14):1–32.

Slobodchikoff, C. N., & H. V. Daly 1971. Systematic and evolutionary implications of parthenogenesis in the Hymenoptera. *Amer. Zool.* 11: 273–282.

Slobodkin, L. B., F. E. Smith, & N. G. Hairston 1967. Regulation in terrestial ecosystems, and the implied balance of nature. *Amer. Nat.* 101:109–124.

Smith, A. 1759. *The theory of moral sentiments.* London: A. Millar. xii + 551 pp.

———. 1776. *An inquiry into the nature and causes of the wealth of nations.* London: W. Strahan and T. Cadell. xiv + 1,097 pp.

Smith, A. G. 1966. The larval development of chitons; Amphineura. *Proc. California Acad. Sci.* 32:433–446.

Smith, C. L. 1967. Contribution to a theory of hermaphroditism. *J. Theoret. Biol.* 17:76–90.

Smith, E. H. 1967. The reproductive system of the British Turridae (Gastropoda: Toxoglossa). *Veliger* 10:176–187.

Smith, G. 1903. Metamorphosis and life-history of *Gnathia maxillaris. Mitth. Zool. Stat. Neapel* 16:469–479.

———. 1904. High and low dimorphism. With an account of certain Tanaidae of the Bay of Naples. *Mitth. Zool. Stat. Neapel* 17:312–340.

———. 1906. Rhizocephala. *Fauna Flora G. Neapel* 29:i–viii, 1–123.

Smith, G. M. 1945. The marine algae of California. *Science* 101:188–192.

Smith, R. I. 1950. Embryonic development in the viviparous nereid polychaete, *Neanthes lighti* Hartman. *J. Morphol.* 87:417–466.

———. 1958. On reproductive pattern as a specific characteristic among nereid polychaetes. *Syst. Zool.* 7:60–73.

Smith, S. G. 1971. Parthenogenesis and polyploidy in beetles. *Amer. Zool.* 11:341–349.

Smouse, P. E. 1971. The evolutionary advantages of sexual dimorphism. *Theoret. Pop. Biol.* 2:469–481.

Snow, C. D., & J. R. Neilsen 1966. Premating and mating behavior of the

Dungeness crab (*Cancer magister* Dana). *J. Fish. Res. Board Canada* 23:1319–1323.

Snow, D. W. 1963. The evolution of manakin displays. *Proc. XIII Internat. Ornithol. Congr.* pp. 553–561.

Sommerhoff, G. 1950. *Analytical biology.* Oxford: University Press. viii + 207 pp.

———. 1969. The abstract characteristics of living systems. In F. E. Emery, *Systems thinking: selected readings.* London: Penguin. Pp. 147–202.

Sonneborn, T. M. 1941. Sexuality in unicellular organisms. In G. N. Calkins & F. M. Summers, *Protozoa in biological research.* New York: Columbia University Press. Pp. 666–709.

Soulé, M. 1972. Phenetics of natural populations. III. Variation in insular populations of a lizard. *Amer. Nat.* 106:429–446.

Soulé, M. & B. R. Stuart 1970. The "niche-variation" hypothesis: a test and alternatives. *Amer. Nat.* 104:85–97.

Southward, T. R. E. 1961. The numbers of species of insect associated with various trees. *J. Animal Ecol.* 30:1–8.

Spalding, J. F. 1942. The nature and formation of the spermatophore and sperm plug in *Carcinus maenas. Quart. J. Microscop. Sci.* 83:399–422.

Spencer, H. 1850. (Reprint 1869.) *Social statics; or, the conditions essential to human happiness specified, and the first of them developed.* New York: D. Appleton. xviii + 523 pp.

———. 1852. The development hypothesis. Reprinted in H. Spencer, 1910, *Essays: scientific, political and speculative.* New York: D. Appleton. Vol. I, pp. 1–7.

———. 1857a. Progress: its law and cause. Reprinted in H. Spencer, 1910, *Essays: scientific, political and speculative.* New York: D. Appleton. Vol. I, pp. 8–62.

———. 1857b. The origin and function of music. Reprinted in H. Spencer, 1910, *Essays: scientific, political, and speculative.* New York: D. Appleton. Vol. II, pp. 400–451.

———. 1860a. The social organism. Reprinted in H. Spencer, 1910, *Essays: scientific, political, and speculative.* New York: D. Appleton. Vol. I, pp. 265–307.

———. 1860b. The physiology of laughter. Reprinted in H. Spencer, 1910, *Essays: scientific, political, and speculative.* New York: D. Appleton. Vol. II, pp. 452–466.

———. 1862. *First principles of a new system of philosophy.* [Reprint, 1864. New York: D. Appleton. viii + 508 pp.]

———. 1871a. Specialized administration. Reprinted in H. Spencer, 1910, *Essays: scientific, political and speculative.* New York: D. Appleton. Vol. III, pp. 401–444.

———. 1871b. Morals and moral sentiments. Reprinted in H. Spencer, 1910, *Essays: scientific, political, and speculative.* New York: D. Appleton. Vol. I, pp. 331–350.

———. 1874. *The study of sociology.* New York: D. Appleton. vii + 423 pp.

———. 1876. *Principles of sociology.* Vol. I. New York: D. Appleton. viii + 704 pp.

———. 1882. *Principles of sociology.* Vol. II. New York: D. Appleton. x + 686 pp.

———. 1884. *Principles of biology.* New York: D. Appleton. xvi + 1061 pp.

———. 1886. *The principles of psychology.* New York: D. Appleton. xx + 1,290 pp.

———. 1896. *Principles of sociology.* Vol. III. New York: D. Appleton. viii + 635 pp.

———. 1904. *An autobiography.* London: Williams & Northgate. xxi + 1,098 pp.

Spieth, H. T. 1968. Evolutionary implications of sexual behavior in *Drosophila. Evol. Biol.* 2:157–193.

Spinage, C. A. 1968. Horns and other bony structures of the skull of the giraffe, and their functional significance. *East Afr. Wildl. J.* 6:53–61.

Spradbery, J. P. 1965. The social organization of wasp communities. *Symp. Zool. Soc. London* 14:61–96.

Spurway, H. 1957. Hermaphroditism with self-fertilization, and the monthly extrusion of unfertilized eggs, in the viviparous fish *Lebistes reticulatus. Nature* 180:1248–1251.

Stanley, S. M. 1968. Post-Paleozoic adaptive radiation of infaunal bivalve molluscs—a consequence of mantle fusion and siphon formation. *J. Paleontol.* 42:214–229.

———. 1972. Functional morphology and evolution of byssally attached bivalve mollusks. *J. Paleontol.* 46:165–212.

———. 1973. An explanation for Cope's rule. *Evolution* 27:1–26.

Stasek, C. R. 1961. The ciliation and function of the labial palps of *Aclia castrensis* (Protobranchia, Nuculidae), with an evaluation of the role of the protobranch organs of feeding in the evolution of the Bivalvia. *Proc. Zool. Soc. London* 137:511–538.

———. 1963. Orientation and form in the bivalved Mollusca. *J. Morphol.* 112:195–214.

———. 1965. Feeding and particle-sorting in *Yoldia ensifera* (Bivalvia: Protobranchia), with notes on other Nuculanids. *Malacologia* 2:349–366.

———. 1966. Views on the comparative anatomy of the bivalved Mollusca. *Malacologia* 5:67–68.

Stauber, L. A. 1945. *Pinnotheres ostreum,* parasitic on the American oyster, *Ostrea* (*Gryphaea*) *virginica. Biol. Bull.* 88:269–291.

Stebbing, A. R. D. 1970. Aspects of the reproduction and life cycle of *Rhabdopleura compacta* (Hemichordata). *Mar. Biol.* 5:205–212.

Stebbins, G. L. 1950. *Variation and evolution in plants.* New York: Columbia University Press. xix + 643 pp.

———. 1957. Self fertilization and population variability in the higher plants. *Amer. Nat.* 91:337–354.

———. 1958. Longevity, habitat, and release of genetic variability in the higher plants. *Cold Spring Harbor Symp. Quant. Biol.* 23:365–378.

———. 1960. The comparative evolution of genetic systems. In S. Tax, *Evolution after Darwin.* Chicago: University Press. Vol. 1, pp. 197–226.

———. 1965. Colonizing species of the native California flora. In H. G. Baker & G. L. Stebbins, *The genetics of colonizing species.* New York & London: Academic Press. Pp. 173–191.

————. 1969. *The basis of progressive evolution.* Chapel Hill: University of North Carolina Press. x + 150 pp.

————. 1970. Adaptive radiation of reproductive characteristics in angiosperms, I: pollination mechanisms. *Ann. Rev. Ecol. Syst.* 1:307–326.

————. 1971. Adaptive radiation of reproductive characteristics in angiosperms, II: seeds and seedlings. *Ann. Rev. Ecol. Syst.* 2:237–260.

Steenstrup, J. J. S. 1846. *Untersuchungen über das Vorkommen des Hermaphroditismus in der Natur: ein naturhistorischer Versuch.* Greifswald: Otte. xvi + 130 pp.

Stehli, F. G. 1968. Taxonomic diversity gradients in pole location: the Recent model. In E. T. Drake, *Evolution and environment.* New Haven: Yale University Press.

Steiner, A. 1932. Die Arbeitsteilung der Feldwespe *Polistes dubia* K. *Z. Vergl. Physiol.* 17:101–152.

————. 1934. Neuere Untersuchungen über die Arbeitsteilung bei Insektenstaaten. *Erg. Biol.* 10:156–176.

Stephenson, J. 1930. *The Oligochaeta.* Oxford: University Press. xvi + 978 pp.

Stephenson, T. A. 1929. On methods of reproduction as specific characters. *J. Mar. Biol. Assoc.* 16:131–172.

Steuer, A. 1925. Rassenbildung bei einem marinen Planktoncopepoden. *Z. Wiss. Zool.* 125:91–101.

Steward, J. H. 1958. Problems of cultural evolution. *Evolution* 12:206–210.

————. 1960. Evolutionary principles and social types. S. Tax, *Evolution after Darwin.* Chicago: University Press. 2:169–186.

Stewart, F. H. 1910. Studies in post-larval development and minute anatomy in the genera *Scalpellum* and *Ibla. Mem. Indian Mus.* 3:33–51.

Stiles, F. G. 1971. Time, energy, and territoriality of the Anna hummingbird (*Calypte anna*). *Science* 173:818–821.

Stillingfleet, B. 1791. *Miscellaneous tracts relating to natural history, husbandry, and physick. To which is appended the calendar of flora.* 4th ed. London: Dodsley, Leigh & Sotheby & Payne. xxxi + 391 pp.

Stitz, H. 1926. Mecoptera. *Biol. Tiere Deutschlands* 21(35):1–28.

Stoltzmann. J. 1885. Quelques rémarques sur le dimorphisme sexuel. *Proc. Zool. Soc. London* 28:421–432.

Stone, G. P. 1965. The play of little children. *Quest* 4:23–31.

Storer, R. W. 1966. Sexual dimorphism and food habits in three North American accipiters. *Auk* 83:423–436.

•Stride, G. O. 1958. On the courtship behaviour of a tropical mimetic butterfly, *Hypolimnas missipus* L. (Nymphalidae). *Proc. X Int. Congr. Entomol.* 2:419–424.

Süffert, F. 1932. Phänomene visueller Anpassung. *Z. Morphol. Okol. Tiere* 26:147–316.

Sugiyama, Y. 1967. Social organization of Hanuman langurs. In S. A. Altmann, *Social communication among primates.* Chicago: University Press. Pp. 221–236.

Summers-Smith, D. 1958. Nest-site selection, pair formation and territory in the house-sparrow *Passer domesticus. Ibis* 100:190–203.

Suomalainen, E. 1950. Parthenogenesis in animals. *Adv. Genet.* 3:193–253.

———. 1962. Significance of parthenogenesis in the evolution of insects. *Ann. Rev. Ent.* 7:349–366.

Sutherland, A. 1898. *The origin and growth of the moral instinct.* London: Longmans, Green & Co. xix + 797 pp.

Sutherland, J. P. 1970. Dynamics of high and low populations of the limpet, *Acmaea scabra* (Gould). *Ecol. Monogr.* 40:169–188.

Sutton-Smith, B. 1966. Piaget on play: a critique. *Psychol. Rev.* 73:104–110.

———. 1967. The role of play in cognitive development. *Young Children* 6:364–369.

Sverdrup, H. U. 1955. The place of physical oceanography in oceanographic research. *J. Marine Res.* 14:287–294.

Swarth, H. S. 1934. The bird fauna of the Galapagos Islands in relation to species formation. *Biol. Rev.* 9:213–234.

Swedmark, B. 1964. The interstitial fauna of marine sand. *Biol. Rev.* 39: 1–42.

Szidat, L. 1964. Sobre la evolucion del dimorfismo sexual secundario en Isopodes parasitos de la familia Cymothoidae (Crust. Isop.). *Anais Congr. Lat. Amer. Zool.* 2:83–87.

Takahasi, S. 1935. Ecological notes on the ocypodian crabs (Ocypodidae) in Formosa, Japan. *Annot. Zool. Japon.* 15:78–87.

Tansley, A. G. 1921. The classification of vegetation and the concept of development. *J. Ecol.* 8:118–149.

———. 1929. Succession: the concept and its values. *Proc. Internat. Congr. Plant Sci., Ithaca* 1926, 1:677–686.

———. 1935. The use and abuse of vegetational concepts and terms. *Ecology* 16:284–307.

———. 1947. The early history of modern plant ecology in Britain. *J. Ecol.* 35:130–137.

Tartar, V. 1967. Morphogenesis in Protozoa. *Res. Protozool.* 2:1–116.

Tate Regan, C. 1925. Dwarfed males parasitic on the females in oceanic angler-fishes (Pediculati Ceratioidea). *Proc. Roy. Soc. London* (B) 97:386–400.

———. 1926. The pediculate fishes of the suborder Ceratioidea. *Oceanogr. Rep. Ed. "Dana"-Committee* 1(2):1–45.

Tavogla, W. N. 1970. Levels of interaction in animal communication. In L. R. Aronson, E. Tobach, D. S. Lehrman, & J. S. Rosenblatt, *Development and evolution of behavior: essays in memory of T. C. Schneirla.* San Francisco: Freeman. Pp. 281–302.

Tchernavin, V. 1938. Changes in the salmon skull. *Trans. Zool. Soc. London* 24:103–178.

Teichert, C. 1964. Morphology of hard parts. *Treatise on Invertebrate Paleontology,* K:13–53.

Tevis, L., Jr. 1950. Summer behavior of a family of beavers in New York State. *J. Mam.* 31:40–65.

Thallwitz, J. 1916. Über Dimorphismus des Männchen bei einem Süsswasser-harpacticiden. *Zool. Anz.* 46:238–240.

Thenius, E. 1962. Die Grosssäugetiere des Pleistozäns von Mitteleuropa: eine Übersicht. Z. *Säugetierkunde* 27:65–83.

Thienemann, A. 1950. Wesen und Bedeutung der Limnologie. *Oikos* 2: 149–161.

Thompson, D. W. 1909. Pycnogonida. *The Cambridge Natural History* 4:501–542.

Thompson, W. F., & J. .B Thompson 1919. The spawning of the grunion (*Leuresthes tenuis*). *California Fish Bull.* (3):1–29.

Thompson, W. F., "and Associates" 1936. The spawning of the silver smelt, *Hypomesus pretiosus*. *Ecology* 17:158–168.

Thompson, W. R. 1958. Social Behavior. In A. Roe & G. G. Simpson, *Behavior and evolution*. New Haven: Yale University Press. Pp. 291–310.

Thorpe, W. H. 1966. Ritualization in ontogeny: I. Animal play. *Phil. Trans. Roy. Soc. London* (B)251:311–319.

Thorson, G. 1946. Reproduction and larval development of Danish marine bottom invertebrates. *Medd. Kom. Havunders. Kobenhavn* (Plankton) 4:1–523.

———. 1950. Reproductive and larval ecology of marine bottom invertebrates. *Biol. Rev.* 25:1–45.

———. 1951. Zur jetzigen Lage der marinen Bodentier-Ökologie. *Verh. Deutsch. Zool. Ges.* 16:276–327.

———. 1955. Modern aspects of marine level-bottom animal communities. *J. Marine Res.* 14:387–397.

———. 1956. Marine level-bottom communities of recent seas, their temperature adaptation and their "balance" between predators and food animals. *Trans. New York Acad. Sci.* (2)18:693–700.

———. 1965. A neotenous dwarf-form of *Capulus ungaricus* (L.) (Gastropoda, Prosobranchia) commensalistic on *Turritella communis* Risso. *Ophelia* 2:175–210.

Tiger, L. 1970. Dominance in human societies. *Ann. Rev. Ecol. Syst.* 1:287–306.

Tiger, L., & R. Fox 1966. The zoological perspective in social science. *Man* 1:75–81.

Tinbergen, L. 1937. Zur Fortpflanzungsethologie von *Sepia officinalis* L. *Arch. Neerl. Zool.* 3:323–364.

Tinbergen, N. 1935. Field observations of East Greenland birds. I. The behaviour of the red-necked phalarope (*Phalaropus lobatus* L.) in spring. *Ardea* 24:1–42.

———. 1936. The function of sexual fighting in birds; and the problem of the origin of "territory." *Bird-Banding* 7:1–8.

———. 1968. On war and peace in animals and man. *Science* 160:1411–1418.

Tobach, E., & L. R. Aronson 1970. T. C. Schneirla: a biographical note. In L. R. Aronson, E. Tobach, D. S. Lehrman, & J. S. Rosenblatt, *Development and evolution of behavior: essays in memory of T. S. Schneirla*. San Francisco: Freeman. Pp. xi–xviii.

Tolman, E. C. 1932. *Purposive behavior in animals and men.* New York: Century. xiv + 463 pp.

Tomlinson, J. T. 1966. The advantages of hermaphroditism and partheno-genesis. *J. Theoret. Biol.* 11:54–58.

———. 1969. The burrowing barnacles (Cirripedia: order Acrothoracica). *United States Nat. Mus. Bull.* (296):vi + 162 pp.

Topoff, H. 1971. Polymorphism in army ants related to division of labor and colony cyclic behavior. *Amer. Nat.* 105:529–548.

———. 1972. Theoretical issues concerning the evolution and development of behavior in social insects. *Amer. Zool.* 12:385–394.

Towe, K. M. 1970. Oxygen-collagen priority and the early metazoan fossil record. *Proc. Nat. Acad. Sci.* 65:781–788.

Triplett, E. L. 1960. Notes on the life history of the barred surfperch, *Amphistichus argenteus* Agassiz, and a technique for culturing embiotocid embryos. *California Fish & Game* 46:433–439.

Trivers, R. L. 1971. The evolution of reciprocal altruism. *Quart. Rev. Biol.* 46:35–57.

———. 1972. Parental investment and sexual selection. In B. Campbell, *Sexual selection and* The Descent of Man *1871–1971.* Chicago: Aldine. Pp. 136–179.

Tschislenko, L. L. 1964. On the sex ratio in marine free-living Copepoda. *Zool. Zhurnal* 43:1400–1402.

Tullock, G. 1971. The coal tit as a careful shopper. *Amer. Nat.* 105:77–80.

Tylor, E. B. 1865. *Researches into the early history of mankind and the development of civilization.* 2nd ed. London: John Murray. vi + 378 pp.

———. 1871. *Primitive culture: researches into the development of mythology, philosophy, religion, art, and custom.* London: John Murray. xviii + 879 pp.

Tyrväinen, H. 1969. The breeding biology of the redwing (*Turdus iliacus* L.). *Ann. Zool. Fenn.* 6:1–46.

Uljanin, M. 1878. Sur le genre *Sagitella* (N. Wagner). *Arch. Zool. Exp. Gén.* 7:1–32.

Unwin, E. E. 1920. Notes upon the reproduction of *Asellus aquaticus. J. Linn. Soc. London (Zool.)*34:335–343.

Uphof, J. C. T. 1938. Cleistogamic flowers. *Bot. Rev.* 4:21–49.

Vale, J. R., & C. A. Vale 1969. Individual differences and general laws in psychology: a reconciliation. *Amer. Psychol.* 24:1093–1108.

Valentine, J. W. 1967. The influence of climatic fluctuations on species diversity within the Tethyan provincial system. *Syst. Assoc. Publ.* (7): 153–166.

———. 1969. Niche diversity and niche size patterns in marine fossils. *J. Paleontol.* 43:905–915.

———. 1971. Resource supply and species diversity patterns. *Lethaia* 4:51–61.

———. 1972. Conceptual models of ecosystem evolution. In T. J. M. Schopf, *Models in paleobiology.* San Francisco: Freeman, Cooper. Pp. 192–215.

Valentine, J. W., & E. M. Moores 1972. Global tectonics and the fossil record. *J. Geol.* 80:167–184.

Van Valen, L. 1971. Group selection and the evolution of dispersal. *Evolution* 25:591–598.

Vandel, A. 1925. Recherches sur la sexualité des Isopodes. Les conditions naturelles de la reproduction chez les Isopodes terrestres. *Bull. Biol. France Belgique* 59:317–371.

———. 1927. La cytologie de la parthénogenèse naturelle. *Bull. Biol. France Belgique* 61:93–125.

———. 1928. La parthénogenèse géographique: contribution a l'étude biologique et cytologique de la parthénogenèse naturelle. *Bull. Biol. France Belgique* 62:164–281.

———. 1931. *La parthénogenèse.* Paris: Doin. xix + 412 pp.

———. 1937. Chromosome number, polyploidy and sex in the animal kingdom. *Proc. Zool. Soc. London* (A)107:519–541.

———. 1940. La parthénogenèse géographique. IV. Polyploidie et distribution géographique. *Bull. Biol. France Belgique* 74:94–100.

———. 1941. Recherches sur la génétique et la sexualité des Isopodes terrestres. VI. Les phènomènes de monogénie chez les Oniscoïdes. *Bull. Biol. France Belgique* 75:316–363.

Verner, J. 1964. Evolution of polygamy in the long-billed marsh wren. *Evolution* 13:252–261.

———. 1965. Selection for sex ratio. *Amer. Nat.* 99:419–421.

———. 1966. The influence of habitats on mating systems of North American birds. *Ecology* 47:143–147.

Verner, J., & M. F. Willson 1966. The influence of habitats on mating systems of North American birds. *Ecology* 47:143–147.

Verwey, J. 1930. Einiges über die Biologie ost-indischer Mangrovenkrabben. *Treubia* 12:167–261.

Vine, I. 1970. Communication by facial-visual signals. In J. H. Crook, *Social behaviour in birds and mammals.* New York: Academic Press. Pp. 279–354.

Volkmann-Rocco, B. 1972. The effect of delayed fertilization in some species of the genus *Tisbe* (Copepoda, Harpacticoida). *Biol. Bull.* 142:520–529.

Vuilleumier, B. S. 1967. The origin and evolutionary development of heterostyly in angiosperms. *Evolution* 21:210–226.

Waddington, C. J. 1967. Palaeomagnetic field reversals and cosmic radiation. *Science* 158:913–915.

Wallace, A. R. 1876. *The geographical distribution of animals: with a study of the relations of living and extinct faunas as elucidating the past changes of the earth's surface.* London: Macmillan. xxxv + 1,110 pp.

———. 1895. *Natural selection and tropical nature: essays on descriptive and theoretical biology.* London: Macmillan. xii + 492 pp.

———. 1912. *Darwinism: an exposition of the theory of natural selection with some of its applications.* 3rd ed. London: Macmillan. xx + 494 pp.

Wallace, W. L. 1969. *Sociological theory: an introduction.* Chicago: Aldine. xiv + 296 pp.

Wallis, D. I. 1965. Division of labour in ant colonies. *Symp. Zool. Soc. London* 14:97–112.

Warburton, F. E. 1961. Inclusion of parental somatic cells in sponge larvae. *Nature* 191:1317.

Washburn, S. L., & C. S. Lancaster 1968. The evolution of hunting. R. B. Lee & I. DeVore, *Man the hunter*. Chicago: Aldine. Pp. 293–303.

Wasmann, E. 1900. *Termitoxenia*, ein neues flügelloses, physogastres Dipterengenus aus Termitennestern. I. Theil. Äussere Morphologie und Biologie. Z. *Wiss. Zool.* 67:599–617.

———. 1902. Zur näheren Kenntnis der termitophilen Dipterengattung *Termitoxenia* Wasm. *Verh. V. Internat. Zool. Congr.* pp. 852–873.

Watson, J. D. 1968. *The double helix: a personal account of the discovery of the structure of DNA.* New York: Atheneum. xvi + 226 pp.

Watson, R. A. 1966. Is geology different: a critical discussion of "The Fabric of Geology." *Phil. Sci.* 33:172–185.

Watt, K. E. F. 1966. *Systems analysis in ecology.* New York: Academic Press. xiii + 276 pp.

Weber, N. A. 1958. Evolution in fungus-growing ants. *Proc. X Int. Congr. Ent.* 2:459–473.

Webster, F. M. 1904. Studies of the life history, habits, and taxonomic relations of a new species of *Oberea* (*Oberea ulmicola* Chittenden). *Bull. Illinois State Lab. Nat. Hist.* 7:1–14.

Weisensee, H. 1916. Die Geschlechtsverhältnisse und der Geschlechtsapparat bei Anodonta. Z. *Wiss. Zool.* 115:262–335.

Weismann, A. 1889. *Essays upon heredity and kindred biological problems.* Vol. I. Oxford: University Press. xii + 455 pp.

———. 1891. *Essays upon heredity and kindred biological problems.* Vol. II. Oxford: University Press. viii + 226 pp.

Weiss, R. F., W. Buchanan, L. Altstatt, & J. P. Lombardo. 1971. Altruism is rewarding. *Science* 171:1262–1263.

Wells, H. W., & M. J. Wells 1961. Observations on *Pinnaxodes floridensis*, a new species of pinnotherid crustacean commensal in holothurians. *Bull. Mar. Sci., Gulf & Caribbean* 11:267–279.

Wells, W. W. 1928. Pinnotheridae of Puget Sound. *Publ. Puget Sound Biol. Sta.* 6:283–314.

———. 1940. Ecological studies on the pinnotherid crabs of Puget Sound. *Univ. Washington Publ. Oceanogr.* 2:19–50.

Wenner, A. M. 1972. Sex ratio as a function of size in marine Crustacea. *Amer. Nat.* 106:321–350.

Wenrich, D. H. 1954a. Sex in Protozoa: a comparative review. D. H. Wenrich, *Sex in microorganisms.* Washington: American Association for the Advancement of Science. Pp. 134–265.

———. 1954b. Comments on the origin and evolution of "sex." D. H. Wenrich, *Sex in microorganisms.* Washington: American Association for the Advancement of Science. Pp. 335–346.

Wertheimer, M. 1945. *Productive thinking.* New York: Harper. xi + 224 pp.

Wesenberg-Lund, C. 1923. Contributions to the biology of the Rotifera. I.

The males of the Rotifera. *D. Kgl. Danske Vidensk. Selsk. Skrifter, naturvidensk. og mathem. Afd.* (8)4(3):191–345.

———. 1926. Contributions to the biology and morphology of the genus *Daphnia:* with some remarks on heredity. *D. Kgl. Danske Vidensk. Selsk. Skrifter, naturvidensk. og mathem. Afd.* (8)11(2):91–250.

———. 1930. Contributions to the biology of the Rotifera. Part II. The periodicity and sexual periods. *D. Kgl. Danske Vidensk. Selsk. Skrifter, naturvidensk. og mathem. Afd.* (9)2(1):3–230.

West, M. J. 1967. Foundress associations in polistine wasps: dominance hierarchies and the evolution of social behavior. *Science* 157:1584–1585.

Westermann, G. E. G., ed., 1969. *Sexual dimorphism in fossil Metazoa and taxonomic implications.* Stuttgart: Schweizerbart. iv + 251 pp.

Weyer, F. 1930. Beobachtungen über die Entstehung neuer Kolonien bei tropishchen Termiten. *Zool. Jahrb. (Syst.)*60:327–380.

Weyrauch, W. 1928. Beitrag zur Biologie von *Polistes. Biol. Centralblatt* 48:407–427.

Wheeler, W. M. 1894. Protandric hermaphroditism in *Myzostoma. Zool. Anz.* 17:177–182.

———. 1897. Sexual phases of *Myzostoma. Mitth. Zool. Stat. Neapel* 12:227–302.

———. 1911. The ant-colony as an organism. *J. Morphol.* 22:307–325.

———. 1923. *Social life among the insects.* New York: Harcourt, Brace. vii + 375 pp.

———. 1924. The courtship of the calobatas. *J. Heredity* 15:484–495.

White, L. A. 1959. *The evolution of culture: the development of civilization to the fall of Rome.* New York: McGraw-Hill. xiv + 378 pp.

White, M. J. D. 1954. *Animal cytology and evolution.* 2nd ed. Cambridge: University Press. xi + 454 pp.

Whitehead, A. N. 1925. *Science and the modern world.* New York: Macmillan. xii + 212 pp.

Whiting, P. W. 1945. The evolution of male haploidy. *Quart. Rev. Biol.* 20:231–260.

Whittaker, R. H. 1969. Evolution of diversity in plant communities. *Brookhaven Symp. Biol.* 22:178–196.

Whittaker, R. H., & P. P. Feeny 1971. Allelochemics: chemical interactions between species. *Science* 171:757–770.

Wickler, W. 1960. Aquarienbeobachtungen an *Aspidontus,* einem ektoparasitischen Fisch. *Z. Tierpsychol.* 17:277–292.

———. 1967. Socio-sexual signals and their intra-specific imitation among primates. In D. Morris, *Primate ethology.* Chicago: Aldine, pp. 69–147.

Wickstead, J. H., & Q. Bone 1959. Ecology of acraniate larvae. *Nature* 184: 1849–1851.

Wiebe, J. P. 1968. The reproductive cycle of the viviparous seaperch, *Cymatogaster aggregata* Gibbons. *Canad. J. Zool.* 46:1221–1234.

Wilbert, H. 1969. Die Ursachen der geschlechtlichen Grössendifferenzen bei *Aphelinus asychis* Walker (Hym., Aphelinidae). *Z. Parasitenk.* 32:220–236.

Willey, A. 1897. On *Ctenoplana*. *Quart. J. Microscop. Sci.* 39:323–342.

———. 1898. The anatomy and development of Peripatus Novae Britanniae. Cambridge: University Press. p. 52.

Williams, C. B. 1957. Insect migration. *Ann. Rev. Ent.* 2:163–180.

Williams, E. E. 1969. The ecology of colonization as seen in the zoogeography of anoline lizards on small islands. *Quart. Rev. Biol.* 44:345–389.

Williams, G. C. 1957. Pleiotropy, natural selection, and the evolution of senescence. *Evolution* 11:398–411.

———. 1966. *Adaptation and natural selection: a critique of some current evolutionary thought.* Princeton: University Press. x + 307 pp.

Williams, G. C., & D. C. Williams 1957. Natural selection of individually harmful social adaptations among sibs with special reference to social insects. *Evolution* 11:32–39.

Williamson, G. B., & C. E. Nelson 1972. Fitness set analysis of mimetic adaptive strategies. *Amer. Nat.* 106:525–537.

Willoughby, E. J., & T. J. Cade 1964. Breeding behavior of the American kestrel (sparrow hawk). *Living Bird* 3:75–96.

Willson, M. F., & E. R. Pianka 1963. Sexual selection, sex ratio and mating system. *Amer. Nat.* 97:405–407.

Wilson, E. O. 1953. The origin and evolution of polymorphism in ants. *Quart. Rev. Biol.* 28:136–156.

———. 1958. The beginnings of nomadic and group-predatory behavior in the ponerine ants. *Evolution* 12:24–31.

———. 1959. Some ecological characteristics of ants in New Guinea rain forests. *Ecology* 40:437–447.

———. 1963. The social biology of ants. *Ann. Rev. Ent.* 8:345–368.

———. 1968a. The superorganism concept and beyond. *Colloques Internat. Centre Nat. Rec. Sci.* 173:27–39.

———. 1968b. The ergonomics of caste in the social insects. *Amer. Nat.* 102:41–66.

———. 1971. *The insect societies.* Cambridge: Harvard University Press. x + 548 pp.

Winn, H. E. 1958. Comparative reproductive behavior and ecology of fourteen species of darters (Pisces—Percidae). *Ecol. Monogr.* 28:155–191.

Winterbourn, M. 1970. The New Zealand species of *Potamopyrgus* (Gastropoda: Hydrobiidae). *Malacologia* 10:283–321.

Withycombe, C. L. 1922. Notes on the biology of some British Neuroptera (Planipennia). *Trans. Ent. Soc. London* 70:501–594.

Woesler, A. 1935. Zur Zwergmännchenfrage bei *Leucobryum glaucum* Schpr. I. *Planta* 24:1–13.

Wolf, E. R. 1970. The study of evolution. In S. N. Eisenstadt, *Readings in social evolution and development.* London: Pergamon Press. Pp. 179–191.

Wolf, L. L. 1969. Female territoriality in a tropical hummingbird. *Auk* 86:490–504.

Wolff, T. 1956. Crustacea Tanaidacea from depths exceeding 6,000 meters. *Galathea Rep.* 2:187–241.

————. 1962. The systematics and biology of bathyal and abyssal Isopoda Asellota. *Galathea Rep.* 6:7–320.

Wollman, E. L., F. Jacob, & W. Hayes 1956. Conjugation and genetic recombination in *Escherichia coli* K–12. *Cold Spring Harbor Symp. Quant. Biol.* 21:141–162.

Wourms, J. P. 1972. The developmental biology of annual fishes III. Preembryonic and embryonic diapause of variable duration in the eggs of annual fishes. *J. Exper. Zool.* 182:389–414.

Wright, S. 1938. Size of population and breeding structure in relation to evolution. *Science* 87:430–431.

Wynne-Edwards, V. C. 1962. *Animal dispersion in relation to social behaviour.* New York: Hafner. xii + 653 pp.

Yaldwyn, J. C. 1966. Protandrous hermaphroditism in decapod prawns of the families Hippolytidae and Campylonotidae. *Nature* 209:1366.

Yanagimachi, R. 1961. The life-cycle of *Peltogasterella* (Cirripedia, Rhizocephala). *Crustaceana* 2:183–186.

Yonge, C. M. 1938. The prosobranchs of Lake Tanganyika. *Nature* 142:464–466.

————. 1939. The protobranchiate Mollusca; a functional interpretation of their structure and evolution. *Phil. Trans. Roy. Soc. London* (B)230:79–147.

————. 1949. On the structure and adaptations of the Tellinacea, deposit-feeding Eulamellibranchia. *Phil. Trans. Roy. Soc. London* (B)234:29–76.

————. 1953. The monomyarian condition in the Lamellibranchia. *Trans. Roy. Soc. Edinburgh* 62:443–478.

Yosii, N. 1931. Note on the organisation of *Baccalaureus japonicus*. *Annot. Zool. Japon.* 13:169–187.

Zenkevitch, L. A. 1966. The systematics and distribution of abyssal and hadal (ultra-abyssal) Echiuroidea. *Galathea Rep.* 8:175–183.

INDEX